Mathematical aspects of boundary element methods

CHAPMAN & HALL/CRC
Research Notes in Mathematics Series

Main Editors

H. Brezis, *Université de Paris*
R.G. Douglas, *Texas A&M University*
A. Jeffrey, *University of Newcastle upon Tyne (Founding Editor)*

Editorial Board

H. Amann, *University of Zürich*
R. Aris, *University of Minnesota*
G.I. Barenblatt, *University of Cambridge*
H. Begehr, *Freie Universität Berlin*
P. Bullen, *University of British Columbia*
R.J. Elliott, *University of Alberta*
R.P. Gilbert, *University of Delaware*
R. Glowinski, *University of Houston*
D. Jerison, *Massachusetts Institute of Technology*
K. Kirchgässner, *Universität Stuttgart*

B. Lawson, *State University of New York at Stony Brook*
B. Moodie, *University of Alberta*
S. Mori, *Kyoto University*
L.E. Payne, *Cornell University*
D.B. Pearson, *University of Hull*
I. Raeburn, *University of Newcastle*
G.F. Roach, *University of Strathclyde*
I. Stakgold, *University of Delaware*
W.A. Strauss, *Brown University*
J. van der Hoek, *University of Adelaide*

Submission of proposals for consideration

Suggestions for publication, in the form of outlines and representative samples, are invited by the Editorial Board for assessment. Intending authors should approach one of the main editors or another member of the Editorial Board, citing the relevant AMS subject classifications. Alternatively, outlines may be sent directly to the publisher's offices. Refereeing is by members of the board and other mathematical authorities in the topic concerned, throughout the world.

Preparation of accepted manuscripts

On acceptance of a proposal, the publisher will supply full instructions for the preparation of manuscripts in a form suitable for direct photo-lithographic reproduction. Specially printed grid sheets can be provided. Word processor output, subject to the publisher's approval, is also acceptable.

Illustrations should be prepared by the authors, ready for direct reproduction without further improvement. The use of hand-drawn symbols should be avoided wherever possible, in order to obtain maximum clarity of the text.

The publisher will be pleased to give guidance necessary during the preparation of a typescript and will be happy to answer any queries.

Important note

In order to avoid later retyping, intending authors are strongly urged not to begin final preparation of a typescript before receiving the publisher's guidelines. In this way we hope to preserve the uniform appearance of the series.

CRC Press UK

Chapman & Hall/CRC Statistics and Mathematics
Pocock House
235 Southwark Bridge Road
London SE1 6LY
Tel: 0171 407 7335

M Bonnet
Ecole Polytechnique, Palaiseau Cedex, France

A-M Sändig
Universität Stuttgart, Germany

W L Wendland
Universität Stuttgart, Germany

(Editors)

Mathematical aspects of boundary element methods

dedicated to Vladimir Maz'ya
on the occasion of his 60th birthday

CHAPMAN & HALL/CRC

Boca Raton   London   New York   Washington, D.C.

Library of Congress Cataloging-in-Publication Data

Mathematical aspects of boundary element methods : dedicated to
  Vladimir Maz'ya on the occasion of his 60th birthday / M. Bonnet, A.
  -M. Sändig, W.L. Wendland, (editors).
        p. cm. (alk. paper)
     Includes bibliographical references.
     ISBN 1-58488-006-6
      1. Boundary value problems—Numerical solutions. 2. Boundary
  element methods. I. Bonnet, Marc, 1960–   . II. Sändig,
  Anna-Margarete, 1944–   III. Wendland, W. L. (Wolfgang
  L.), 1936–   .
  QA379.M385 1999
  515'35—dc21   99-38269
                                                                              CIP

This book contains information obtained from authentic and highly regarded sources. Reprinted material is quoted with permission, and sources are indicated. A wide variety of references are listed. Reasonable efforts have been made to publish reliable data and information, but the author and the publisher cannot assume responsibility for the validity of all materials or for the consequences of their use.

Neither this book nor any part may be reproduced or transmitted in any form or by any means, electronic or mechanical, including photocopying, microfilming, and recording, or by any information storage or retrieval system, without prior permission in writing from the publisher.

All rights reserved. Authorization to photocopy items for internal or personal use, or the personal or internal use of specific clients, may be granted by CRC Press LLC, provided that $.50 per page photocopied is paid directly to Copyright Clearance Center, 222 Rosewood Drive, Danvers, MA 01923 USA. The fee code for users of the Transactional Reporting Service is ISBN 1-58488-006-6/99/$0.00+$.50. The fee is subject to change without notice. For organizations that have been granted a photocopy license by the CCC, a separate system of payment has been arranged.

The consent of CRC Press LLC does not extend to copying for general distribution, for promotion, for creating new works, or for resale. Specific permission must be obtained in writing from CRC Press LLC for such copying.

Direct all inquiries to CRC Press LLC, 2000 N.W. Corporate Blvd., Boca Raton, Florida 33431.

Trademark Notice: Product or corporate names may be trademarks or registered trademarks, and are used only for identification and explanation, without intent to infringe.

© 2000 by Chapman & Hall/CRC

No claim to original U.S. Government works
International Standard Book Number 1-58488-006-6
Library of Congress Card Number 99-38269
Printed in the United States of America   1 2 3 4 5 6 7 8 9 0
Printed on acid-free paper

Dedication

This volume is dedicated to our esteemed colleague and friend Vladimir Maz'ya, who celebrated his 60th birthday on $31^{\underline{st}}$ December 1997.

Vladimir Maz'ya was born in St. Petersburg (Russia) — at that time Leningrad. His childhood and youth were marked by the cruelties of the second world war and the very hard post–war time in Leningrad. His father perished at the frontline in the winter 1941 and his grandparents died during the siege of Leningrad. Then the solicitude for young Volodja was in the hands of his loving mother alone.

Already in his school years Vladimir won first prizes in mathematics and physics. During his studies at the Department of Mathematics at the Leningrad University, beginning in 1955, he had shown his great mathematical talent and continued with winning prizes. So Professor S.G. Mikhlin became aware of this bright student and then supported and encouraged him. Vladimir Maz'ya's first paper on second order elliptic equations appeared in 1959 when he was just a third–year student. At the age of 22 he graduated from the Leningrad University. Already two years later he received the Ph.D from the Moscow University with his thesis "Classes of sets and imbedding theorems for function spaces". In 1965 he got the Doctors of Science degree from the Leningrad University with the thesis "Dirichlet and Neumann problems in domains with irregular boundaries".

Vladimir Maz'ya was employed from 1960 until 1990 in Leningrad at different Research Institutes, at the Academy of Sciences and at the Leningrad University. In 1971 he became a Professor in Applied Mathematics. This time was not easy for him; in particular, due to Soviet ruling it was almost impossible to publish in international journals and to care for international connections and cooperations. Then, in 1990, he emigrated to Sweden together with his mother, his beloved wife and cooperator Tatyana Shaposhnikova and their son Michael, those persons who are and have always been his source of solicitude, strength and inspiration. Now he is Professor at the Linköping University in Sweden.

Vladimir Maz'ya is extremely hard working – you cannot read so fast as he is writing. Up to now he produced more than 300 papers and 17 books, some together with his colleagues and friends. He made significant and deep contributions to different mathematical fields: function spaces, theory of capacities and nonlinear potential theory, elliptic boundary value problems, theory of multipliers, asymptotic analysis, surface waves, iterative procedures and, last but not least, boundary integral equations (see e.g. [1,8]). Hence, the minisymposium "Mathematical Aspects of Boundary Element Methods" during the IABEM 98 conference in Paris was dedicated to Vladimir Maz'ya on the occasion of his 60th birthday and dealt with constructive and numerical methods for boundary integral equations which to a great extent are based on his work.

Vladimir Maz'ya's work began with the analysis of classical boundary integral equations on nonsmooth boundaries extending J. Radon's fundamental functional analysis of boundary potentials operating on continuous boundary functions from two to several dimensions and proving with Yu. Burago and V. Sapozhnikova that the essential norm of the double layer potential operator generated by the maximum norm equals the quotient of the total variation of the local solid angle and the halfsphere surface, a result which also is due to J. Král. With G. Kresin, Vladimir Maz'ya then generalized these results to hydrodynamic and elastic potentials and discovered that even for polyhedral boundaries, the Fredholm radius can be strictly larger than the inverse of the essential norm generated by the maximum norm. D. Medkova and J. Král show in their contribution to this volume that for the harmonic double layer potential there always exists some weighted supremum norm for which the associated weighted Borel measure of the local solid angle is equal to the corresponding essential norm and also equals the inverse of the Fredholm radius. These theoretical results are decisive for stability and convergence of the classical panel method on nonsmooth boundaries which is one of the "working horses" in potential flow used for airplane and ship design.

Near to corners and edges solutions to elliptic partial differential equations might become singular but admit asymptotic expansions defining, e.g., intensity factors in the stress field corresponding to linear elasticity. V. Maz'ya with B.A. Plamenevskii and recently with V. Kozlov and J. Rossmann developed the corresponding analysis extending V.A. Kondratiev's approach. With S. Nazarov and B.A. Plamenevskii, the asymptotics of solutions was combined with singular pertubation analysis imposed by changing geometry. During the last decade, V. Maz'ya with V. Kozlov extended decisively functional calculus related to operator–valued coefficients differential equations.

The explicit Maz'ya–Plamenevskii formulae for stress intensity factors are nowadays indispensable in crack analysis, related problems of mathematical elasticity and numerical methods. V. Maz'ya and his coworkers brought the theory of boundary integral equations on surfaces with polyhedral angles, edges and cusps to a rather complete level so that computational methods can take great advantage of the detailed knowledge of the solutions's behaviour at the geometrical discontinuities.

In his early work in the sixties, V. Maz'ya applied boundary integral equations also to oblique derivative problems in higher dimensions where the directional field on the boundary might become tangential on lower–dimensional boundary parts. Then the pseudodifferential operator of the integral equation is not elliptic anymore. Here V. Maz'ya with B. Paneyah first unified the theory for the case of so–called transversal degeneration and then V. Maz'ya obtained in 1972 the most general result yet known for general generic degeneration.

In most of the singular boundary integral equations, the principal part of the operators is convolutional; and corresponding analysis in Sobolev spaces requires detailed study of the associated multipliers. Together with his wife Tatyana Shaposhnikova, Vladimir Maz'ya developed a new theory of multipliers in Sobolev spaces which allows under weakest assumptions to find essential norms, compactness properties, smoothing of commutators (see [4]). Recently their wonderful book on Jaques Hadamard [11] came out — describing his personality, his life and his mathematical creations — J. Hadamard is one of the great forefathers of boundary integral equations!

Besides the overwhelming results on basic theory, V. Maz'ya also contributed substantially to problems in mechanics of continua such as submerged bodies in fluids, hydrodynamic surface waves, flows around irregularly bounded obstacles, the Cosserat spectrum in elasticity, thermoelastic problems, the Babuška paradox. He presented an algorithm for solving the Lamé equations with data on some part of the boundary. With G. Schmidt he developed an extremely efficient method for evaluating Newton potentials by combining wavelike decompositions with special basis functions in the so–called method of approximate approximations.

We have met "Volodja and Tanya" at several conferences and we have enjoyed their hospitality and warmness at their homes in Leningrad and Linköping. We are gratefully and deeply impressed by the inexhaustible energy of Vladimir, by his profoundly rich mathematical knowledge and by his incredible creativity. He was and is a promoter for numerous young mathematicians and is very stimulating for the scientific community! On the occasion of his 60th birthday we wish to express our heartfelt thanks to him for his contributions to mathematics and to our lives and we wish him plenty of new mathematical creations and a happy and long life.

The Editors

Books by Vladimir Maz'ya:

[1] Potential Theory and Function Theory for Irregular Regions (with Yu. Burago), Seminars in Mathematics, Steklov Institute, Leningrad, Vol. 3, Consultants Bureau, New York, 1969 (Russian Version 1967).

[2] Einbettungssätze für Sobolevsche Räume, I.; II., Teubner Verlag Leipzig 1979; 1980.

[3] Zur Theorie Sobolevscher Räume, Teubner Verlag Leipzig 1981.

[4] Abschätzungen für Differentialoperatoren im Halbraum (with I. Gelman), Akademie Verlag Berlin 1981 and Birkhäuser Basel 1982.

[5] Sobolev Spaces, Springer–Verlag Berlin 1985 (Russian: Leningrad Univ. Press 1985).

[6] Theory of Multipliers in Spaces of Differentiable Functions (with T. Shaposhnikova), Pitman, London 1985 (Russian: Leningrad Univ. Press 1986).

[7] Elliptic Boundary Value Problems (with N. Morozov, B. Plamenevskii, L. Stupyalis), American Mathematical Society Translations, AMS Providence 1984.

[8] I. Linear Integral Equations, by S. Prössdorf; II. Boundary Integral Equations, by V. Maz'ya in Analysis IV, Vol. 27 of the Encyclopedia of Mathematical Sciences (eds. V. Maz'ya, S.H. Nikolski) Springer–Verlag Berlin 1991.

[9] Asymptotische Theorie elliptischer Randwertaufgaben in singulär gestörten Gebieten, I;II (with S. Nazarov and B. Plamenevskii), Akademie Verlag Berlin 1991; 1992.

[10] Elliptic Boundary Value Problems in Domains with Point Singularities (with V. Kozlov and J. Rossman), American Mathematical Society, Providence 1997.

[11] Theory of a Higher–Order Sturm–Liouville Equation (with V. Kozlov), Springer–Verlag Berlin, Lecture Notes in Mathematics, 1997.

[12] Differentiable Functions on Bad Domains (with S. Poborchi), World Scientific Singapore 1997.

[13] Jacques Hadamard, a Universal Mathematician (with T. Shaposhnikova), American Mathematical Society and London Mathematical Society, 1998.

[14] Asymptotic Analysis of Fields in Multistructures (with V. Kozlov and A. Movchan), to appear at Oxford University Press 1999.

[15] Differential Equations with Operator Coefficients (with V. Kozlov), to appear at Springer–Verlag Berlin 1999.

[16] Conical Singularities of Solutions to Elliptic Equations (with V. Kozlov and and J. Rossman), submitted to the American Mathematical Society.

[17] Linear Time–Harmonic Water Waves. A Mathematical Approach (with N. Kuznetsov and B. Vainberg), to appear at Cambridge University Press.

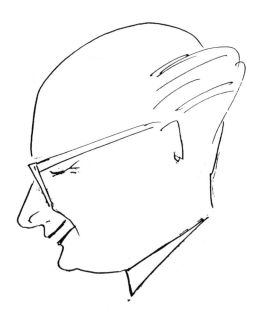

Vladimir Maz'ya's spirit floating over the auditorium

Contents

Preface

Coupling Integral Equations Method and Finite Volume Elements for the Resolution of the Leontovich Boundary Value Problem for the Time-Harmonic Maxwell Equations in Three Dimensional Heteregeneous Media
H. AMMARI and J.-C. NEDELEC 11

Smoothness Properties of Solutions to Variational Inequalities Describing Propagation of Mode-1 Cracks
M. BACH and S.A. NAZAROV 23

Edge Singularities and Kutta Condition for 3D Unsteady Flows in Aerodynamics
P. BASSANINI, C.M. CASCIOLA, M.R. LANCIA and R. PIVA 33

Approximation using Diagonal-Plus-Skeleton Matrices
M. BEBENDORF, S. RJASANOW and E. E. TYRTYSHNIKOV 45

Variational integral formulation in the problem of elastic scattering by a buried obstacle
M. BEN TAHAR, C. GRANAT and T. HA-DUONG 53

Sensitivity Analysis for Elastic Fields in Non Smooth Domains
M. BOCHNIAK and A.-M. SÄNDIG 66

A Formulation for Crack Shape Sensitivity Analysis Based on Galerkin BIE, Domain Differentiation and Adjoint Variable
M. BONNET 78

Periodic and Stochastic BEM for Large Structures Embedded in an Elastic Half-Space
D. CLOUTEAU, D. AUBRY, M. L. ELHABRE and E. SAVIN 91

Self-regularized hypersingular BEM for Laplace's equation
T. A. CRUSE and J. D. RICHARDSON 103

An Adaptive Boundary Element Method for Contact Problems
C. ECK and W.L. WENDLAND 116

Fast Summation Methods and Integral Equations
Y. FU, J. R. OVERFELT and G. J. RODIN 128

Hybrid Galerkin Boundary Elements on Degenerate Meshes
I.G. GRAHAM, W. HACKBUSCH and S.A. SAUTER 140

The Poincaré-Steklov Operator within Countably Normed Spaces
N. HEUER and E. P. STEPHAN 152

Boundary Layer Approximate Approximations for the Cubature of Potentials
T. IVANOV, V. MAZ'YA and G. SCHMIDT 165

A Simplified Approach to the Semi-discrete Galerkin Method for the Single-layer Equation for a Plate
D. MAUERSBERGER and I.H. SLOAN 178

Construction of Basis Functions for High Order Approximate Approximations
V. MAZ'YA and G. SCHMIDT 191

L_p–Theory of Direct Boundary Integral Equations on a Contour with Peak
V. MAZ'YA and A. SOLOVIEV 203

Essential Norms of the Integral Operator Corresponding to the Neumann Problem for the Laplace Equation
D. MEDKOVÁ and J. KRÁL 215

Polynomial Collocation Methods for 1D Integral Equations with Nonsmooth Solutions
G. MONEGATO and L. SCUDERI 227

Singularities in Discretized BIEs for Laplace's Equation; Trailing-Edge Conditions in Aerodynamics
L. MORINO and G. BERNARDINI 240

Fluid-Structure Interaction Problems
D. NATROSHVILI, A-.M.SÄNDIG and W.L.WENDLAND 252

Extraction, Higher Order Boundary Element Methods and Adaptivity
H. SCHULZ, Ch. SCHWAB and W.L. WENDLAND 263

Asymptotic Solution of Boundary Integral Equations
A. SELLIER 275

Sobolev Multipliers in the Theory of Integral Convolution Operators
T. SHAPOSHNIKOVA 285

Stable Boundary Element Approximations of Steklov–Poincaré Operators
O. STEINBACH 296

Preface

This volume, dedicated to Vladimir Maz'ya on the occasion of his 60th birthday, contains 25 papers based on lectures given at the two–days Minisymposium *Mathematical Aspects of Boundary Element Methods* of the IABEM 98 conference in Palaiseau, Ecole Polytechnique, 25 - 29 May 1998.

The International Association for Boundary Element Methods, in short IABEM, organizes international conferences about every 2 years in different countries and aims to bring together engineers, mathematicians and practitioners. The IABEM 98 in Paris was attended by about 110 participants from 24 countries and all continents. The contributions are devoted to computational techniques, mathematical analyses as well as engineering applications. The latter involve fluid flow, heat transfer, elasticity, non-homogeneous media, nonlinear solid mechanics, plate theory, wave propagation, sensitivity analysis, fracture mechanics, geomechanics and acoustics. The mathematical and numerical issues for the treatment of the resulting integral equations have reflected new results as such on nonsmooth boundaries, and new methods such as domain decomposition and parallelization, preconditioned iterative techniques, multipole expansions, higher order boundary elements and approximate approximations. Of course, having Vladimir Maz'ya as a participant, most of the stimulating contributions and discussions were due to him.

Within the minisymposium most of the papers are devoted to the mathematical and numerical analysis of boundary integral operators which correspond to different boundary, transmission, contact and crack problems in solid and fluid mechanics. Almost all presentations of the minisymposium and a few additional papers of the general conference are presented here.

Marc Bonnet	Anna-Margarete Sändig	Wolfgang L. Wendland
Palaiseau, France	*Stuttgart, Germany*	*Stuttgart, Germany*

Acknowledgements: The editors want to thank kindly J. Elschner and Tatyana Shaposhnikova for conveying manuscripts on V. Maz'ya's work, the pictures and the help with the dedication. They also want to thank Ingrid Bock and Gisela Wendland for collecting manuscripts and for the technical preparation of this book, Michael Bach for his valuable technical advice and the Collaborative Research Center SFB 404 "Field Interaction Problems" of the German Research Foundarion DFG at the University of Stuttgart for its support.

His first Teubner book just appeared, 1979

Speaking at the commemoration of S. Mikhlin at the Accademia dei Lincei 1994

H. AMMARI and J.-C. NEDELEC

Coupling Integral Equations Method and Finite Volume Elements for the Resolution of the Leontovich Boundary Value Problem for the Time-Harmonic Maxwell Equations in Three Dimensional Heteregeneous Media

1. Introduction

In this paper, we describe and analyze a method for computing an approximation to the time-harmonic electromagnetic field scattered by a bounded dielectric material Ω^d surrounding a lossy highly conductive body Ω^c. At high conductivity, the lossy body Ω^c can be replaced by a Leontovich boundary condition applied at its boundary Γ^c. This approximate boundary condition provides an approximate relationship between the electric and magnetic field on Γ^c. The method is to couple a finite element scheme in the dielectric medium Ω^c with an integral equation on its boundary Γ^d. This is a promising technique to compute scattering from complex objects. The finite element method is used to compute scattering from complex scatterers including several material characteristics which may be anisotropic and inhomogeneous. The integral equations account for the propagation in the unbounded space that surrounds the scatterer providing an exact boundary condition. This exact boundary condition is written directly on the surface of the scatterer, no mesh of the surrounding medium is required.

By proving that a saddle point structure holds for the continuous as well as the discretized problem, we extend the error estimates proved by Bendali [5] for the scattering problem by a conducting object to the more general problem of electromagnetics, where we also have dielectric and metallic objects, not necessary perfect conductors.

Our method is based on the use of a Hodge decomposition of the electric field given in [4] and an integral representation formula for the tangential component of the magnetic field on the boundary of the dielectric medium Γ^d which was originally developed in our paper [1].

Two different coupling formulations are presented here. They may be considered as the generalization of the ones discussed in our recent papers [2] and [3] to the treatment of a Leontovich boundary condition.

Our present work as well as [2] and [3] is related to several other works in the literature. Levillain [14] in his thesis has implemented several versions of this coupling procedure. Kirsch and Monk [12] [13] have used exact nonlocal boundary conditions on an artificial sphere to study the scattering problem by a dielectric object. De La Bourdonnaye [6] has analyzed the properties of the electromagnetic operator. We also refer the reader to Wendland [17] and Gatica and Wendland [10] for a survey of asymptotic error estimates for symmetric and nonsymmetric coupling of finite and boundary element methods and to Nédélec [16] for a recent survey of the use and numerical analysis of integral equations in electromagnetics.

2. Formulation of the problem

Let Ω^i be a bounded smooth domain in $\mathbb{R}^3$ and let Ω^e be the complement of $\overline{\Omega^i}$ in $\mathbb{R}^3$. Let $\Omega^c \subset \Omega^i$ be a lossy highly conductive body and Γ^c its boundary. $\Omega^d = \Omega^i \setminus \overline{\Omega^c}$ is the dielectric coating. By Γ^d we denote the dielectric interface between Ω^d and the exterior medium Ω^e and by ε and μ we denote the electric permittivity and the magnetic permeability in $\mathbb{R}^3 \setminus \overline{\Omega^c}$, respectively. We assume that these functions are $\mathcal{C}^2$ in $\overline{\Omega^d}$ and they are such that $\Re e\,\varepsilon > \varepsilon'$, $\Re e\,\mu > \mu'$, $\varepsilon = \varepsilon_0$, and $\mu = \mu_0$ in the exterior domain Ω^e where $\varepsilon', \mu', \varepsilon_0$ and μ_0 are positive constants. Throughout this paper, for any smooth vector field $\mathbf{u}$, we denote by $\mathbf{u}_{\Gamma^d}$ (resp. $\mathbf{u}_{\Gamma^c}$) its tangential component on Γ^d (resp. Γ^c). By div_{Γ^d} (respectively $\mathbf{grad}_{\Gamma^d}$), we denote the surface divergence of a field (respectively the tangential gradient of a function). We also denote by $\mathbf{curl}_{\Gamma^d}$ (respectively curl_{Γ^d}) the vector rotational of a function (respectively the scalar rotational of a vector), and by Δ_{Γ^d} the Laplace-Beltrami operator on Γ^d. To simplify the presentation, we use the same notation $\mathbf{n}$ to denote the outward pointing normal at Γ^c or Γ^d. The meaning of $\mathbf{n}$ should always be clear from the contexts. We shall also use the following notations for the usual functional spaces: $\mathrm{L}^2(\Omega^d)$ is the space of complex square integrable functions defined in Ω^d, $\mathrm{H}^s(\Gamma^d)$ the usual Sobolev space of order s of functions defined on Γ^d, $\mathrm{H}^s(\Omega^d)$ the usual Sobolev space of order s of functions defined in Ω^d,

$$
\begin{aligned}
\mathrm{H}(\mathrm{curl}, \Omega^d) &= \left\{ \mathbf{u} \in (\mathrm{L}^2(\Omega^d))^3, \mathbf{curl}\,\mathbf{u} \in (\mathrm{L}^2(\Omega^d))^3 \right\}, \\
\mathrm{TH}^s(\Gamma^d) &= \left\{ \mathbf{u} \in (\mathrm{H}^s(\Gamma^d))^3, \mathbf{u}.\mathbf{n} = 0 \right\}, \\
\mathrm{TL}^2(\Gamma^c) &= \left\{ \mathbf{u} \in (\mathrm{L}^2(\Gamma^c))^3, \mathbf{u}.\mathbf{n} = 0 \right\}, \\
\mathrm{TH}^s(\mathrm{div}, \Gamma^d) &= \left\{ \mathbf{u} \in \mathrm{TH}^s(\Gamma^d), \mathrm{div}_{\Gamma^d}\,\mathbf{u} \in \mathrm{H}^s(\Gamma^d) \right\}, \\
\mathrm{TH}^s(\mathrm{curl}, \Gamma^d) &= \left\{ \mathbf{u} \in \mathrm{TH}^s(\Gamma^d), \mathrm{curl}_{\Gamma^d}\,\mathbf{u} \in \mathrm{H}^s(\Gamma^d) \right\}, \\
\mathbb{H}(\Omega^d) &= \left\{ \mathbf{u} \in \mathrm{H}(\mathrm{curl}, \Omega^d), \mathbf{u}_{\Gamma^c} \in \mathrm{TL}^2(\Gamma^c) \right\}, \\
\mathbb{P}(\Omega^d) &= \left\{ p \in \mathrm{H}^1(\Omega^d), p|_{\Gamma^c} \in \mathrm{H}^1(\Gamma^c) \right\}, \\
\mathbb{M}(\Omega^d) &= \left\{ \mathbf{u} \in (\mathrm{H}^1(\Omega^d))^3, \mathrm{div}\,\mu\mathbf{u} = 0 \text{ in } \Omega^d, \mathbf{u}.\mathbf{n} = 0 \text{ on } \Gamma^d, \underset{\Gamma^c}{\mathrm{div}}\,\mathbf{u}_{\Gamma^c} = 0 \right\}.
\end{aligned}
$$

Finally, by

$$\Phi(\mathbf{x} - \mathbf{y}) = e^{i\omega\sqrt{\varepsilon_0 \mu_0}|\mathbf{x}-\mathbf{y}|}/4\pi|\mathbf{x}-\mathbf{y}|,$$

we denote the outgoing fundamental solution of the Helmholtz equation.

Our main purpose in this paper is to analyze symmetric and nonsymmetric coupling BEM/FEM procedures for solving the heteregeneous Maxwell equations in $\mathbb{R}^3 \setminus \overline{\Omega^c}$ with a Leontovich boundary condition on Γ^c. Some of the material presented here was already discussed in [2], [3] and [4]. We will refer to these papers for complete proofs. However, details of the results presented here for the first time are given.

3. The continuous problem

We shall analyze a numerical method coupling BEM/FEM for solving the heterogeneous Maxwell equations:

$$\begin{aligned}
\mathbf{curl}\,\mathbf{E} &= i\omega\mu\,\mathbf{H}, \\
\mathbf{curl}\,\mathbf{H} &= -i\omega\varepsilon\,\mathbf{E} \quad \text{in } \mathbb{R}^3 \setminus \overline{\Omega^c}, \\
\mathbf{E}_{\Gamma^c} &+ i\lambda\,\mathbf{H}\wedge\mathbf{n} = 0 \text{ on } \Gamma^c, \\
\lim_{|\mathbf{x}|\to+\infty} &\left(\sqrt{\mu_0}\,(\mathbf{H}-\mathbf{H}^{in})\wedge\frac{\mathbf{x}}{|\mathbf{x}|} - \sqrt{\varepsilon_0}\,(\mathbf{E}-\mathbf{E}^{in})\right) = 0,
\end{aligned} \tag{1}$$

where $(\mathbf{E}^{in},\mathbf{H}^{in})$ is an incident plane wave with frequency ω and λ is the impedance of the lossy body Ω^c. We assume that λ is a smooth function and it satisfies $\Im m(\lambda)\leq 0$ and $\Re e(\lambda) < \lambda_0 < 0$.

Results on existence and uniqueness of a solution of the boundary value problem (1) are established in [4]. Here our aim is to present coupling integral equations method and finite volume elements for the resolution of this boundary value problem. This provides a quite new proof of the existence of a solution to (1).

In order to obtain symmetric and nonsymmetric coupling formulations, we will follow the same procedure as for the derivation of a nonsymmetrical formulation in our previous paper [2].

Lemma 1. *a) The space $\mathbb{H}(\Omega^d)$ is the direct sum of $\mathbb{M}(\Omega^d)$ and $\mathbf{grad}\,\mathbb{P}(\Omega^d)$:*

$$\mathbb{H}(\Omega^d) = \mathbb{M}(\Omega^d) \oplus \mathbf{grad}\,\mathbb{P}(\Omega^d).$$

b) The Hodge decomposition:

$$\mathrm{TH}^{-1/2}(\mathrm{div},\Gamma^d) = \mathbf{grad}_{\Gamma^d}\,\mathrm{H}^{3/2}(\Gamma^d)/C \oplus \mathbf{curl}_{\Gamma^d}\,\mathrm{H}^{1/2}(\Gamma^d)/C,$$

is orthogonal with respect to the duality product between $\mathrm{TH}^{1/2}(\Gamma^d)$ and $\mathrm{TH}^{-1/2}(\Gamma^d)$.

Proof. Decomposition a) was proved in [4]. Here we give a slightly different proof. Since $\mathbf{H}_{\Gamma^c} \in \mathrm{TL}^2(\Gamma^c) \cap \mathrm{TH}^{-1/2}(\mathrm{curl},\Gamma^c)$ there exists a tangential field $\mathbf{a}_{\Gamma^c} \in \mathrm{TH}^{1/2}(\Gamma^c)$ such that $\mathrm{div}_{\Gamma^c}\,\lambda\,\mathbf{a}_{\Gamma^c} = 0$ and $\mathbf{curl}_{\Gamma^c}\,\mathbf{a}_{\Gamma^c} = \mathbf{curl}_{\Gamma^c}\,\mathbf{H}_{\Gamma^c} \in \mathrm{TH}^{-1/2}(\Gamma^c)$. Furthermore, there exists a unique function $f \in \mathrm{H}^1(\Omega^d)$ such that

$$\operatorname*{div}_{\Gamma^c}\lambda\,\mathbf{grad}_{\Gamma^c}f = \operatorname*{div}_{\Gamma^c}\lambda\,\mathbf{H}_{\Gamma^c} \in \mathrm{H}^{-1}(\Gamma^c).$$

Let $\mathbf{u} \in \mathrm{H}^1(\Omega^d)$ be defined by

$$\begin{cases}
\mathbf{curl}\,\mathbf{u} = \mathbf{curl}\,\mathbf{H} \in (L^2(\Omega^d))^3 & \text{in } \Omega^d \\
\mathrm{div}\,\mu\,\mathbf{u} = 0 & \text{in } \Omega^d, \\
\mathbf{u}\cdot\mathbf{n} = 0 & \text{on } \Gamma^d, \\
\mathbf{u}_{\Gamma^c} = \mathbf{a}_{\Gamma^c} \in \mathrm{TH}^{1/2}(\Gamma^c) & \text{on } \Gamma^c.
\end{cases} \tag{2}$$

There exists a unique $p \in \mathbb{P}(\Omega^d)$ such that
$$\begin{cases} \operatorname{div} \mu \operatorname{\mathbf{grad}} p = \operatorname{div} \mu \mathbf{H} & \text{in } \Omega^d, \\ p = \Delta_{\Gamma^d}^{-1}(\operatorname*{div}_{\Gamma^d} \mathbf{H}_{\Gamma^d} - \operatorname*{div}_{\Gamma^d} \mathbf{u}_{\Gamma^d}) \in H^{1/2}(\Gamma^d) & \text{on } \Gamma^d, \\ p = f \in H^1(\Gamma^c) & \text{on } \Gamma^c. \end{cases} \quad (3)$$

It is easy to see that (2) admits only one solution. Indeed, $\mathbf{H} - \mathbf{u} = \operatorname{\mathbf{grad}} p$ where p satisfies (3). Since the boundary value problem (3) is uniquely solvable in $\mathbb{P}(\Omega^d)$ the assertion a) holds.

b) holds under the assumption that Γ^d is simply connected. We refer to [6] for its proof. □

Now by making use of the above *Hodge decomposition lemma* together with an *integral representation formula* for the magnetic field $\mathbf{H}$, we obtain the following result:

Theorem 1. *The following variational formulation, where the unknowns are* $\mathbf{u} \in \mathbb{IM}(\Omega^d), p \in \mathbb{P}(\Omega^d)$ *and* $\mathbf{m} = \operatorname{\mathbf{grad}}_{\Gamma^d} \varphi + \operatorname{\mathbf{curl}}_{\Gamma^d} \psi \in \mathrm{TH}^{-1/2}(\operatorname{div}, \Gamma^d)$:

$$\begin{cases} \displaystyle\int_{\Omega^d} \frac{1}{\varepsilon} \operatorname{\mathbf{curl}} \mathbf{u} . \operatorname{\mathbf{curl}} \mathbf{u}^t - \omega^2 \int_{\Omega^d} \mu \mathbf{u} . \mathbf{u}^t + i\omega \int_{\Gamma^d} \mathbf{m} . \mathbf{u}_{\Gamma^d}^t \\ \displaystyle+ \omega \int_{\Gamma^c} \lambda \mathbf{u}_{\Gamma^c} . \mathbf{u}_{\Gamma^c}^t = 0, \quad \forall \, \mathbf{u}^t \in \mathbb{IM}(\Omega^d), \end{cases} \quad (4)$$

$$\begin{cases} \displaystyle -\omega^2 \int_{\Omega^d} \mu \operatorname{\mathbf{grad}} p . \operatorname{\mathbf{grad}} p^t + i\omega \int_{\Gamma^d} \mathbf{m} . \operatorname{\mathbf{grad}}_{\Gamma^d} p^t \\ \displaystyle + \omega \int_{\Gamma^c} \lambda \operatorname{\mathbf{grad}}_{\Gamma^c} p . \operatorname{\mathbf{grad}}_{\Gamma^c} p^t = 0, \quad \forall \, p^t \in \mathbb{P}(\Omega^d), \end{cases} \quad (5)$$

$$\begin{cases} \displaystyle -\frac{i}{\omega\mu_0} \int_{\Gamma^d}\int_{\Gamma^d} \Phi \Delta_{\Gamma^d} \varphi \, \Delta_{\Gamma^d} \varphi^t + i\omega\varepsilon_0 \int_{\Gamma^d}\int_{\Gamma^d} \Phi \operatorname{\mathbf{grad}}_{\Gamma^d} \varphi . \operatorname{\mathbf{grad}}_{\Gamma^d} \varphi^t \\ \displaystyle +i\omega\varepsilon_0 \int_{\Gamma^d}\int_{\Gamma^d} \Phi \operatorname{\mathbf{curl}}_{\Gamma^d} \psi . \operatorname{\mathbf{grad}}_{\Gamma^d} \varphi^t = -\int_{\Gamma^d} \mathbf{H}^{in} . \operatorname{\mathbf{grad}}_{\Gamma^d} \varphi^t \\ \displaystyle + \int_{\Gamma^d}\int_{\Gamma^d} \operatorname{\mathbf{grad}}_{\mathbf{x}} \Phi \wedge ((\mathbf{u} + \operatorname{\mathbf{grad}} p) \wedge \mathbf{n}) . \operatorname{\mathbf{grad}}_{\Gamma^d} \varphi^t \\ \displaystyle +\frac{1}{2} \int_{\Gamma^d} (\mathbf{u}_{\Gamma^d} + \operatorname{\mathbf{grad}}_{\Gamma^d} p) . \operatorname{\mathbf{grad}}_{\Gamma^d} \varphi^t, \quad \forall \, \varphi^t \in H^{3/2}(\Gamma^d)/\mathbb{C}, \end{cases} \quad (6)$$

$$\begin{cases} \displaystyle -i\omega\varepsilon_0 \int_{\Gamma^d}\int_{\Gamma^d} \Phi \operatorname{\mathbf{curl}}_{\Gamma^d} \psi . \operatorname{\mathbf{curl}}_{\Gamma^d} \psi^t - i\omega\varepsilon_0 \int_{\Gamma^d}\int_{\Gamma^d} \Phi \operatorname{\mathbf{grad}}_{\Gamma^d} \varphi . \operatorname{\mathbf{curl}}_{\Gamma^d} \psi^t \\ \displaystyle = \int_{\Gamma^d} \mathbf{H}^{in} . \operatorname{\mathbf{curl}}_{\Gamma^d} \psi^t - \frac{1}{2}\int_{\Gamma^d} \mathbf{u}_{\Gamma^d} . \operatorname{\mathbf{curl}}_{\Gamma^d} \psi^t \\ \displaystyle -\int_{\Gamma^d}\int_{\Gamma^d} \operatorname{\mathbf{grad}}_{\mathbf{x}} \Phi \wedge ((\mathbf{u} + \operatorname{\mathbf{grad}} p) \wedge \mathbf{n}) . \operatorname{\mathbf{curl}}_{\Gamma^d} \psi^t, \quad \forall \, \psi^t \in H^{1/2}(\Gamma^d)/\mathbb{C}, \end{cases} \quad (7)$$

is such that

$$\mathbf{H} = \mathbf{u} + \operatorname{grad} p \text{ and } \mathbf{E} = \frac{i}{\omega \varepsilon} \operatorname{curl} \mathbf{u}$$

are solutions (locally in $H(\operatorname{curl}, \mathbf{R}^3 \setminus \overline{\Omega^c}) \times H(\operatorname{curl}, \mathbf{R}^3 \setminus \overline{\Omega^c})$*) of the Maxwell equations* (1) *and* $\mathbf{m} = \mathbf{E} \wedge \mathbf{n}$ *on* Γ^d *if*

$$\begin{aligned}
\frac{1}{2} \mathbf{m} \wedge \mathbf{n} &= \mathbf{E}^{in} \wedge \mathbf{n} - (\operatorname{curl} \int_{\Gamma^d} \Phi \, \mathbf{m}) \wedge \mathbf{n} \\
&\quad - \frac{i}{\omega \varepsilon_0} (\operatorname{grad} \int_{\Gamma^d} \Phi \operatorname*{div}_{\Gamma^d} ((\mathbf{u} + \operatorname{grad} p) \wedge \mathbf{n})) \wedge \mathbf{n} \qquad (8) \\
&\quad - i\omega \mu_0 (\int_{\Gamma^d} \Phi ((\mathbf{u} + \operatorname{grad} p) \wedge \mathbf{n})) \wedge \mathbf{n}.
\end{aligned}$$

Furthermore, if $\mathbf{E}$ *and* $\mathbf{H}$ *are solutions of the boundary value problem* (1) *then* $\mathbf{u}, p$ *and* $\mathbf{m}$ *satisfy* (4) − (7) *and* (8).

We refer to [2] for the proof of this theorem.

Now we show that the variational formulation (4) − (7) under the identity (8) is of Fredholm type and we derive a symmetric coupling formulation. To do so, we need the following identities, deduced from an integral representation formula on Γ^d of the magnetic field $\mathbf{H}$. From [2] the following holds:

$$\begin{aligned}
\frac{1}{2} \int_{\Gamma^d} \mathbf{m} \cdot \mathbf{u}^t_{\Gamma^d} &= -\frac{i}{\omega \varepsilon_0} \int_{\Gamma^d} \int_{\Gamma^d} \Phi \operatorname{curl}_{\Gamma^d} \mathbf{u}_{\Gamma^d} \operatorname{curl}_{\Gamma^d} \mathbf{u}^t_{\Gamma^d} \\
&\quad + i\omega \mu_0 \int_{\Gamma^d} \int_{\Gamma^d} \Phi \operatorname{curl}_{\Gamma^d} p \cdot \mathbf{u}^t_{\Gamma^d} \\
&\quad - \int_{\Gamma^d} \int_{\Gamma^d} \left[\partial_{n_x} \Phi \, \mathbf{m} - \operatorname{grad}_x \Phi \, (\mathbf{n_x} - \mathbf{n_y}) \cdot \mathbf{m} \right] \cdot \mathbf{u}^t_{\Gamma^d} \qquad (9) \\
&\quad + \int_{\Gamma^d} \mathbf{E}^{in} \wedge \mathbf{n} \cdot \mathbf{u}^t_{\Gamma^d} + i\omega \mu_0 \int_{\Gamma^d} \int_{\Gamma^d} \Phi \mathbf{u} \wedge \mathbf{n} \cdot \mathbf{u}^t_{\Gamma^d},
\end{aligned}$$

$$\begin{aligned}
\frac{1}{2} \int_{\Gamma^d} \mathbf{m} \cdot \operatorname{grad}_{\Gamma^d} p^t &= i\omega \mu_0 \int_{\Gamma^d} \int_{\Gamma^d} \Phi \operatorname{curl}_{\Gamma^d} p \cdot \operatorname{grad}_{\Gamma^d} p^t \\
&\quad - \int_{\Gamma^d} \int_{\Gamma^d} \left[\partial_{n_x} \Phi \, \mathbf{m} - \operatorname{grad}_x \Phi \, (\mathbf{n_x} - \mathbf{n_y}) \cdot \mathbf{m} \right] \cdot \operatorname{grad}_{\Gamma^d} p^t \\
&\quad + \int_{\Gamma^d} \mathbf{E}^{in} \wedge \mathbf{n} \cdot \operatorname{grad}_{\Gamma^d} p^t \\
&\quad + i\omega \mu_0 \int_{\Gamma^d} \int_{\Gamma^d} \Phi \mathbf{u} \wedge \mathbf{n} \cdot \operatorname{grad}_{\Gamma^d} p^t,
\end{aligned} \qquad (10)$$

and
$$\int_{\Gamma^d}\int_{\Gamma^d} \Phi \, \mathbf{curl}_{\Gamma^d} \, p \cdot \mathbf{grad}_{\Gamma^d} \, p^t = \int_{\Gamma^d}\int_{\Gamma^d} (\mathbf{n_x} - \mathbf{n_y}) \wedge \mathbf{grad}_y \, \Phi \cdot \mathbf{grad}_{\Gamma^d} \, p^t \, p. \quad (11)$$

Now to derive a symmetrical coupling formulation we replace, by using identity (9), the term
$$i\omega \int_{\Gamma^d} \mathbf{m} \cdot \mathbf{u}^t_{\Gamma^d} = i\omega \int_{\Gamma^d} (\mathbf{grad}_{\Gamma^d}\varphi + \mathbf{curl}_{\Gamma^d}\psi) \cdot \mathbf{u}^t_{\Gamma^d}$$
by
$$\frac{i}{2}\omega \int_{\Gamma^d} (\mathbf{grad}_{\Gamma^d}\varphi + \mathbf{curl}_{\Gamma^d}\psi) \cdot \mathbf{u}^t_{\Gamma^d} + \frac{1}{\varepsilon_0} \int_{\Gamma^d}\int_{\Gamma^d} \Phi \, \mathbf{curl}_{\Gamma^d} \, \mathbf{u}_{\Gamma^d} \, \mathbf{curl}_{\Gamma^d} \, \mathbf{u}^t_{\Gamma^d}$$
$$-\omega^2 \mu_0 \int_{\Gamma^d}\int_{\Gamma^d} \Phi \, \mathbf{curl}_{\Gamma^d} \, p \cdot \mathbf{u}^t_{\Gamma^d}$$
$$-i\omega \int_{\Gamma^d}\int_{\Gamma^d} \left[\partial_{n_x} \Phi \, \mathbf{m} - \mathbf{grad}_{\mathbf{x}}\Phi \, (\mathbf{n_x} - \mathbf{n_y}) \cdot \mathbf{m}\right] \cdot \mathbf{u}^t_{\Gamma^d}$$
$$+i\omega \int_{\Gamma^d} \mathbf{E}^{in} \wedge \mathbf{n} \cdot \mathbf{u}^t_{\Gamma^d} - \omega^2 \mu_0 \int_{\Gamma^d}\int_{\Gamma^d} \Phi \, \mathbf{u} \wedge \mathbf{n} \cdot \mathbf{u}^t_{\Gamma^d},$$

in the variational formulation (4) – (7). We do the same for the term
$$i\omega \int_{\Gamma^d} \mathbf{m} \cdot \mathbf{grad}_{\Gamma^d} \, p^t.$$

That is from identity (10)
$$i\omega \int_{\Gamma^d} \mathbf{m} \cdot \mathbf{grad}_{\Gamma^d} \, p^t = \frac{i}{2}\omega \int_{\Gamma^d} \mathbf{m} \cdot \mathbf{grad}_{\Gamma^d} \, p^t - \omega^2 \mu_0 \int_{\Gamma^d}\int_{\Gamma^d} \Phi \, \mathbf{curl}_{\Gamma^d} \, p \cdot \mathbf{grad}_{\Gamma^d} \, p^t$$
$$-i\omega \int_{\Gamma^d}\int_{\Gamma^d} \left[\partial_{n_x} \Phi \, \mathbf{m} - \mathbf{grad}_{\mathbf{x}}\Phi \, (\mathbf{n_x} - \mathbf{n_y}) \cdot \mathbf{m}\right] \cdot \mathbf{grad}_{\Gamma^d} \, p^t$$
$$+i\omega \int_{\Gamma^d} \mathbf{E}^{in} \wedge \mathbf{n} \cdot \mathbf{grad}_{\Gamma^d} \, p^t$$
$$-\omega^2 \mu_0 \int_{\Gamma^d}\int_{\Gamma^d} \Phi \, \mathbf{u} \wedge \mathbf{n} \cdot \mathbf{grad}_{\Gamma^d} \, p^t.$$

Doing this we obtain:

Theorem 2. *Assuming (8), the variational formulation (4) – (7) can be rewritten in the following form:*
$$\begin{cases} a(\mathbf{u}, \mathbf{u}^t) + \frac{i}{2}\omega \int_{\Gamma^d} (\mathbf{grad}_{\Gamma^d}\varphi + \mathbf{curl}_{\Gamma^d}\psi) \cdot \mathbf{u}^t_{\Gamma^d} + \omega \int_{\Gamma^c} \lambda \mathbf{u}_{\Gamma^c} \cdot \mathbf{u}^t_{\Gamma^c} \\ + \Big(\text{compact terms in } \mathbf{u}\Big) + \Big(\text{coupling compact terms}\Big) = \int_{\Gamma^d} \mathbf{g}^{in} \cdot \mathbf{u}^t_{\Gamma^d}, \\ \forall \, \mathbf{u}^t \in \mathrm{IM}(\Omega^d), \end{cases} \quad (12)$$

$$\begin{cases} \hat{a}(p, p^t) + \dfrac{i}{2}\omega \displaystyle\int_{\Gamma^d} \mathbf{grad}_{\Gamma^d}\varphi \cdot \mathbf{grad}_{\Gamma^d} p^t + \Big(\text{compact terms in } p\Big) \\ + \Big(\text{coupling compact terms}\Big) = \displaystyle\int_{\Gamma^d} \mathbf{g}^{in} \cdot \mathbf{grad}_{\Gamma^d} p^t, \quad \forall\, p^t \in \mathbb{P}(\Omega^d), \end{cases} \quad (13)$$

$$\begin{cases} -\dfrac{1}{\mu_0} \displaystyle\int_{\Gamma^d}\int_{\Gamma^d} \Phi_0\, \Delta_{\Gamma^d}\varphi\, \Delta_{\Gamma^d}\varphi^t + \dfrac{i}{2}\omega \displaystyle\int_{\Gamma^d} (\mathbf{u}_{\Gamma^d} + \mathbf{grad}_{\Gamma^d} p) \cdot \mathbf{grad}_{\Gamma^d}\varphi^t \\ + \Big(\text{compact terms in } \varphi\Big) + \Big(\text{coupling compact terms in } \psi\Big) \\ = i\omega \displaystyle\int_{\Gamma^d} \mathbf{H}^{in} \cdot \mathbf{grad}_{\Gamma^d} \varphi^t, \quad \forall\, \varphi^t \in H^{3/2}(\Gamma^d)/C, \end{cases} \quad (14)$$

$$\begin{cases} \omega^2 \varepsilon_0 \displaystyle\int_{\Gamma^d}\int_{\Gamma^d} \Phi_0\, \mathbf{curl}_{\Gamma^d} \psi \cdot \mathbf{curl}_{\Gamma^d} \psi^t + \dfrac{i}{2}\omega \displaystyle\int_{\Gamma^d} \mathbf{u}_{\Gamma^d} \cdot \mathbf{curl}_{\Gamma^d}\psi^t \\ + \Big(\text{compact terms in } \psi\Big) + \Big(\text{coupling compact terms}\Big) \\ = i\omega \displaystyle\int_{\Gamma^d} \mathbf{H}^{in} \cdot \mathbf{curl}_{\Gamma^d} \psi^t, \quad \forall\, \psi^t \in H^{1/2}(\Gamma^d)/C, \end{cases} \quad (15)$$

where

$$a(\mathbf{u}, \mathbf{u}^t) = \int_{\Omega^d} \frac{1}{\varepsilon} \mathbf{curl}\, \mathbf{u} \cdot \mathbf{curl}\, \mathbf{u}^t - \omega^2 \int_{\Omega^d} \mu\, \mathbf{u} \cdot \mathbf{u}^t$$
$$+ \frac{1}{\varepsilon_0} \int_{\Gamma^d}\int_{\Gamma^d} \Phi_0\, \mathbf{curl}_{\Gamma^d} \mathbf{u}_{\Gamma^d}\, \mathbf{curl}_{\Gamma^d} \mathbf{u}^t_{\Gamma^d},$$

$$\hat{a}(p, p^t) = -\omega^2 \int_{\Omega^d} \mu\, \mathbf{grad}\, p \cdot \mathbf{grad}\, p^t + \omega \int_{\Gamma^c} \lambda\, \mathbf{grad}_{\Gamma^c} p \cdot \mathbf{grad}_{\Gamma^c} p^t,$$

and

$$\mathbf{g}^{in} = -i\omega\, \mathbf{E}^{in} \wedge \mathbf{n} \quad \text{on } \Gamma^d.$$

This gives the following

Theorem 3. *The variational formulation* (12) − (15) *is of Fredholm type.*

Proof. Since the proof of this theorem is similar to that of theorem 3.2 in [2], we only discuss the details which are unlike those found in [2]. Only the terms

$$\frac{i}{2}\omega \int_{\Gamma^d} \mathbf{curl}_{\Gamma^d}\psi \cdot \mathbf{u}^t, \quad \frac{i}{2}\omega \int_{\Gamma^d} \mathbf{u}_{\Gamma^d} \cdot \mathbf{curl}_{\Gamma^d}\psi^t, \quad (16)$$

17

$$\frac{i}{2}\omega \int_{\Gamma^d} \mathbf{grad}_{\Gamma^d} p \cdot \mathbf{grad}_{\Gamma^d} \varphi^t, \quad \frac{i}{2}\omega \int_{\Gamma^d} \mathbf{grad}_{\Gamma^d}\varphi \cdot \mathbf{grad}_{\Gamma^d} p^t, \qquad (17)$$

and

$$\omega \int_{\Gamma^c} \lambda \mathbf{u}_{\Gamma^c} \cdot \mathbf{u}^t_{\Gamma^c}, \qquad (18)$$

have to be carefully studied. The treatment of the others terms follows from [2]. The term (18) is compact on $\mathbb{M}(\Omega^d) \times \mathbb{M}(\Omega^d)$. The term

$$+\omega \int_{\Gamma^c} \lambda \, \mathbf{grad}_{\Gamma^c} p \cdot \mathbf{grad}_{\Gamma^c} p^t$$

has the right sign: the bilinear form $\hat{a}(p, p^t)$ on $\mathbb{P}(\Omega^d) \times \mathbb{P}(\Omega^d)$ is coercive. But the other's new terms (16)−(17) are not compact. However, they are paired if we associate equation (12) to (15) and (13) to (14). To obtain the a priori estimates we add (12) to (15) (with $\mathbf{u}^t = \overline{\mathbf{u}}$ and $\varphi^t = \overline{\varphi}$). Taking the real part the terms (16) cancelled:

$$\Re e \left(\frac{i}{2}\omega \int_{\Gamma^d} \mathbf{curl}_{\Gamma^d}\psi \cdot \overline{\mathbf{u}_{\Gamma^d}} + \frac{i}{2}\omega \int_{\Gamma^d} \mathbf{u}_{\Gamma^d} \cdot \overline{\mathbf{curl}_{\Gamma^d}\psi} \right) = 0.$$

It is similar with (13) and (14) for the treatment of the terms (17). Doing that the coercive terms yield the desired a priori estimate as in [2]. □

Now if we replace the terms $i\omega \int_{\Gamma^d} \mathbf{m} \cdot \mathbf{u}^t_{\Gamma^d}$ and $i\omega \int_{\Gamma^d} \mathbf{m} \cdot \mathbf{grad}_{\Gamma^d} p^t$ by their expressions derived from identities (9) and (10) we arrive at a nonsymmetric coupling variational formulation. The following holds.

Theorem 4. *Assuming (8), the variational formulation (4) − (7) can be rewritten in the following form:*

$$\begin{cases} a(\mathbf{u}, \mathbf{u}^t) + \Big(\text{compact terms in } \mathbf{u}\Big) \\ \qquad + \Big(\text{coupling compact terms}\Big) = \int_{\Gamma^d} \mathbf{g}^{in} \cdot \mathbf{u}^t_{\Gamma^d}, \quad \forall\, \mathbf{u}^t \in \mathbb{M}(\Omega^d), \end{cases} \qquad (19)$$

$$\begin{cases} \hat{a}(p, p^t) + \Big(\text{compact terms in } p\Big) \\ \qquad + \Big(\text{coupling compact terms}\Big) = \int_{\Gamma^d} \mathbf{g}^{in} \cdot \mathbf{grad}_{\Gamma^d} p^t, \quad \forall\, p^t \in \mathbb{P}(\Omega^d), \end{cases} \qquad (20)$$

$$\begin{cases} -\dfrac{1}{\mu_0} \displaystyle\int_{\Gamma^d}\int_{\Gamma^d} \Phi_0\, \Delta_{\Gamma^d}\varphi\, \Delta_{\Gamma^d}\varphi^t + \Big(\text{compact terms in } \varphi\Big) \\ \qquad + \Big(\text{coupling compact terms in } \psi\Big) = i\omega \displaystyle\int_{\Gamma^d} \mathbf{H}^{in} \cdot \mathbf{grad}_{\Gamma^d}\varphi^t, \\ \forall\, \varphi^t \in H^{3/2}(\Gamma^d)/C\,, \end{cases} \qquad (21)$$

$$\begin{cases} \omega^2 \varepsilon_0 \int_{\Gamma^d} \int_{\Gamma^d} \Phi_0 \, \mathbf{curl}_{\Gamma^d} \, \psi \cdot \mathbf{curl}_{\Gamma^d} \, \psi^t + \Big(\text{compact terms in } \psi\Big) \\ + \Big(\text{coupling terms}\Big) = -i\omega \int_{\Gamma^d} \mathbf{H}^{in} \cdot \mathbf{curl}_{\Gamma^d} \, \psi^t, \quad \forall \, \psi^t \in \mathrm{H}^{1/2}(\Gamma^d)/C \,. \end{cases} \quad (22)$$

Furthermore, the nonsymmetric variational formulation (19) − (22) *is also of Fredholm type.*

It has to be noted that in (22) the coupling terms are not compact but this does not affect the Fredholm type of the nonsymmetric formulation (19) − (22).

Finally, the equivalence between (12) − (15) and the boundary value problem (1) yields the following statement.

Theorem 5. *The symmetric variational formulation* (12) − (15) *admits a unique solution in the space* $\mathcal{X} = \mathrm{IM}(\Omega^d) \times \mathrm{IP}(\Omega^d) \times \mathrm{H}^{3/2}(\Gamma^d)/C \times \mathrm{H}^{1/2}(\Gamma^d)/C$.

The same result holds for the nonsymmetric variational formulation (19) − (22).

In what follows we rewrite the variational formulation (12) − (15) (or (19) − (22)) in the more suitable form:

$$\mathcal{A}((\mathbf{u}, p, \varphi, \psi), (\mathbf{u}^t, p^t, \varphi^t, \psi^t)) = ((\mathbf{g}^{in}, \mathbf{u}^t), (g^{in}, p^t), (g^{in}, \varphi^t, \psi^t)),$$

for all $(\mathbf{u}^t, p^t, \varphi^t, \psi^t) \in \mathcal{X}$. We can prove the following inf-sup condition on $\mathcal{A}$:

Lemma 2. *There is a positive constant C such that*

$$\sup_{(\mathbf{u}^t, p^t, \varphi^t, \psi^t) \in \mathcal{X} \setminus \{0\}} \frac{|\mathcal{A}((\mathbf{u}, p, \mathbf{m}), (\mathbf{u}^t, p^t, \varphi^t, \psi^t))|}{|||(\mathbf{u}^t, p^t, \varphi^t, \psi^t)|||} \geq C \, |||(\mathbf{u}, p, \varphi, \psi)|||, \quad (23)$$

for all $(\mathbf{u}, p, \varphi, \psi) \in \mathcal{X}$ *where*

$$|||(\mathbf{u}, p, \varphi, \psi)|||^2 = ||\mathbf{u}||^2_{\mathrm{IM}(\Omega^d)} + ||p||^2_{\mathrm{IP}(\Omega^d)} + ||\varphi||^2_{\mathrm{H}^{3/2}(\Gamma^d)/C} + ||\psi||^2_{\mathrm{H}^{1/2}(\Gamma^d)/C} \,.$$

4. The discrete problem

We shall discretize (12) − (15) (or (19) − (22)) using a family of finite element subspaces $\mathbb{H}_h$ in $\mathbb{H}(\Omega^d)$ where the parameter h measures the maximum diameter of the elements in the associated finite element mesh. We shall assume that the family $\mathbb{H}_h$ satisfies the Hodge decomposition given in Lemma 1. Any vector $\mathbf{H}_h \in \mathbb{H}_h$ can be written as $\mathbf{H}_h = \mathbf{u} + \mathbf{grad}\, p$, where $\mathbf{u} \in \mathrm{IM}(\Omega^d)$ and $p \in \mathrm{IP}(\Omega^d)$. We also require that

$$\frac{1}{||\mathbf{H}_h||_{\mathrm{H}(\mathrm{curl}, \Omega^d)}} ||R_h \mathbf{u} - \mathbf{u}||_{(L^2(\Omega^d))^3} \to 0, \quad h \to 0,$$

where $R_h : (\mathrm{H}^2(\Omega^d))^3 \mapsto \mathbb{H}_h$ is an interpolation operator. The family used to discretize the vector unknown $\mathbf{u}$ is then the projection of $\mathbb{H}_h$ on $\mathrm{IM}(\Omega^d)$. We may use $\mathbb{H}_h = P_1$ Lagrange finite element since $\mathbf{grad}\, \mathrm{H}^2(\Omega^d)$ is dense in $\mathrm{IP}(\Omega^d)$ (even for Ω^d a nonconvex polyhedra).

Natural approximations for φ and ψ are
- P_1 Lagrange finite element approximation for ψ and then $\mathbf{curl}_{\Gamma^d}\psi$ is the space of Raviart-Thomas;
- any $\mathcal{C}^1$ finite element for φ.

In fact, we only need $\mathbf{m}$ and $\mathrm{div}_{\Gamma^d}\,\mathbf{m} = \Delta_{\Gamma^d}\,\varphi$.

Now, let $\mathcal{X}_h \subset \mathcal{X}$ be the dicretized subspace associated to h. We have the following basic Babuska-Brezzi condition.

Lemma 3. *There is a positive constant C independent of h such that*

$$\sup_{(\mathbf{u}_h^t,p_h^t,\varphi_h^t,\psi_h^t)\in \mathcal{X}_h\setminus\{0\}} \frac{|\mathcal{A}((\mathbf{u}_h,p_h,\varphi_h,\psi_h),(\mathbf{u}_h^t,p_h^t,\varphi_h^t,\psi_h^t))|}{|||(\mathbf{u}_h^t,p_h^t,\varphi_h^t,\psi_h^t)|||} \geq$$
$$C\,|||(\mathbf{u}_h,p_h,\varphi_h,\psi_h)|||, \quad \forall\,(\mathbf{u}_h,p_h,\varphi_h,\psi_h) \in \mathcal{X}_h.$$

This lemma yields to the desired result. We have the following theorem.

Theorem 6. *There exists $h_0 > 0$ such that the discret solution*

$$\mathbf{H}_h = \mathbf{u}_h + \mathrm{grad}\,p_h$$

is well defined provided $0 < h < h_0$ and $\mathbf{H}_h$ satisfies the error estimate

$$\|\mathbf{H} - \mathbf{H}_h\|_{\mathrm{H}(\mathrm{curl},\Omega^d)} \leq C \inf_{\mathbf{F}_h \in \mathcal{X}_h} \|\mathbf{H} - \mathbf{F}_h\|_{\mathrm{H}(\mathrm{curl},\Omega^d)},$$

where C is a positive constant independent of h.

Finally, if $\mathbf{H} \in (\mathrm{H}^2(\Omega^d))^3$ (this holds provided $\varepsilon \in \mathcal{C}^2(\overline{\Omega^d}), \mu \in \mathcal{C}^2(\overline{\Omega^d}), \lambda \in \mathcal{C}^2(\Gamma^c)$ and the boundaries Γ^d and Γ^c are of class $\mathcal{C}^3$), we can prove the following result.

Theorem 7. *There exist $h_0 > 0$ and $C > 0$ such that for $0 < h < h_0$, the following optimal error estimate holds*

$$\|\mathbf{H} - \mathbf{H}_h\|_{\mathrm{H}(\mathrm{curl},\Omega^d)} \leq C\,h\,\|\mathbf{H}\|_{(\mathrm{H}^2(\Omega^d))^3}.$$

5. Conclusion

We have analyzed a symmetric and a nonsymmetric method of computing an approximation to the time-harmonic electromagnetic fields by a dielectric material surrounding a highly conducting body. The conducting body is replaced by a Leontovich boundary condition applied at its boundary. Our method is to couple a finite element scheme on the dielectric material with a boundary element method on the interface between the dielectric material and the exterior medium. Our main result is that these proposed numerical methods attain unique solutions with optimal approximation properties. The method is very general, particularly, it can be extended to treat the generalized impedance boundary conditions used to approximate the effect of thin dielectric layers on electromagnetic scattering. An interesting future of the method is to decouple the finite element and integral equations solutions by using an iterative procedure. The integral equations could be solved by a fast technique (the fast mutipole methods) independently of the finite element solutions, making the integral equations part of the code negligible in terms of time.

References

[1] H. Ammari, J.-C. Nédélec: Time-harmonic electromagnetic fields in thin chiral curved layers, *SIAM J. Math. Anal.* **29 (2)** (1998), 395-423.

[2] H. Ammari, J.-C. Nédélec: Couplage éléments finis équations intégrales pour la résolution des équations de Maxwell en milieu hétérogène. In: *Equations aux Dérivées Partielles et Applications. Articles dédiés à Jacques-Louis Lions*, Elsevier, 1998.

[3] H. Ammari, J.-C. Nédélec: *Coupling of Finite and Boundary Element Methods for the Time-harmonic Maxwell Equations. Part II: A Symmetric Formulation*, Preprint.

[4] H. Ammari, C. Latiri-Grouz, J.-C. Nédélec: Scattering of Maxwell's equations with a Leontovich boundary condition in an inhomogeneous medium: A singular perturbation problem, *SIAM J. Appl. Math.* **59**, to be published (1999).

[5] A. Bendali: Numerical analysis of the exterior boundary value problem for the time-harmonic Maxwell equations by a boundary finite element, Part I: The continuous problem, *Math. Comp.* **43** (1984), 29-46.

[6] A. de La Bourdonnaye: Some formulations coupling finite element and integral equation method for Helmholtz equation and electromagnetism, *Numer. Math.* **69** (1995), 257-268.

[7] D. Colton, R. Kress: Time harmonic electromagnetic waves in inhomogeneous medium, *Proc. Royal Soc. Edinburgh* **116 A** (1990), 279-293.

[8] M. Costabe, E. P. Stephan: Strongly elliptic boundary integral equations for electromagnetic transmission problems, *Proc. Royal Soc. Edinburgh* **109 A** (1988), 271-296.

[9] M. Costabel, E. P. Stephan: Coupling of finite and boundary element methods for an elastoplastic interface problem, *SIAM J. Numer. Anal.*, **27** (1990), 1212-1226.

[10] G. N. Gatica, W. L. Wendland: Coupling of mixed finite elements and boundary elements for linear and nonlinear elliptic problems, *Appl. Anal.* **63** (1996), 39-75.

[11] C. Johnson, J.-C. Nédélec: On the coupling of boundary integral and finite element methods, *Math. Comp.* **35** (1980), 1063-1079.

[12] A. Kirsch, P. Monk: A Finite element /spectral method for approximating the time-harmonic Maxwell system in R^3, *SIAM J. Appl. Math.* **55** (1995), 1324-1344.

[13] A. Kirsch, P. Monk: Corrigendum to "A Finite element /spectral method for approximating the time-harmonic Maxwell system in R^3, *SIAM J. Appl. Math.* **55** (1995), 1324-1344," *SIAM J. Appl. Math.* **58** (1998), 2024-2028.

[14] V. Levillain: Ph.D. Thesis, Ecole Polytechnique, 1991.

[15] J.-C. Nédélec: A new family of mixed finite elements in $\mathbf{R}^3$, *Numer. Math.* **50** (1986), 57-81.

[16] J.-C. Nédélec: New trends in the use and analysis of integral equations, *Proc. Symp. Appl. Math.* **48** (1994), 151-176.

[17] W. L. Wendland: On asymptotic error estimates for combined BEM and FEM, in finite element and boundary element techniques from mathematical and engineering point of view, 273-333, *CISM Courses and Lectures*, 301, Springer, Viennna, 1988.

H. Ammari and J.-C. Nédélec
Centre de Mathématiques Appliquées
UMR CNRS 7641, Ecole Polytechnique
91128 Palaiseau Cedex, France.
Email: ammari@polytechnique.fr

M. BACH and S.A. NAZAROV

Smoothness Properties of Solutions to Variational Inequalities Describing Propagation of Mode-1 Cracks

1. Introduction

Let G be a domain in the plane bounded by a simple closed smooth contour Γ; besides, we avoid to distinguish between 2-D sets on $\partial \mathbb{R}^3_+$ and their immersions into $\mathbb{R}^3$. We consider the elasticity problem in the elastic isotropic space $\mathbb{R}^3$ containing the plane crack G. We assume that there are no mass forces and that the symmetric normal loading P applied to the crack surfaces $G^\pm$ opens the crack and therefore the stress intensity factor (SIF) K_1 is a positive function on the crack front Γ (i.e. we consider the crack of a normal rupture with mode–1 stress–strain state). Note that $K_1 \in C^\infty(\Gamma)$ if the function P is smooth in the vicinity of Γ. Thus the displacement field $u = (u_1, u_2, u_3)$ should satisfy the homogeneous Lamé system, vanish at infinity, and the components of the stress tensor $\sigma(u)$ should satisfy the conditions

$$\sigma_{13}(u, y, \pm 0) = 0, \quad \sigma_{23}(u, y, \pm 0) = 0, \quad \sigma_{33}(u, y, \pm 0) = \mp P(y) \qquad y \in G.$$

Due to the Papkovich–Neuber representation (see, e.g. [9])

$$u_j = v_j + x_3 \frac{\partial v_0}{\partial x_j} \qquad j = 1 \ldots 3$$

($v_j, j = 0 \ldots 3$ are harmonic functions), it is possible to reduce the 3–D elasticity problem in $\mathbb{R}^3 \setminus \overline{G}$ to the following scalar problem in the half–space:

Find the harmonic function v in $\mathbb{R}^3_+ = \{x = (y, z) : z > 0\}$ which vanishes at infinity and satisfies homogeneous Dirichlet conditions on $\mathbb{R}^2_+ \setminus \overline{G}$ and inhomogeneous Neumann conditions on G:

$$\begin{aligned}
\Delta v(y, z) &= 0 & (y, z) &\in \mathbb{R}^3_+, \\
v(y, 0) &= 0, & y &\in \mathbb{R}^2_+ \setminus \overline{G}, \\
-\partial_z v(y, 0) &= p(y) = \alpha^{-1} P(y), & y &\in G.
\end{aligned} \qquad (1)$$

Here $\alpha = (1 - \nu)^{-1} \mu$, where μ stands for the shear modulus and ν for the Poisson ratio. The harmonic functions $v_j, j = 0 \ldots 2$, can be calculated from v_3 as

$$\begin{aligned}
v_0 &= -\frac{\lambda + \mu}{\lambda + 2\mu} v_3, \\
\frac{\partial v_j}{\partial x_3} &= -\frac{\lambda + \mu}{\lambda + 2\mu} \frac{\partial v_3}{\partial x_j} & j = 1, 2.
\end{aligned}$$

This work was partially supported by the INTAS - Association, INTAS project 96-876 on 'Mathematical Theory of Cracks and their Propagation'.

Since the stress σ_{33} coincides on $\partial \mathbb{R}^3_+$ with $a\partial_z v$, one has the local representation

$$a\partial_z v(y,0) = (2\pi r)^{\frac{1}{2}}\big(K_1(s) + r k_1(s)\big) - \alpha p_\Gamma(s) + \mathcal{O}(r^2), \quad y \in \mathbb{R}^2_+ \setminus \overline{G},$$

where $k_1(s)$ denotes the 'junior' SIF, i.e., the factor at the low-order singularity; p_Γ implies the trace of p on $\Gamma = \partial G$, s is the arc–length on Γ and (r,φ) are the polar coordinates in the planes perpendicular to Γ ($r = \text{dist}\{x,\Gamma\}$ and $\varphi \in (-\pi,\pi)$).

2. Rupture criteria

Let P be the critical loading, i.e. a positive increment of the loading leads to the propagation of the crack. The aim is to find the shape of the growing crack $G(t)$ with the front

$$\Gamma(t) = \{y \in \partial \mathbb{R}^3_+ \cap U : r = t\,h(s)\},$$

where $t\,h(s)$ describes the depth of the propagation along the normal to Γ at the point s and $t \geq 0$ is a fixed time–like parameter (supposed to be small) describing the increment of the loading

$$P(t;y) = P(y) + tP'(y). \tag{2}$$

In other words, $\Gamma(t)$ is regarded as a perturbation of the initial front $\Gamma := \Gamma(0)$ and, as $t \to +0$, the asymptotics of the SIF $K_1(t;s)$ on $\Gamma(t)$ for the loading $P(t;y)$ must be found.

2.1. Irwin criterion

The Irwin criterion is a stress criterion and it is based on the comparison of the SIF K_1 with the critical SIF K_{1c} (a material parameter). First of all, due to the irreversibility of the fracture process we observe that

$$h(s) \geq 0 \quad \forall s \in \Gamma.$$

In the case of 'stretching' loading P (i.e. $P(y) \geq 0, y \in G$) the maximum principle furnishes the inequality $K_1 \geq 0$ on Γ. We assume K_1 to be positive. Denoting by $K_1(t;s)$ the SIF at $\Gamma(t)$ for the loading $P(t;y)$ on the crack surfaces $G^\pm(t)$, and assuming the quasistatic propagation of the crack, i.e. neglecting dynamical effects, we apply the Irwin criterion which leads to the relations

$$\begin{aligned} h(s) = 0 &\implies K_1(t;s) \leq K_{1c}, \\ h(s) > 0 &\implies K_1(t;s) = K_{1c}. \end{aligned} \tag{3}$$

These relations mean that the crack $G(t)$ appears to be in equilibrium, i.e. $K_1(t;s) \leq K_{1c}$, at all points on its front $\Gamma(t)$. Moreover, in virtue of (3) the SIF $K_1(t;s)$ takes the critical value on the new parts of the crack front (otherwise, the inequalities $h(t;s) > 0, K_1(t;s) < K_{1c}$ would imply that the propagation of the crack at the point s might stop for a smaller t).

Reformulating (3) as a variational inequality we get the following variational problem (see [1, 3, 7, 8]):

Find the non–negative function $H(s) = K_1(s)\,h(s)$ such that for every $X \in C^\infty(\Gamma)$ with $X \geq 0$ the following inequality is valid:

$$\langle K_1(t;s) - K_{1c}, H(s) - X(s)\rangle_\Gamma \geq 0. \tag{4}$$

2.2. Griffith criterion

The Griffith criterion is an energy criterion and it is based on the calculation of the total energy. It implies that the crack-front variation will be in such a way that the total energy of the cracked body will be minimized. Denoting by Π_t the total energy, by U_t the potential energy and by S_t the surface energy at the time t, we have to solve the minimization problem

$$\Pi_t(h) := U_t(h) + S_t(h) \longrightarrow \min. \tag{5}$$

Therefore we get for the first variation the condition

$$\frac{d}{d\delta}\Pi_t(H + \delta(H - X))\big|_{\delta=0} \geq 0 \tag{6}$$

with $H = K_1 h$ and $X \in C^\infty(\Gamma)$. (6) defines a variational inequality a solution H of which describes propagation of the crack (see [3, 8]).

3. Asymptotic expansions

To describe the variational inequalities (4) and (6) more precisely, which means to calculate $K_1(t;s)$ and Π_t, it is necessary to indicate the coefficients in the asymptotics of problem (1) near the crack edge. Introducing curvilinear coordinates (r, φ, s) in the neighborhood U of the contour Γ, we can write down the asymptotic formula for the energetic solution v of problem (1):

$$v(y,z) = \frac{1}{\alpha}\left(\frac{2}{\pi}\right)^{\frac{1}{2}}\left\{K_1(s)r^{\frac{1}{2}}\sin\left(\frac{\varphi}{2}\right) + \frac{1}{3}k_1(s)r^{\frac{3}{2}}\sin\left(\frac{3\varphi}{2}\right) + \right.$$

$$\left. \kappa(s)K_1(s)r^{\frac{3}{2}}\left(\frac{1}{4}\sin\left(\frac{\varphi}{2}\right) - \frac{1}{12}\sin\left(\frac{3\varphi}{2}\right)\right)\right\} - p_\Gamma(s)r\sin\varphi + \mathcal{O}(r^2),$$

(see [7]) where $\kappa(s)$ denotes the curvature of Γ at the point s.

In addition, we need the asymptotic form

$$V(H, x) = (2\pi r)^{-\frac{1}{2}} H(s) \sin\left(\frac{\varphi}{2}\right) + \alpha\left(\frac{2r}{\pi}\right)^{\frac{1}{2}} MH(s)\sin\left(\frac{\varphi}{2}\right)$$

$$- \kappa(s)\left(\frac{r}{2\pi}\right)^{\frac{1}{2}} H(s)\left(\frac{1}{4}\sin\left(\frac{3\varphi}{2}\right) - \frac{3}{4}\sin\left(\frac{\varphi}{2}\right)\right) + \mathcal{O}(r^2),$$

of the weight function V (see [2, 5, 6]), which is a (non-energetic) solution of problem (1) with homogeneous Neumann conditions, and has the square root singularity distributed with the density H over $\Gamma(t)$. The 'SIF' $MH(s)$ of $V(H, x)$, of course, depends on H while according to [7] this dependence is described by the integral operator

$$MH(s) = \int_\Gamma (H(\tilde{s}) - H(s))Z(\tilde{s}, s)\, d\tilde{s} + \tilde{Z}(s)H(s)$$

with the symmetric positive kernel Z of the form

$$Z(\tilde{s}, s) = (2\pi)^{-1}|\tilde{s} - s|^{-2} + \mathcal{O}(|\ln|\tilde{s} - s||).$$

Both the kernel Z and the factor $\tilde{Z}$ depend on the shape of the crack G.

In the special case of the penny–shaped crack $G = \{y : |y| < 1/2\}$ (see [1]), the operator $\boldsymbol{M}$ takes the form

$$\boldsymbol{M}H(s) = \int_\Gamma \frac{H(\tilde{s}) - H(s)}{2\pi |sin(\tilde{s} - s)|^2}\, d\tilde{s}.$$

4. Variational inequalities

In accordance with the method of matched asymptotic expansions the outer and inner representations of the solution $v(t; x)$ of problem (1), corresponding to the crack G_t and the loading $P(t; y)$, can be constructed (using the asymptotics of v and V). The inner representation gives the two-term asymptotic of the SIF $K_1(t; s)$ and the outer representation the three-term asymptotic of the energy Π_t. Inserting these asymptotics in the variational inequalities (4) and (6) and denoting by the index 'I'(G) the solution/data in case of the Irwin (Griffith) criterion, we have to solve the variational problem(s) (see [1, 3, 7, 8])

Find $H_{I(G)} \in W_{2,+}^{\frac{1}{2}}(\Gamma)$ such that $\forall X \in W_{2,+}^{\frac{1}{2}}(\Gamma)$

$$\langle \beta_{I(G)} H_{I(G)},\, X - H_{I(G)} \rangle - \langle \boldsymbol{B} H_{I(G)},\, X - H_{I(G)} \rangle \geq \langle F_{I(G)},\, X - H_{I(G)} \rangle, \quad (7)$$

where the set $W_{2,+}^{\frac{1}{2}}(\Gamma)$ stands for the convex cone of non-negative functions in $W_2^{\frac{1}{2}}(\Gamma)$, $\langle \cdot, \cdot \rangle$ for the inner product in $L_2(\Gamma)$ and $\boldsymbol{B}$ for the integral operator

$$\boldsymbol{B} H_{I(G)}(s) := \int_\Gamma \big(H_{I(G)}(\tilde{s}) - H_{I(G)}(s) \big) Z(\tilde{s}, s)\, d\tilde{s}. \quad (8)$$

$\beta_{I(G)}$ and $F_{I(G)}$ are function constructed from the data of the problem for the crack in the initial position and have the form

$$\beta_I := -\frac{1}{2}\frac{k_1(s)}{K_1(s)},\quad \beta_G := -\frac{1}{2}\frac{k_1(s)}{K_1(s)} - \frac{\kappa(s)}{2}\left(1 - \frac{K_{1c}^2}{K_1^2}\right),$$

$$F_I := t^{-1}(K_1 - K_{1c}) + K_1^*(s),\quad F_G := t^{-1}(K_1 - K_{1c})\frac{K_1 + K_{1c}}{2K_1} + K_1^*(s).$$

(K_1^* is the first SIF of the solution to the problem (1) on the crack in the initial position with the loading $\partial_t P(0; s)$ as Neumann condition).

Theorem 1. $\beta \in \{\beta_I, \beta_G\}, F \in \{F_I, F_G\}, H \in \{H_I, H_G\}$. *Let $\beta > 0$ on Γ. Then for every $F \in L_2(\Gamma)$ there exists a unique solution $H \in W_{2,+}^{\frac{1}{2}}(\Gamma)$ of the problem (1) and the estimate*

$$\|H; W_{2,+}^{\frac{1}{2}}(\Gamma)\| \leq c\, \|F; L_2(\Gamma)\| \quad (9)$$

is valid with a constant c independent of F.

Proof. Using the symmetry of the kernel Z we obtain

$$-\langle \boldsymbol{B}H, H\rangle = \frac{1}{2}\int_\Gamma \int_\Gamma |H(s) - H(\tilde{s})|^2 Z(s,\tilde{s})\, ds\, d\tilde{s}\,.$$

Together with the other properties of Z and the assumption $\beta > 0$ it follows that the expression

$$\langle \beta H, H\rangle - \langle \boldsymbol{B}H, H\rangle$$

implies a norm in the space $W_2^{\frac{1}{2}}(\Gamma)$. Therefore we conclude the theorem by referring to the general results in [4]. □

Theorem 2. $\beta \in \{\beta_I, \beta_G\}, F \in \{F_I, F_G\}, H \in \{H_I, H_G\}$. Let $\beta > 0$ on Γ and $F \in L_p(\Gamma), 2 \le p < +\infty$. Then the solution $H \in W_2^{\frac{1}{2}}(\Gamma)$ of the problem (1) belongs to the Sobolev space $W_p^1(\Gamma)$ and satisfies the estimate

$$\|H; W_p^1(\Gamma)\| \le c\|F; L_p(\Gamma)\| \tag{10}$$

with a constant c independent of F.

Proof. Let us first consider the equation

$$\beta(s)H(s) - \boldsymbol{B}[H](s) = \mathcal{F}(s) \qquad s \in \Gamma\,. \tag{11}$$

Similar to Theorem 1, for every $\mathcal{F} \in L_2(\Gamma)$ this equation has a unique solution in $W_2^{\frac{1}{2}}(\Gamma)$. Due to the integral representation of $\boldsymbol{B}$ and the properties of the kernel Z, $\boldsymbol{B}$ is an elliptic pseudodifferential operator with the principal symbol $\frac{1}{2}|\xi|$. This means in particular, that for every $\mathcal{F} \in L_p(\Gamma)$ the solution H of the equation (11) belongs to $W_p^1(\Gamma)$ (see, e.g. [11]), and there holds the estimate

$$\|H; W_p^1(\Gamma)\| \le c(\|\mathcal{F}; L_p(\Gamma)\| + \|H; L_2(\Gamma)\|)\,.$$

Recalling estimate (9), we achieve

$$\|H; W_p^1(\Gamma)\| \le c_p\|\mathcal{F}; L_p(\Gamma)\|\,. \tag{12}$$

We proceed with the penalty equation

$$\beta(s)H^\varepsilon(s) - \boldsymbol{B}[H^\varepsilon](s) - \varepsilon^{-1}H_-^\varepsilon(s) = F(s) \qquad s \in \Gamma\,, \tag{13}$$

where T_- stands for the negative part of the real number T,

$$T_- := \frac{1}{2}(|T| - T)\,.$$

In other words, $H_-^\varepsilon(s)$ vanishes if $H^\varepsilon \ge 0$, otherwise we have $H_-^\varepsilon(s) = -H^\varepsilon(s) > 0$. Thus,

$$-\langle H_-^\varepsilon, H^\varepsilon\rangle = \langle H_-^\varepsilon, H_-^\varepsilon\rangle$$

and, referring to [4], we get the existence of a unique solution $H^\varepsilon \in W_2^{\frac{1}{2}}(\Gamma)$ to the equation (13) with $F \in L_2(\Gamma)$. Moreover, as shown in [4] for general penalization methods, H^ε converges for $\varepsilon \to +0$ to the solution H of the variational inequality (7) weakly in $W_2^{\frac{1}{2}}(\Gamma)$ and, therefore, strongly in $L_p(\Gamma)$ with $p \in [2, +\infty)$.

If $F \in L_p(\Gamma)$, then $\mathcal{F} = F + \varepsilon^{-1} H_-^\varepsilon \in L_p(\Gamma)$ and, according to (12), we obtain the inclusion $H^\varepsilon \in W_p^1(\Gamma)$ and the estimate

$$\|H^\varepsilon; W_p^1(\Gamma)\| \leq c_p(\|F; L_p(\Gamma)\| + \varepsilon^{-1}\|H_-^\varepsilon; L_p(\Gamma)\|), \tag{14}$$

with a constant c_p independent of F and $\varepsilon \in (0,1)$.

Multiplying (13) by $\varepsilon^{1-p} H_-^\varepsilon(s)^{p-1}$ and integration over Γ leads to

$$\begin{aligned}\|\varepsilon^{-1} H_-^\varepsilon; L_p(\Gamma)\|^p &= -\langle F, \varepsilon^{1-p}(H_-^\varepsilon)^{p-1}\rangle + \varepsilon^{1-p}\langle \beta H^\varepsilon, (H_-^\varepsilon)^{p-1}\rangle \\ &\quad - \varepsilon^{1-p}\langle \mathbf{B} H^\varepsilon, (H_-^\varepsilon)^{p-1}\rangle \\ &=: I_1 + \varepsilon^{1-p} I_2 + \varepsilon^{1-p} I_3. \end{aligned} \tag{15}$$

Obviously, there holds

$$I_1 \leq \|F; L_p(\Gamma)\| \, \|\varepsilon^{-1} H_-^\varepsilon; L_p(\Gamma)\|^{p-1}, \tag{16}$$

and

$$I_2 = -\langle \beta H_-^\varepsilon, (H_-^\varepsilon)^{p-1}\rangle \leq 0. \tag{17}$$

Employing again the properties of the kernel Z of the integral operator $\mathbf{B}$, we get

$$\begin{aligned} I_3 &= -\frac{1}{2}\int_\Gamma H_-^\varepsilon(s)^{p-1} \int_\Gamma (H^\varepsilon(\tilde{s}) - H^\varepsilon(s)) Z(s,\tilde{s}) \, d\tilde{s}\, ds \\ &\quad -\frac{1}{2}\int_\Gamma H_-^\varepsilon(\tilde{s})^{p-1} \int_\Gamma (H^\varepsilon(s) - H^\varepsilon(\tilde{s})) Z(\tilde{s},s) \, ds\, d\tilde{s} \\ &= \frac{1}{2}\int_\Gamma\int_\Gamma (H^\varepsilon(\tilde{s}) - H^\varepsilon(s))(H_-^\varepsilon(\tilde{s})^{p-1} - H_-^\varepsilon(s)^{p-1}) Z(s,\tilde{s}) ds\, d\tilde{s} \\ &= -\frac{1}{2}\int_\Gamma\int_\Gamma (H_-^\varepsilon(\tilde{s}) - H_-^\varepsilon(s))(H_-^\varepsilon(\tilde{s})^{p-1} - H_-^\varepsilon(s)^{p-1}) Z(s,\tilde{s}) ds\, d\tilde{s} \\ &\leq 0. \end{aligned} \tag{18}$$

Here we have used the identity

$$(b-a)(a_-^{p-1} - b_-^{p-1}) = (a_- - b_-)(a_-^{p-1} - b_-^{p-1}) \quad \forall a,b \in \mathbb{R}, \tag{19}$$

and the inequality

$$(a-b)(a^{p-1} - b^{p-1}) \geq |a-b|^p \quad \forall a,b \in [0, +\infty). \tag{20}$$

Since $p \geq 2$, (19) is evident. (20) follows from the inequality

$$1 - t^q \geq (1-t)^q \qquad \forall t \in [0,1], \qquad (21)$$

where $q = p - 1 \geq 1$. The change $t \mapsto 1 - t$ does not influence (21), hence we can restrict (21) for $t \in [0, \frac{1}{2}]$. To prove (21) and (20), it remains to mention that the derivatives of the both sides of (21) are in the relationship

$$-qt^{q-1} \geq -q(1-t)^{q-1}$$

as $t \leq 1 - t$ and $q - 1 \geq 0$.

Due to (17) and (18), we derive from (15) together with (16)

$$\|\varepsilon^{-1} H_-^\varepsilon; L_p(\Gamma)\|^p \leq I_1 \leq \|F; L_p(\Gamma)\| \, \|\varepsilon^{-1} H_-^\varepsilon; L_p(\Gamma)\|^{p-1},$$

and therefore with (14)

$$\|H^\varepsilon; W_p^1(\Gamma)\| \leq 2c_p \|F; L_p(\Gamma)\|. \qquad (22)$$

Since H^ε converges to H, the estimate (22) holds true for the solution H of the variational inequality (7). The estimate (10) and, therefore, theorem 2 are proven. □

Remark 1. The authors do not know if it is possible to extend theorem 2 for the case $p = +\infty$, i.e. to prove that the derivative $H'(s)$ is bounded. At the same time all asymptotics and numerical solutions predict that $H \in W_\infty^1(\Gamma)$. To prove the latter inclusion is still an open question.

5. Numerical realization and example

Under the assumptions of theorem 1 and theorem 2 we now determine the solution of the variational inequality (7). For a corresponding algorithm we use the penalization method (compare (13)) in order to transform the variational inequality into a nonlinear variational equation of the form:

Find $H^\varepsilon \in W_2^{\frac{1}{2}}(\Gamma)$ such that

$$\langle \beta H^\varepsilon - BH^\varepsilon, X \rangle - \varepsilon^{-1} \langle H_-^\varepsilon, X \rangle = \langle F, X \rangle \quad \forall X \in W_2^{\frac{1}{2}}(\Gamma) \qquad (23)$$

(note that this is the same equation as in (13)).

Using the orthogonal projection of $W_2^0(\Gamma)$ onto spline spaces S_N (for example the space of piecewise continuous, linear splines where N denotes the degree of freedom) we determine the solution H_N^ε of the nonlinear discrete equations

$$\langle \beta H_N^\varepsilon - BH_N^\varepsilon + \varepsilon^{-1} H_N^\varepsilon, X_N \rangle = \langle F, X_N \rangle \quad \forall X_N \in S_N. \qquad (24)$$

It can be shown ([1]) that the sequence of solutions H_N^ε of (24) converges to the solution of the variational inequality (7). The nonlinear discrete scheme can be solved with an appropriate relaxation method.

The depth $th(s)$ of the crack propagation is determined by

$$H(s) = K_1(s)h(s)$$

via the solution of the variational inequality (7). This solution can vanish on parts of Γ, so that the crack may grow only on some part of its front.

As an example of such a situation, we present the calculations for the penny–shaped crack

$$G = \{y : |y| < 1/2\},$$

whose surfaces are loaded by the stretching force P concentrated at the point $Q = (0.007, 0)$. The material constants are $\mu = 77.52$, $\nu = 0.29$ and $K_{1c} = 0.06$. The calculations are done for the variational inequality based on the Griffith criterion (the reference configuration is always the unperturbed penny–shaped crack).

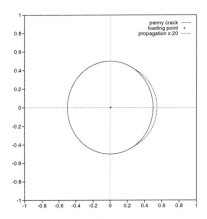

Figure 1:
Asymmetrical propagation of a penny–shaped crack under non–centered loading, $t = 0.01$.

Figure 1 shows the propagation of the penny–shaped crack under the loading $P = 0.116$ at the 'time' $t = 0.01$. We emphasize that the propagated crack is presented with a magnification factor of 20 to make the new crack front visible. The crack grows only on some parts of the front.

Figure 2 shows the propagation of the penny–shaped crack under the loading $P = 0.121$ at the 'time' $t = 0.05$. The crack grows in every part of its front except of one point.

Figure 3 shows the propagation of the penny–shaped crack under the loading $P = 0.122$ at the 'time' $t = 0.06$. There is everywhere crack propagation, and it is circular around the loading point (note the special magnification factor).

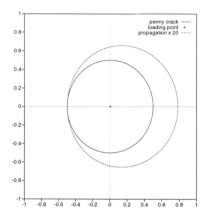

Figure 2:
Asymmetrical propagation of a penny–shaped crack under non–centered loading, $t = 0.05$.

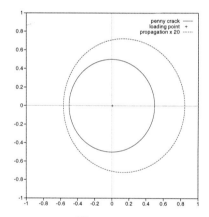

Figure 3:
Asymmetrical propagation of a penny–shaped crack under non–centered loading, $t = 0.06$.

References

[1] M. Bach, S.A. Nazarov, W.L. Wendland: Propagation of a penny shaped crack under the Irwin criterion. In: Bach, M., Constanda, C., Hsiao, G.C., Sändig, A.-M. and Werner, P. (eds.), *Analysis, Numerics and Applications of Differential and Integral Equations*, Pitman Research Notes in Mathematics Series No. 379, Addison Wesley Longman Ltd., Harlow, UK, 1998, 17-21.

[2] H.F. Bueckner: A novel principle for the computation of stress intensity factors, *ZAMM*, **50** (1970), 529–546.

[3] L.H. Kolton, S.A. Nazarov : Variation of the shape of the front of plane mode-one crack which is not in equilibrium locally. *Mekhanika tverd. tela*, **3** (1997), 125-133. (Russian).

[4] J.-L. Lions : *Quelques Méthods de Résolution des Problémes aux Limites non Linéaires*. Dunod Gauthier-Villars, Paris, 1969.

[5] V.G. Mazya, B.A. Plamenewskij: On the coefficients in the asymptotics of solutions of elliptic boundary value problems near the edge, *Dokl. Akad. Nauk SSSR*, **229** (1976), 970–974.

[6] V.G. Mazya, B.A. Plamenewskij: On the coefficients in the asymptotics of solutions of elliptic boundary value problems in domains with conical points, *Math. Nachr.*, **76** (1977), 29–60.

[7] S.A. Nazarov : Derivation of variational inequality for shape of small increment of an mode-1 crack. *Mekhanika tverd. tela*, **2** (1989), 152-160. (English transl.: *Mech. Solids*, **24** (1989), 145-152).

[8] S.A. Nazarov, O.R. Polyakova: On the equivalence of the fracture criteria for a mode-one crack in an elastic space. *Mekhanika tverd. tela*, **2** (1992), 101-113. (Russian).

[9] Y.N. Rabotnov : *Mechanics of Deformable Solids*. Nauka, Moscow, 1979.

[10] J.R. Rice: Weight function theory for three-dimensional crack problems by a finite perturbation method. In: Wei, R.P., Gangloff, R.P. (eds.), *Fracture Mechanics: perspectives and directions (twentieth symposium)*, American Society for Testing and Materials, 1989, 29-57.

[11] M.E. Taylor: *Pseudodifferential Operators*, New Jersey - Princeton University Press, Princeton, 1981.

M. Bach, University of Stuttgart, Math. Institute A.,
Pfaffenwaldring 57, 70569 Stuttgart, Germany;
S.A. Nazarov, Laboratory of Mathematical Modelling of Wave Phenomena,
Institute of Mechanical Engineering problems,
V.O. Bolshoi pr. 61, 199178 St. Petersburg, Russia.
E-mail: michael@mathematik.uni-stuttgart.de

P. BASSANINI, C.M. CASCIOLA, M.R. LANCIA and R. PIVA
Edge Singularities and Kutta Condition for 3D Unsteady Flows in Aerodynamics

1. Introduction

The role of the trailing edge of a wing in releasing a vortical wake and generating lift is crucial in aerodynamics. In the limit of very high Reynolds numbers the wing is modeled by a smooth surface Γ with a sharp trailing edge TE and the wake by a vortex sheet $\mathcal{W}$ detaching from TE, where vorticity is shed into the fluid in the form of a jump in tangential velocity. The flow is almost everywhere irrotational and can be described in terms of a potential $\Phi = \Phi(\mathbf{x}, t)$, a function of space and time which is harmonic outside the body + wake and, together with its tangential gradient, suffers a discontinuity jump across $\mathcal{W}$.

We consider here forced motions of a wing, or, equivalently, unsteady flow with free stream velocity $\mathbf{U}(t)$, a given vector function of time, past a fixed lifting body. For simplicity we will suppose that the trailing edge is straight and unswept (see section 2 for details). Mathematically, the problem splits up into four distinct steps, namely, to determine (i) the harmonic function Φ satisfying a Neumann boundary condition on the body and discontinuous across the wake $\mathcal{W} = \mathcal{W}(t)$, (ii) the vorticity $\boldsymbol{\omega}$ shed into the wake at TE and hence the wake configuration at detachment at any given time, (iii) the time evolution of the wake, and (iv) the pressure distribution, by means of Bernoulli's theorem. Obviously, these four steps are not independent since, for example, solving the first requires to know the wake configuration and location at the given time moment and hence solve step (iii). This, in turn, requires to know the shedding of vorticity and the wake configuration at the trailing edge, i.e. solve step (ii).

We will address here the first two steps only, and admit without further discussion that, once they are solved, the time evolution of the wake follows from the weak form of Euler equations (cf [1]). The problem becomes then essentially one of potential theory, at any given time instant, and a crucial role is played by the behavior of the harmonic function Φ at the sharp edge, which is intimately related to the choice of $\boldsymbol{\omega}$ in point (ii). This in turn raises a uniqueness issue: one can construct an infinity of possible flows characterized by their behavior at the trailing edge and corresponding to each choice of $\boldsymbol{\omega}$ (section 2). For example, choosing $\boldsymbol{\omega} = 0$ yields an irrotational flow, without a wake, and the gradient of the harmonic function Φ (the "velocity") has an unbounded singularity of "power-law type" at the trailing edge, where the dihedral angle seen from the flow is obtuse, [6, 8]. In the presence of a vortex sheet, an additional weaker velocity singularity at TE of logarithmic type may also appear, induced by the vortex sheet, and depending on the value of $\boldsymbol{\omega}$ (sections 2 and 3). Furthermore, the wake always implies a jump in the velocity component tangent to TE.

The shed vorticity $\boldsymbol{\omega}$ must be chosen in such a way that the velocity field $\mathbf{u} = grad\Phi$ is *bounded,* and therefore all unbounded singularities, both power-law and logarithmic, must be removed by means of a suitable cancellation process. As in our previous papers devoted to steady flows, this turns out to be easier to do if a formulation in terms of velocity is adopted (section 3). Then, for unsteady flows, this "unbounded singularity

---

The research of P.B. was partially supported by CNR under Grants 96.03842.01, 97.00857.01.

removal" principle (*Kutta condition*), coupled with Bernoulli's and Kelvin's theorems, yields a full set of equations for the determination of the wake configuration and of the vorticity shedding $\boldsymbol{\omega}$ at the trailing edge. As a result, the wake must separate tangentially to the upper or lower wing face in each wing section, according to the sign of the time derivative of the circulation and the local flow field; for flat plate wings, the wake must be aligned to Γ at separation (see section 4).

In the limit of steady flows, we recover the description of the wake at detachment predicted by Mangler and Smith [11]. The relation of steady flow to the asymptotic limit of the unsteady flow as t goes to infinity is briefly discussed at the end of section 4.

2. The flow problem

We consider unsteady incompressible inviscid fluid flow with given free stream velocity $\mathbf{U} = \mathbf{U}(t)$ past a three-dimensional conventional wing, represented geometrically by a smooth compact surface Γ with a sharp trailing edge, TE. For simplicity, the trailing edge is assumed to be straight and orthogonal to $\mathbf{U}$ (unswept wing), and all effects due to the possible presence of conic points at the wing tips are neglected.

We adopt a cartesian coordinate system $O(X, Y, Z)$, with origin on (the center of) TE, Y-axis along TE, with unit vector $\mathbf{T}$, and X-axis oriented into the unbounded flow domain, Ω. We denote by $\mathbf{n}$ the normal to Γ oriented towards Ω, and by $\alpha(Y)$, $\pi < \alpha(Y) < 2\pi$, the outer dihedral angle of Γ at TE, i.e. the angle between the upper and lower tangents $\mathbf{t}^+, \mathbf{t}^-$ to each Y-section of Γ at TE. The unit vectors $\mathbf{t}^+, \mathbf{t}^-$ are oriented away from TE, left part of fig. 1. We will denote the flow velocity in this coordinate system by $\mathbf{v} = \mathbf{v}(\mathbf{x},t)$, $\mathbf{x} = (X, Y, Z)$, and the perturbation velocity by $\mathbf{u} = \mathbf{v} - \mathbf{U}$.

A simple, putative flow field could be constructed by assuming that $\mathbf{v}$ is irrotational, i.e. that the vorticity $\boldsymbol{\zeta} = curl\,\mathbf{u}$ is identically zero in all of Ω. Then $\mathbf{u} = grad\Phi$, and the perturbation potential $\Phi = \Phi(\mathbf{x},t)$ satisfies the exterior Neumann problem at any given time t

$$\begin{aligned} \Delta\Phi &= 0 \quad , \quad \mathbf{x} \in \Omega \\ \frac{\partial\Phi}{\partial n} &= -\mathbf{U}\cdot\mathbf{n} \quad , \quad \mathbf{x} \in \Gamma\backslash TE \\ \Phi &= O(1/|\mathbf{x}|^2) \quad , \quad |\mathbf{x}| \longrightarrow \infty \end{aligned} \quad (1)$$

The solution of this problem is regular in all Ω (precisely, $\Phi \in C^\infty(\Omega) \cap C^o(\overline{\Omega}) \cap H^1(\Omega)$) but its gradient is singular at the trailing edge. Denoting by $\rho = \sqrt{X^2 + Z^2} = dist(\mathbf{x}, TE)$ the distance from the point $\mathbf{x}$ to TE in the Y-section, the normal component of $grad\Phi$ has an unbounded singularity at TE of power-law type [6, 10]

$$\frac{\partial\Phi}{\partial\rho} \sim c(Y)\rho^{-\beta(Y)} \quad , \quad \rho \to 0 \quad (2)$$

where $c(Y)$ is a smooth function of Y along the trailing edge, and $\beta(Y) = 1 - \pi/\alpha(Y)$ satisfies $0 < \beta(Y) < 1/2$.

This irrotational solution is unique, but is not the "actual" solution. In order to have non-zero lift, the potential must be discontinuous in Ω. In fact, the actual velocity field is rotational and, under the present modeling assumptions for the flow,

u is characterized by being bounded everywhere and having vorticity ζ concentrated on an infinitely thin wake $\mathcal{W}$ detaching from TE, which evolves in time according to the Euler equations. The potential Φ has a jump discontinuity across $\mathcal{W}$, with two different limiting values Φ^+, Φ^- on the upper and lower face of $\mathcal{W}$, respectively, and the tangential velocity also jumps across $\mathcal{W}$, while the pressure p and the normal component of velocity are continuous [12]. We will denote $[f]_S = f^+ - f^-$ the jump of a quantity f across a surface S, by a bar the closure of a set (e.g., $\overline{S}$), and by $\wedge$ the cross product in $\mathbb{R}^3$. The wake can be viewed either as a *vortex sheet* with density

$$\gamma = [\mathbf{N} \wedge \mathbf{u}]_\mathcal{W} \equiv [\mathbf{N} \wedge grad\Phi]_\mathcal{W} \equiv \mathbf{N} \wedge grad[\Phi]_\mathcal{W}$$

(**N** the normal to $\mathcal{W}$) or as a *double layer* distribution with density $[\Phi]_\mathcal{W} = \Phi^+ - \Phi^-$. The two descriptions of the wake are equivalent under the assumption that $[\Phi]_\mathcal{W}$ vanishes at the free edges of $\mathcal{W}$, see the Appendix. Actually, using a formulation in terms of the velocity variable and thinking of $\mathcal{W}$ as a vortex layer turns out to be more convenient for the present purposes (see [2, 3]).

We denote $\omega = \gamma|_{TE}$ the vorticity shed into the wake at the trailing edge. Then, we may think of γ as being determined as a function of $\mathbf{x} \in \mathcal{W}$ and t by its "initial value" $\omega = \omega(Y, \tau)$ (for all Y along TE and all times $\tau \leq t$) and by the Euler equations, so that $\gamma = \gamma[\omega]$ will be a known functional of ω. For details in this connection see [1]. For unsteady flows, the shed vorticity can be decomposed in the form

$$\omega(Y, t) = \omega_1(Y, t)\boldsymbol{\nu} + \omega_2(Y, t)\mathbf{T} \tag{3}$$

where **T** is the unit tangent to TE, $\mathbf{N}_o = \mathbf{N}_o(Y, t)$ is the "initial" value of **N** at TE, oriented upwards, and $\boldsymbol{\nu} = \mathbf{T} \wedge \mathbf{N}_o(Y, t)$ is a unit vector normal to TE and tangent to $\mathcal{W}$, oriented away from Γ. The velocity field induced by the presence of the vortex sheet can be written in terms of a vector potential $\mathbb{W}\gamma[\omega]$ in the form

$$\mathbf{u}_\mathcal{W}(\mathbf{x}, t) = curl \int_\mathcal{W} \frac{\gamma[\omega]}{4\pi r} dS_y \equiv curl\,(\mathbb{W}\gamma[\omega]) \tag{4}$$

where $r = |\mathbf{x}-\mathbf{y}|$ (see [3] and section 3). Since many years, it is well understood (see e.g. [7]), that $\mathbf{N} \cdot \mathbf{u}_\mathcal{W}$ may present a logarithmic singularity as $\mathbf{x}$ approaches the boundary of $\mathcal{W}$; in particular, an easy calculation shows that, as $\mathbf{x}$ approaches the trailing edge,

$$curl\,(\mathbb{W}\gamma[\omega]) = -\frac{1}{2\pi}\mathbf{N}_o\omega_2(Y, t)\log\rho + O(1) \quad , \rho \to 0 \tag{5}$$

3. Integral formulation in terms of velocity and analysis of the edge singularities

The flow perturbation velocity can be represented by the Poincaré formula [2, 3]

$$\mathbf{u}(\mathbf{x}, t) = grad \int_\Gamma \frac{\mathbf{n} \cdot \mathbf{U}}{4\pi r} dS + curl \int_\Gamma \frac{\mathbf{n} \wedge \mathbf{u}}{4\pi r} dS + curl\,(\mathbb{W}\gamma[\omega]) \quad , \quad \mathbf{x} \in \Omega \tag{6}$$

where $\mathbf{n}\wedge\mathbf{u}$ is unknown on Γ and, according to (4), we have assumed the flow vorticity $\boldsymbol{\zeta}$ to be concentrated on a vortex sheet $\mathcal{W}$,

$$\boldsymbol{\zeta} = \gamma[\omega]\delta_{\mathcal{W}}(\mathbf{x}), \qquad \mathbf{x} \in \Omega \tag{7}$$

where $\delta_{\mathcal{W}}(\mathbf{x})$ is the Dirac distribution supported by $\mathcal{W}$. For unsteady flows, the wake $\mathcal{W} = \mathcal{W}(t)$ is represented geometrically by a smooth bounded surface and (6) implies the asymptotic behavior

$$\mathbf{u} = O(1/|\mathbf{x}|^2), \; \Phi = O(1/|\mathbf{x}|) \quad \text{as} \; |\mathbf{x}| \to \infty$$

By introducing a surface potential ϕ and a surface stream function Ψ on Γ, we can enforce the Hodge-type decomposition

$$\mathbf{n} \wedge \mathbf{u} = \mathbf{n} \wedge grad\phi - grad_\Gamma \Psi \tag{8}$$

for $\mathbf{n}\wedge\mathbf{u}$ on Γ, where $grad_\Gamma$ denotes the surface gradient[2]. Then (8) yields $\Delta_\Gamma \Psi = \mathbf{n}\cdot\boldsymbol{\zeta}$ where $\Delta_\Gamma = div_\Gamma grad_\Gamma \equiv \mathbf{n}\cdot curl(\mathbf{n}\wedge grad)$ is the Laplace-Beltrami operator on Γ, and $\mathbf{n}\cdot\boldsymbol{\zeta}$ is concentrated at the trailing edge according to (3) and (7). Using for $\mathbf{n}$ at TE either the value $\mathbf{n}^+$ or the value $\mathbf{n}^-$ for the normal to the upper or lower face of Γ, respectively, from well-known properties of the Dirac function we obtain

$$\mathbf{n}\cdot\boldsymbol{\zeta} = \mathbf{n}^\pm\cdot(\omega_1\boldsymbol{\nu} + \omega_2\mathbf{T})\delta_{\mathcal{W}}(\mathbf{x}) = \omega_1\delta_{TE}(\mathbf{x}) \qquad , \mathbf{x} \in \Gamma$$

As ω_1 has average value zero on TE, the inverse operator Δ_Γ^{-1} is well-defined on Γ and the surface stream function on Γ

$$\Psi = \Delta_\Gamma^{-1}(\omega_1\delta_{TE}) \tag{9}$$

is known from ω_1 as in the steady case (see [2]). On the contrary, the scalar function ϕ is unknown on Γ. Letting $\mathbf{x}$ approach Γ and taking the scalar product $\mathbf{n}\cdot\mathbf{u}$ in (6) yields

$$\begin{aligned}\frac{1}{2}\mathbf{n}\cdot\mathbf{u}(\mathbf{x},t) &= \frac{\partial}{\partial n}\int_\Gamma \frac{\mathbf{n}_y\cdot\mathbf{U}}{4\pi|\mathbf{x}-\mathbf{y}|}dS_y + \mathbf{n}\cdot curl\int_\mathcal{W} \frac{\gamma[\omega]}{4\pi|\mathbf{x}-\mathbf{y}|}dS_y \\ &+\mathbf{n}\cdot curl\int_\Gamma \frac{\mathbf{n}_y\wedge grad\phi}{4\pi|\mathbf{x}-\mathbf{y}|}dS_y - \mathbf{n}\cdot curl\int_\Gamma \frac{grad_\Gamma\Delta_\Gamma^{-1}(\omega_1\delta_{TE})}{4\pi|\mathbf{x}-\mathbf{y}|}dS_y\end{aligned} \tag{10}$$

for $\mathbf{x} \in \Gamma$, where $\mathbf{n}\cdot\mathbf{u} = -\mathbf{n}\cdot\mathbf{U}$. We thus obtain the boundary equation for ϕ

$$D\phi = F[\mathbf{U},\boldsymbol{\omega}]$$

where D is the pseudodifferential operator of order 1 on Γ defined by

$$D\phi = -\mathbf{n}\cdot curl\int_\Gamma \frac{\mathbf{n}_y\wedge grad\phi}{4\pi r}dS$$

and $F[\mathbf{U},\boldsymbol{\omega}]$ can be decomposed into the sum $F_\infty + F_\omega$ of two source terms, such that

$$F_\infty = \left(\frac{1}{2}\mathbf{n}\cdot\mathbf{U} + \frac{\partial}{\partial n}\int_\Gamma \frac{\mathbf{n}\cdot\mathbf{U}}{4\pi r}dS\right)$$

depends (linearly) only on $\mathbf{U} = \mathbf{U}(t)$, while

$$F_\omega = -\mathbf{n}\cdot curl \int_W \frac{\boldsymbol{\gamma}[\boldsymbol{\omega}]}{4\pi r} dS_y - \mathbf{n}\cdot curl \int_\Gamma \frac{grad_\Gamma \Delta_\Gamma^{-1}(\omega_1 \delta_{TE})}{4\pi r} dS_y$$

depends (linearly) on the arbitrary vector function $\boldsymbol{\omega} = \omega_1 \boldsymbol{\nu} + \omega_2 \mathbf{T}$. This suggests the similar decomposition for ϕ

$$\phi = \phi_\infty + \phi_\omega \tag{11}$$

so that ϕ_∞, ϕ_ω satisfy the two boundary equations on Γ

$$D\phi_\infty = F_\infty \quad , \quad D\phi_\omega = F_\omega$$

For each choice of the "control vector function" $\boldsymbol{\omega}$, these equations have unique solutions $\phi_\infty, \phi_\omega \in H^{1/2}(\Gamma)/\mathbb{R}$ [4], [5], [8], [13], [2] so that the tangent trace $\mathbf{n} \wedge \mathbf{u}$ is uniquely determined as a functional of $\boldsymbol{\omega}$, and a unique flow $\mathbf{u} = \mathbf{u}[\boldsymbol{\omega}]$ follows by (6). For example, choosing $\boldsymbol{\omega} \equiv 0$ yields $\boldsymbol{\gamma}[\boldsymbol{\omega}] \equiv 0, F_\omega \equiv 0$, hence $\phi_\omega \equiv 0$ (up to an inessential additive constant) and we get the irrotational flow field corresponding to the solution of the Neumann problem (1). Thus, there are an infinity of possible flows $\mathbf{u}[\boldsymbol{\omega}]$, parameterized by the scalar control functions ω_1, ω_2 and the unit vector $\boldsymbol{\nu}$ (or $\mathbf{N}_o$), which are at this stage arbitrary and must be chosen appropriately in order to get the physically relevant solution.

It may be worth stressing the fact that the surface potential ϕ does not coincide with the trace of the potential Φ on Γ unless Ψ is constant, i.e., $\omega_1 \equiv 0$. Indeed, $\phi \in H^{1/2}(\Gamma)$ cannot have jump discontinuities, whereas Φ jumps at TE in the presence of a wake.

We need now a more detailed analysis of the behavior of $\mathbf{u}$ on Γ near TE, already anticipated in section 2. From (8) and (11) we find for the trace of $\mathbf{u}$ on $\Gamma \backslash TE$

$$\mathbf{u}|_\Gamma = grad_\Gamma(\phi_\infty + \phi_\omega) + \mathbf{n} \wedge grad_\Gamma \Psi - (\mathbf{n} \cdot \mathbf{U})\mathbf{n} \tag{12}$$

where $\Psi = \Delta_\Gamma^{-1}(\omega_1 \delta_{TE})$. As discussed in [2], the term $\mathbf{n} \wedge grad_\Gamma \Psi$ is *bounded* on Γ and only provides a jump in the tangential velocity

$$-[\mathbf{T} \cdot \mathbf{u}]_{TE} = \omega_1(Y, t) \tag{13}$$

at the trailing edge. Similarly, $(\mathbf{n} \cdot \mathbf{U})\mathbf{n}$ is bounded on Γ and jumps at TE. In contrast, the first two terms in (12) contribute an unbounded singularity at the trailing edge whose leading term is of power-law type

$$grad_\Gamma \phi_\infty \sim \mathbf{t}^\pm c_\infty(Y) \rho^{-\beta(Y)} \quad , \quad grad_\Gamma \phi_\omega \sim \mathbf{t}^\pm c_\omega(Y) \rho^{-\beta(Y)}$$

as $\rho \to 0$ (cf (2)). Here

$$c_\infty(Y) = \pm \lim_{\rho \to 0} \rho^{\beta(Y)} \mathbf{t}^\pm \cdot grad_\Gamma \phi_\infty \,, \, c_\omega(Y) = \pm \lim_{\rho \to 0} \rho^{\beta(Y)} \mathbf{t}^\pm \cdot grad_\Gamma \phi_\omega \tag{14}$$

are smooth functions of Y, whose linear functional dependence on $\mathbf{U}$ and $\boldsymbol{\omega}$, respectively, is known. (The limits in (14) can be taken indifferently along the upper or lower wing side, by choosing the appropriate signs.) Let $c(Y) = c_\infty(Y) + c_\omega(Y)$. Then this power law singularity will be removed by taking $c(Y) \equiv 0$, see (20) below. After doing this, we are left with a "regular" part ϕ_r of the function $\phi = \phi[\boldsymbol{\omega}]$, continuous on Γ, with $grad_\Gamma \phi_r$ Holder continuous on $\Gamma \setminus TE$, and such that *finite limits* $\mathbf{t}^\pm \cdot grad_\Gamma \phi_r$ *on the upper and lower face of* Γ *exist at* TE and are continuous in Y. Similarly, Ψ is continuous on Γ, $grad_\Gamma \Psi$ is Holder continuous on $\Gamma \setminus TE$, so that the tangential derivative $\partial \Psi / \partial Y |_{TE}$ is finite and continuous along TE. The functions γ and ω will be assumed bounded and smooth: the validity of this assumption can be checked after the fact. From (6), rewritten here according to (8), we obtain the expression

$$\mathbf{u}[\boldsymbol{\omega}] = grad \int_\Gamma \frac{\mathbf{n} \cdot \mathbf{U}}{4\pi r} dS + curl \int_\Gamma \frac{\mathbf{n} \wedge grad\phi[\boldsymbol{\omega}]}{4\pi r} dS \qquad (15)$$
$$- curl \int_\Gamma \frac{grad_\Gamma \Psi[\omega_1]}{4\pi r} dS + curl (\mathbb{W}\boldsymbol{\gamma}[\boldsymbol{\omega}])$$

and we can seek the asymptotic behavior of $\mathbf{u}(\mathbf{x}, t) = \mathbf{u}[\boldsymbol{\omega}]$ as $\mathbf{x}$ approaches the trailing edge. As $\rho \to 0$, each integral in (15) gives rise to an "orthogonal" log singularity at the edge TE: indeed, an easy calculation yields

$$grad \int_\Gamma \frac{\mathbf{n} \cdot \mathbf{U}}{4\pi r} dS = -\frac{1}{2\pi}(\mathbf{n}^+ \cdot \mathbf{U}\, \mathbf{t}^+ + \mathbf{n}^- \cdot \mathbf{U}\, \mathbf{t}^-)_{TE} \log \rho + O(1) \qquad (16)$$

$$curl \int_\Gamma \frac{\mathbf{n} \wedge grad\phi}{4\pi r} dS = \frac{-1}{2\pi}(\mathbf{t}^+ \cdot grad_\Gamma \phi\, \mathbf{n}^+ + \mathbf{t}^- \cdot grad_\Gamma \phi\, \mathbf{n}^-)_{TE} \log \rho$$
$$+ O(1) \qquad (17)$$

$$curl \int_\Gamma \frac{grad_\Gamma \Psi[\omega_1]}{4\pi r} dS = \frac{1}{2\pi}(\mathbf{n}^+ - \mathbf{n}^-) \frac{\partial \Psi}{\partial Y} |_{TE} \log \rho + O(1)$$

in addition to (5). Thus the precise asymptotic behavior of $\mathbf{u}(\mathbf{x}, \mathbf{t})$ as $\mathbf{x}$ approaches TE along Γ is given by

$$\mathbf{u}(\mathbf{x}, t) = \mathbf{t}\, c(Y) \rho^{-\beta(Y)} - \frac{1}{2\pi} \{ (\mathbf{t}^+ \cdot grad_\Gamma \phi) \mathbf{n}^+ + (\mathbf{t}^- \cdot grad_\Gamma \phi) \mathbf{n}^-$$
$$+ \mathbf{N}_o \omega_2 + \frac{1}{2\pi}(\mathbf{n}^+ - \mathbf{n}^-)\mathbf{T} \cdot grad_\Gamma \Psi + (\mathbf{n}^+ \cdot \mathbf{U}) \mathbf{t}^+ \qquad (18)$$
$$+ (\mathbf{n}^- \cdot \mathbf{U}) \mathbf{t}^- \}_{TE} \log \rho + O(1)$$

where $\mathbf{t}$ is either $\mathbf{t}^+$ or $\mathbf{t}^-$, and $\mathbf{T} \cdot grad_\Gamma \Psi = \partial \Psi / \partial Y |_{TE}$. Only the "regular" part ϕ_r of ϕ contributes to the log term, which disappears altogether when $\boldsymbol{\omega} \equiv 0$, i.e., in the absence of the wake, or more generally whenever $\omega_2 \equiv 0$ (see the Appendix).

4. Kutta condition and trailing edge behavior of the wake

As already mentioned, the physical flow field is bounded and the control function $\boldsymbol{\omega}$ (i.e., ω_1, ω_2 and $\mathbf{N}_o$) must be chosen in such a way that the unbounded singularities of $\mathbf{u} = \mathbf{u}[\boldsymbol{\omega}]$ at TE cancel out. This boundedness requirement can be written in the form

$$\sup_{\mathbf{x} \in \Gamma} |\mathbf{u}(\mathbf{x})| < \infty \tag{19}$$

and, as $\mathbf{n} \cdot \mathbf{u} = -\mathbf{n} \cdot \mathbf{U}$ is bounded on Γ, it actually applies to $\mathbf{n} \wedge \mathbf{u}$. By (18), the $\rho^{-\beta(Y)}$ singularity will be canceled by taking

$$c(Y) \equiv c_\infty(Y) + c_\omega(Y) = 0 \tag{20}$$

for all Y varying along TE, and the log singularity will be eliminated by setting equal to zero the brackets term in (18). After some simple manipulations, we find the two scalar equations

$$\begin{aligned} v_2^+ \sin\alpha(Y) &= -\mathbf{t}^- \cdot \mathbf{N}_o \omega_2(Y, t) \\ v_2^- \sin\alpha(Y) &= -\mathbf{t}^+ \cdot \mathbf{N}_o \omega_2(Y, t) \end{aligned} \tag{21}$$

which involve $\omega_2(Y, t), \mathbf{N}_o$ and the limiting values $\mathbf{t}^\pm \cdot \mathbf{v}^\pm$ at TE along the upper and lower face of Γ, respectively, of the component of the total velocity $\mathbf{v} = \mathbf{u} + \mathbf{U}$ tangential to Γ and orthogonal to TE

$$v_2^\pm \equiv \mathbf{t}^\pm \cdot \mathbf{v}^\pm = \mathbf{t}^\pm \cdot \mathbf{U} + \lim_{\rho \to 0} (\mathbf{t}^\pm \cdot grad_\Gamma \phi \pm \mathbf{T} \cdot grad_\Gamma \Psi)$$

The problem is to see whether the Kutta condition (19) as enforced by the two relations (20), (21) uniquely determines the control functions ω_1 and ω_2 *and* the vector $\mathbf{N}_o$, i.e. the tangent $\boldsymbol{\nu}$ to the wake $\mathcal{W}$ at separation from TE. To answer this question we need further information. First of all, since $\mathbf{v}$ is bounded, the wake must detach in such a way that, with the orientations chosen in section 2, $v_2^\pm \leq 0$ and

$$\mathbf{n}^\pm \cdot \boldsymbol{\nu} \geq 0 \tag{22}$$

at each location Y (see [11]). As $\pi < \alpha(Y) < 2\pi$, $\mathbf{n}^+ \cdot \boldsymbol{\nu} > 0$ implies $v_2^+ = 0$ and $\mathbf{n}^- \cdot \boldsymbol{\nu} > 0$ implies $v_2^- = 0$, so that in every case

$$v_2^+ v_2^- = 0 \tag{23}$$

as schematically illustrated on the right part of fig. 1. It is easy to check that (21), (22) and (23) leave only three possibilities (for a fixed, arbitrary value of Y):

$$(L): \quad \mathbf{N}_o \cdot \mathbf{t}^- = 0, \ v_2^+ = 0, \ \omega_2 = v_2^- < 0$$

and the wake detaches tangent to the lower face of Γ, $\nu = -\mathbf{t}^-$;

(U) : $\mathbf{N}_o \cdot \mathbf{t}^+ = 0$, $v_2^- = 0$, $\omega_2 = -v_2^+ > 0$

and the wake detaches tangent to the upper face of Γ, $\nu = -\mathbf{t}^+$;

(I) : $v_2^- = v_2^+ = 0$, $\omega_2 = 0$

and the wake starts with intermediate tangent ν between $-\mathbf{t}^-$ and $-\mathbf{t}^+$, so that (22) is satisfied.

Note that (20) – (23) imply that $\boldsymbol{\omega} = \omega_1 \boldsymbol{\nu} + \omega_2 \mathbf{T}$ *matches continuously* with $\mathbf{N}_o \wedge [\mathbf{v}]_\Gamma$ at the trailing edge. Indeed, from (13) and the relations above obtained from (21) we find

$$\omega_1(Y,t) = -[\mathbf{T} \cdot \mathbf{u}]_{TE} , \quad \omega_2(Y,t) = [\boldsymbol{\nu} \cdot \mathbf{u}]_{TE} \tag{24}$$

and the assertion follows from the relations $\mathbf{N}_o = \boldsymbol{\nu} \wedge \mathbf{T}$, $[\mathbf{v}]_{TE} = [\mathbf{u}]_{TE}$, and (3). In other words, the derivatives of the potential jump satisfy the matching conditions at TE

$$-\omega_1 \equiv \frac{\partial[\Phi]_\mathcal{W}}{\partial Y} = \frac{\partial[\Phi]_\Gamma}{\partial Y}, \quad \omega_2 \equiv \frac{\partial[\Phi]_\mathcal{W}}{\partial \nu} = [\frac{\partial \Phi}{\partial \nu}]_\Gamma$$

where $[\frac{\partial \Phi}{\partial \nu}]_\Gamma := \pm v_2^\mp$. The matching of the potential jump itself $[\Phi]_\mathcal{W} = [\Phi]_\Gamma$ [9] is a consequence of the continuity properties of Φ (see the Appendix). This common value $[\Phi]_{TE}$ for the jump of Φ can be interpreted as the circulation $\mathcal{K}(Y,t)$ of $\mathbf{v}$ around the planar contour $\mathcal{C}$ obtained by intersecting Γ with a Y-plane, and we immediately have the relation

$$\frac{\partial \mathcal{K}(Y,t)}{\partial Y} \equiv \frac{\partial[\Phi]_{TE}}{\partial Y} = -\omega_1(Y,t). \tag{25}$$

Note that the lift is proportional to the vector circulation [12, 3] and hence to the integral

$$\int_{TE} \mathcal{K}(Y,t)\, dY \equiv \int_{TE} [\Phi]_{TE}\, dY. \tag{26}$$

As the pressure is continuous across $\mathcal{W}$, Bernoulli's theorem yields

$$\frac{\partial \mathcal{K}(Y,t)}{\partial t} = -\frac{1}{2}(|\mathbf{v}^+|^2 - |\mathbf{v}^-|^2).$$

Let us decompose $\mathbf{v}^+, \mathbf{v}^-$ in the form $\mathbf{v}^\pm = v_2^\pm \mathbf{t}^\pm + v_1^\pm \mathbf{T}$, with $v_1^\pm = \mathbf{v}^\pm \cdot \mathbf{T}$, $v_2^\pm \equiv \mathbf{t}^\pm \cdot \mathbf{v}^\pm$, and let

$$\mathbf{V} = \frac{1}{2}(\mathbf{v}^+ + \mathbf{v}^-) =: V_1 \mathbf{T} - V_2 \tag{27}$$

40

denote the "average wake velocity" at separation. Then

$$\omega_1(Y,t) = -[v_1]_{TE} \equiv v_1^- - v_1^+ . \tag{28}$$

Since $|\mathbf{v}^+|^2 - |\mathbf{v}^-|^2 = (v_1^+)^2 - (v_1^-)^2 + (v_2^+)^2 - (v_2^-)^2$, where at least one of v_2^+, v_2^- is zero (see above), the Bernoulli theorem may be rewritten as

$$\dot{\mathcal{K}} - V_1 \omega_1 = \begin{cases} \frac{1}{2}\omega_2^2 & \text{in case (L)}, \ \omega_2 = v_2^- < 0 \\ -\frac{1}{2}\omega_2^2 & \text{in case (U)}, \ \omega_2 = -v_2^+ > 0 \\ 0 & \text{in case (I)}, \ \omega_2 = 0. \end{cases} \tag{29}$$

The relations just obtained imply the L–U alternative for unsteady flows:

(L) If $\dot{\mathcal{K}} - V_1\omega_1 > 0$, then $\omega_2(Y,t) < 0$ and the wake separates tangentially to the lower face of Γ in the Y-section under consideration.

(U) If $\dot{\mathcal{K}} - V_1\omega_1 < 0$, then $\omega_2(Y,t) > 0$, and the wake starts tangent to the upper face of Γ.

Thus the wake configuration at detachment (i.e., the vector $\boldsymbol{\nu}$) is determined by the sign of $\dot{\mathcal{K}} - V_1\omega_1$, while the scalar control functions ω_1, ω_2 must be determined as the fixed points of the equations (20), (29) and (25). A possible iteration process is the following:

(i) Solve (20) for $\omega_1(Y,t)$ as a functional of $\omega_2(Y,t)$. In conjunction with (25), this determines $\mathcal{K} = \mathcal{K}[\omega_2]$.

(ii) Use the boundary integral representation to evaluate V_1. Note that the value of $\dot{\mathcal{K}} - V_1\omega_1$ determines the sign of ω_2 at each location Y.

(iii) Solve the fixpoint equation for $|\omega_2(Y,t)|$ along TE obtained from (29)

$$|\omega_2| = \sqrt{2 \, |\dot{\mathcal{K}}[\omega_2] - V_1[\omega_2]\omega_1[\omega_2]|}. \tag{30}$$

The complete dynamical problem can be solved numerically by extending the time-step scheme previously proposed for 2D flows [1]. On this basis we infer that for unsteady flows the Kutta condition (19) yields a *unique bounded* flow velocity $\mathbf{u}(\mathbf{x},t)$ with a well–determined wake $\mathcal{W} = \mathcal{W}(t)$.

In the case of a flat plate wing, $\alpha(Y) \equiv 2\pi$, and boundedness of $\mathbf{v}$ implies that the wake must start aligned to Γ at the trailing edge. Because $\sin \boldsymbol{\alpha}(Y) \equiv 0$ and $\mathbf{t}^+ \equiv \mathbf{t}^-$ is orthogonal to $\mathbf{N}_0$, (21) is then automatically satisfied, and the Kutta condition reduces to (20), which together with (29) serves to determine a unique control function.

For steady flows, $\dot{\mathcal{K}}= 0$ and relation (29) reduces to

$$-V_1\omega_1 = \begin{cases} \frac{1}{2}\omega_2^2 & \text{in case (L)}, \ \omega_2 = v_2^- < 0 \\ -\frac{1}{2}\omega_2^2 & \text{in case (U)}, \ \omega_2 = -v_2^+ > 0 \\ 0 & \text{in case (I)}, \ \omega_2 = 0, \end{cases} \tag{31}$$

thus, recovering the physical behavior described by Mangler and Smith in [11], implying that the steady state solution is obtained as the limit of a pseudotransient formulation.

5. Appendix

We recall the well-known equivalence of doublet and vortex layers on a smooth surface S in $\mathbb{R}^3$

$$grad \int_S g \frac{\partial}{\partial n}(\frac{1}{4\pi r})dS = curl \int_S \frac{\mathbf{N} \wedge grad\, g}{4\pi r}dS - curl \int_{\partial S} \frac{g\,\mathbf{t}}{4\pi r}ds \qquad (32)$$

($\mathbf{N}$ the normal to S, $\mathbf{t}$ the tangent to ∂S). If $\partial S \neq \emptyset$ and g does not vanish on ∂S, the vortex filament term along ∂S in the right-hand side does not vanish and the gradient of the double layer potential on the left-hand side is not locally square summable in $\mathbb{R}^3$, as it is easy to check (for example, take g =constant). Conversely, $g|_{\partial S} = 0$ and g smooth, say $g \in H_{oo}^{1/2}(S)$, implies that the double layer potential is locally $H^1(\mathbb{R}^3)$. Here $S = \mathcal{W}$ and $g = [\Phi]_\mathcal{W}$, so the requirement is $g|_{\partial S} = [\Phi]_{\partial \mathcal{W}} = 0$. This assumption actually concerns only the "free edges" of $\mathcal{W}$, i.e., the part of $\partial \mathcal{W}$ not including the trailing edge because, due to the matching condition $[\Phi]_\mathcal{W} = [\Phi]_\Gamma$, the integral along TE is cancelled by an opposite contribution coming from the side of Γ. By defining $\Phi|_{\mathcal{W}^\pm} = \Phi^\pm$, Φ is then continuous in the closure of the domain Ω cut by $\mathcal{W}$, $\Phi \in H^1_{loc}(\Omega) \cap C^o(\overline{\Omega \setminus \mathcal{W}})$.

The analysis in section 3, see (16) and (17), shows that "an edge on a smooth surface Γ acts like a border" for a single layer or vortex layer, so that the induced velocities present an "orthogonal" log singularity there. For a velocity vector field $\mathbf{u}[\boldsymbol{\omega}]$ in the presence of a concentrated vorticity wake, the various log singularities add up according to (18). It is easy to prove that, *if* $\omega_2 \equiv 0$, as in the case of an unswept wing in steady flow [2, 3], *the log singularity term in* (18) *disappears*. Indeed, since $\omega_2 \equiv 0$ implies $v_2^\pm \equiv 0$, there is no log singularity coming from the wake, see (5), and the log singularities coming from the two sides of Γ at TE cancel out, as (21) are automatically satisfied.

Let us examine in more detail how the log singularities at the edge cancel out in the case of a harmonic function, i.e., when $\boldsymbol{\omega} \equiv \mathbf{0}$. Consider a function $\Phi = \Phi(\mathbf{x})$, harmonic in Ω, continuous in $\overline{\Omega}$, and of class $H^1(\Omega)$, solution of the Neumann problem (1). We can decompose $\Phi = \Phi_r + \Phi_{sing}$ as the sum of two harmonic functions, such that Φ_r satisfies the non-homogeneous Neumann boundary condition in (1), while Φ_{sing} satisfies a homogeneous Neumann boundary condition on $\Gamma \setminus TE$ and $grad\Phi_{sing}$ has the power-law singularity (2) at TE with an $o(1)$ remainder,

$$\frac{\partial \Phi_{sing}}{\partial n}|_\Gamma = 0 \quad , \quad \frac{\partial \Phi_{sing}}{\partial \rho} = c(Y)\rho^{-\beta(Y)} + o(1) \quad \text{as } \rho \to 0$$

From Green's representation formula for Φ_r we then find

$$grad\Phi_r(\mathbf{x}) = -grad \int_\Gamma \frac{1}{4\pi r}\frac{\partial \Phi_r}{\partial n}dS + grad \int_\Gamma \Phi_r \frac{\partial}{\partial n}(\frac{1}{4\pi r})dS \quad , \quad \mathbf{x} \in \Omega$$

where Φ_r and $\partial\Phi_r/\partial n = -\mathbf{U} \cdot \mathbf{n}$ are smooth on $\Gamma \setminus TE$. By transforming the gradient of the double layer term into a vortex layer as in (32) we obtain, since $[\Phi_r]_{TE} = 0$,

$$grad\Phi_r(\mathbf{x}) = -grad \int_\Gamma \frac{\mathbf{n} \cdot grad\Phi_r}{4\pi r}dS + curl \int_\Gamma \frac{\mathbf{n} \wedge grad\Phi_r}{4\pi r}dS \quad , \quad \mathbf{x} \in \Omega$$

This equation is nothing else but the Poincaré formula (15) for $grad\Phi_r(\mathbf{x})$. As $\rho \to 0$, the relation $\mathbf{t}^\pm \cdot grad_\Gamma \Phi_r \sim -\mathbf{t}^\pm \cdot \mathbf{U}$ holds because $v_2^\pm \equiv 0$, and from (16) and (17) we have

$$grad\Phi_r(\mathbf{x}) = \frac{1}{2\pi}\{(\mathbf{t}^+ \cdot \mathbf{U})\mathbf{n}^+ + (\mathbf{t}^- \cdot \mathbf{U})\mathbf{n}^-$$
$$-(\mathbf{n}^+ \cdot \mathbf{U})\mathbf{t}^+ - (\mathbf{n}^- \cdot \mathbf{U})\mathbf{t}^-\}\log\rho + O(1)$$

Due to the fact that $(\mathbf{n}^+, \mathbf{n}^-)$ and $(\mathbf{t}^-, \mathbf{t}^+)$ are biorthogonal bases, the terms in brackets cancel out, and no log singularity in $\mathbf{u} = grad\Phi$ arises at the edge, TE.

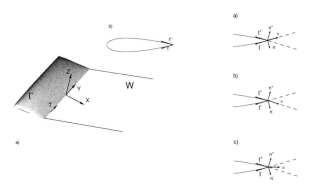

Figure 1: On the left: a) Sketch of the wing. Definition of the coordinate system and notation. Note the unit vector T aligned with the trailing edge. b) Generic section of the wing showing the vectors t^+ and t^-. On the right: the possible configurations of the wake. a) $v_2^+ = 0$, b) $v_2^- = 0$, c) $v_2^+ = v_2^- = 0$.

References

[1] P. Bassanini, C. M. Casciola, M. R. Lancia, R. Piva: A boundary integral formulation for the kinetic field in aerodynamics. II: Applications to unsteady 2D flows. *Eur. J. Mech. B/Fluids* **11**, 65-92, 1992.

[2] P. Bassanini, C. M. Casciola, M. R. Lancia, R. Piva: On the trailing edge singularity and Kutta condition for 3D airfoils. *Eur. J. Mech. B/Fluids* **15**, 809-830, 1996.

[3] P. Bassanini, C. M. Casciola, M. R. Lancia, R. Piva: A theoretical model for multiply connected wings. To appear in *Eur. J. Appl. Math.*, 1998.

[4] M. Costabel: Boundary integral operators on Lipschitz domains: elementary results. *SIAM J. Math. Anal.* **19**, 613-626, 1988.

[5] G. C. Hsiao: Solution of boundary value problems by integral equations of the first kind—An update. In: *Proc. IABEM Symposium on Boundary Integral Methods* (Eds. L. Morino & R. Piva), Rome, Italy, Oct. 15-19, 1990; Springer Verlag, 231-240, 1991.

[6] V. A. Kondrat'ev, O. A. Oleinik: Boundary value problems for partial differential equations in non-smooth domains: *Russian Math. Surveys* **38**, 1-86, 1983.

[7] J. Legras: Influence de l'attauqe oblique et de la non permanence du movement sur une aile de grand allongement. *Le cahiers de l'aerodynamique* **7**, 9-28, 1947.

[8] V. G. Maz'ya: Boundary Integral Equations. *Encyclopedia of Math. Sci.* **27**, Springer Verlag, 127-222, 1991.

[9] L. Morino: Helmholtz decomposition revisited: vorticity generation and trailing edge condition. *Comput. Mech.* **1**, 65-90, 1986.

[10] V. G. Maz'ya, B. A. Plamenevskii: Problems with oblique derivatives in regions with piecewise smooth boundaries. *Funct. Anal. Appl.* **5**, 256-257, 1971.

[11] K. W. Mangler, J. H. B. Smith: Behaviour of the vortex sheet at the trailing edge of a lifting wing. *Aero. J. Royal Aeron. Soc.* **74**, 906-908, 1970.

[12] L. M. Milne-Thomson: *Theoretical Hydrodynamics*, McMillan, London 1966.

[13] W. L. Wendland: Variational methods for BEM' In: *Proc. IABEM Symposium on Boundary Integral Methods* (Eds. L. Morino & R. Piva), Rome, Italy, Oct. 15-19, 1990; Springer Verlag, 15-34, 1991.

P. Bassanini: Universita' di Roma "La Sapienza", Dip. di Matematica,
C.M. Casciola, R. Piva: Universita' di Roma "La Sapienza",
Dip. di Meccanica e Aeronautica,
M.R. Lancia: Universita' di Roma "La Sapienza", Dip. Metodi Modelli Matematici.
Email: bassanini@mercurio.mat.uniroma1.it

M. BEBENDORF, S. RJASANOW and E. E. TYRTYSHNIKOV

Approximation using Diagonal-Plus-Skeleton Matrices

1. Introduction

Large dense matrices coming from integral equations have no explicit structure in the general case. However, it might be a good idea to approximate them by some structured matrices, for example, by circulants or block circulants (see [2, 5, 7]).

Above all, these approximations can be useful as preconditioners [7]. They match some principal part of the integral operator, yet give rather large errors as approximants. The residue, all the same, might be close to a low-rank matrix (for the operator minus its principal part is a compact operator in appropriate functional spaces). Thus, we can decrease the error if we use approximations by structured-plus-low-rank matrices.

On the other hand, large nonsingular matrices frequently contain rather large blocks close to some low-rank matrices [6]. Adding some structured matrices to the latter might lead to much higher approximation accuracy. These mixed approximants are easy to store and handle, and likely to bring about some new fast approximate matrix-vector multiplication algorithms. As yet, however, this approximation has not been well studied.

We are thus led to the following matrix approximation problem.

Given a matrix $A \in \mathbb{C}^{n \times n}$, a class $\mathcal{S}$ of structured matrices, and an integer $r \geq 0$, find an approximant $B = S + R$, where $S \in \mathcal{S}$ and rank $R \leq r$, providing the minimal possible value for $\|A - B\|_F$.

In this paper we first present a more detailed, experimental BEM motivation for this approximation problem. Then we confine ourselves to a model problem, simple enough to produce some theory yet not trivial. Taking scalar matrices as structured matrices we present an algorithm and prove its convergence for normal matrices. We could note that, rather strikingly, the greater the size the faster the convergence. This problem is certainly too particular. However, it is still of some theoretical interest, since it suggests how we could treat more practical cases. We look at this in the final section, where we discuss possible approaches and difficulties. Our main purpose is to attract attention to some intriguing problems and ideas which might be very useful in future research.

2. A BEM motivation

In this section we give a more detailed motivation to the above ideas using the following example. We consider the numerical solution of the single-layer boundary integral equation

$$-\frac{1}{2\pi} \int_\Gamma \ln|x - y| u(x) ds_x = f(y), \ y \in \Gamma.$$

This work was supported by the Volkswagen-Stiftung. The work of the third author was also supported in part by the Russian Fund for Basic Research.

This equation corresponds to the Dirichlet boundary value problem for the two-dimensional Laplace equation. Here $\Gamma \in \mathbb{R}^2$ denotes a given closed smooth two-dimensional boundary curve having a regular parametric representation

$$\Gamma = \{x \in \mathbb{R}^2, \ x = x(t), \ 0 \le t < 1, \ |\dot{x}(t)| \ge \kappa > 0\}.$$

Thus, the boundary integral equation can be rewritten as

$$-\frac{1}{2\pi}\int_0^1 \ln|x(t) - x(\tau)| \tilde{u}(t) dt = f(\tau), \ 0 \le \tau < 1,$$

where $x(\tau) = y$, $f(\tau)$ abbreviates $f(x(\tau))$ and $\tilde{u}(t) = u(x(t))|\dot{x}(t)|$. The following decomposition of the kernel of this integral operator corresponds to the continuous and exact representation of the operator as the sum of the "circulant" operator (constructed on a circle) and low-rank operator (compact operator):

$$-\frac{1}{2\pi}\int_0^1 \ln|\sin(\pi(t-\tau))|\tilde{u}(t)dt - \frac{1}{2\pi}\int_0^1 \ln\frac{|x(t)-x(\tau)|}{|\sin(\pi(t-\tau))|}\tilde{u}(t)dt = f(\tau).$$

This decomposition was used by many authors in particular to construct an efficient circulant preconditioner for this problem. The Galerkin procedure with 1-periodic ansatz and test functions leads to an algebraic system of linear equations

$$Ay = b, \ A \in \mathbb{R}^{N \times N}, \ y, b \in \mathbb{R}^N,$$

where N denotes the number of elements in the Galerkin basis, y contains the unknown coefficients of the Galerkin approximation to the solution $\tilde{u}(t)$ and b denotes the corresponding right-hand side. Due to the above decomposition of the kernel the matrix A decomposes in a sum $A = A_1 + A_2$ of a circulant A_1 and a remainder matrix A_2. In order to illustrate our ideas we first cut the matrix A generated for the ellipse

$$x(t) = (\cos(2\pi t), 0.1\sin(2\pi t))^T, \ 0 \le t < 1$$

into quadratic blocks of size equal to $N/2$ or $N/4$. The corresponding blocks of the matrix A_1 are then obviously of the Toeplitz structure. The blocks of the matrix A_2 can be approximated with the relative accuracy 10^{-6} (in the Frobenius norm) using the following number of skeletons (rank–one matrices):

13	55
55	13

8	5	8	29
5	8	29	8
8	29	8	5
29	8	5	8

The numbers in the above tables are almost independent of the dimension N of the system matrix. For $N = 512$ (which is large enough for two-dimensional problems) the storage required for the compressed matrix will be 25.7% in the first and 19.5% in the second case. For $N = 1024$ this storage will be about 13% and 9%, respectively; for $N = 2048$ it is expected to reduce to 6.5% and 4.5%, and so on (if we stick to the 10^{-6} accuracy).

In the above example, the structured component was known explicitly. In spite of the soundness of our reason for taking it from the circulant accounting for the principal part, we might ask if it is the best choice. The answer is no. For example, we can produce some other Toeplitz approximants to the blocks with less skeletons:

7	31
31	7

8	3	8	16
3	8	16	8
8	16	8	3
16	8	3	8

For the accuracy 10^{-4} (which might be quite sufficient), the Toeplitz approximants taken from the circulant matching the principal part of the integral operator yield the following numbers of skeletons:

8	34
34	8

5	3	5	18
3	5	18	5
5	18	5	3
18	5	3	5

Again, some other Toeplitz approximants could provide us with less skeletons:

5	18
18	5

4	2	5	11
2	4	11	5
5	11	5	2
11	5	2	5

Now the "analytical" Toeplitz approximants are noticeably worse than the "algebraic" ones.

These examples inspire us to study the above algebraic approximation problem in some depth. However, it seems new and not easy to tackle by standard means.

In the attempts to obtain some theory, we now simplify the problem by getting to very simple structured approximants such as scalar matrices. Then we try to extend the findings and inquire further into the difficulties.

3. Model problem

Given $A \in \mathbb{C}^{n \times n}$, let us minimise $\|A - \alpha I - R\|_F$, where $\alpha \in \mathbb{C}$ and R is a skeleton (in other words, a matrix of the form uv^*, $u, v \in \mathbb{C}^n$). The minimum of $\|A - \alpha I - R\|_F$ is always attained at some $\tilde{\alpha}$ and $\tilde{R}$.

Two extreme problems are naturally related to ours. First, the functional $\|A - \alpha I\|_F$ attains its minimal value iff

$$\alpha = \frac{1}{n} \operatorname{trace}(A).$$

Second, we are aware that

$$\min_{\operatorname{rank} R \leq r} \|A - R\|_F = \sqrt{\sum_{i=r+1}^{n} \sigma_i^2},$$

where $\sigma_1 \geq \cdots \geq \sigma_n$ are the singular values of A. If $A = \sum_{i=1}^{n} \sigma_i u_i v_i^*$ is the singular value decomposition of A, then the minimum is attained at $R = \sum_{i=1}^{r} \sigma_i u_i v_i^*$. This is well known and goes back to Eckart and Young (see [4]). Let us write

$$R = \operatorname{skeleton}_r A.$$

Note that R can be computed by the SVD algorithm or by the Lanczos biorthogonalisation procedure (see [1]). Sometimes R is not unique; if so then we mean by skeleton$_r$ A any matrix providing the minimum. The extreme cases suggest that, in the mixed case, we can proceed as follows.

The IR algorithm:

$$R_0 = 0;$$
$$\alpha_i = \frac{1}{n}\,\text{trace}\,(A - R_{i-1}), \quad R_i = \text{skeleton}_1\,(A - \alpha_i I), \quad i = 1, 2, \ldots.$$

It is easy to verify that $\|A - D_i - R_i\|_F$ decreases monotonously in i. However, the IR algorithm can produce a local minimum.

Theorem 1. *For any normal matrix A of order n, the* skeleton$_1$ *operation can be implemented so that the IR algorithm converges to the minimum linearly:*

$$\|(\alpha_i I + R_i) - (\tilde{\alpha} I + \tilde{R})\|_F \leq \frac{c}{\sqrt{n-1}\,n^i}$$

with

$$c = \max_k |\sum_{j \neq k}(\lambda_j - \lambda_k)|,$$

where $\lambda_1, \ldots, \lambda_n$ are a complete set of eigenvalues of A.

Proof. Since A is normal, it can be expressed by its spectral decomposition

$$A = \sum_{j=1}^{n} \lambda_j q_j q_j^*,$$

where $q_1, \ldots, q_n$ are the orthonormal eigenvectors of A and $\lambda_1, \ldots, \lambda_n$ are the corresponding eigenvalues. In line with the IR algorithm,

$$\alpha_1 = \frac{1}{n}\sum_{j=1}^{n}\lambda_j, \quad A - \alpha_1 I = \sum_{j=1}^{n}(\lambda_j - \alpha_1)q_j q_j^*.$$

For any α, denote by $k = k(\alpha)$ any index such that

$$|\lambda_k - \alpha| \geq |\lambda_j - \alpha| \quad \forall\, j.$$

It is not difficult to see that we may set

$$R_1 = \text{skeleton}_1\,(A - \alpha_1 I) = (\lambda_k - \alpha_1)q_k q_k^*, \quad k = k(\alpha_1).$$

Consequently,

$$\alpha_2 = \frac{1}{n}\left(\alpha_1 + \sum_{j \neq k(\alpha_1)} \lambda_j\right).$$

It is important that $k = k(\alpha_1)$ remains among the possible values for $k(\alpha_2)$. To prove this, note that $\lambda_{k(\alpha_2)} = \lambda_{k(\alpha_1)}$:

$$\alpha_1 - \alpha_2 = \frac{\lambda_k - \alpha_1}{n},$$

and if $\lambda_k \neq \lambda_j$,

$$\begin{aligned}|\lambda_j - \alpha_2| &\leq |\lambda_j - \alpha_1| + |\alpha_1 - \alpha_2| \\ &\leq |\lambda_k - \alpha_1| + |\alpha_1 - \alpha_2| \\ &= \frac{n+1}{n}|\lambda_k - \alpha_1| = |\lambda_k - \alpha_2|.\end{aligned}$$

The equal sign in the first estimate holds only if $\lambda_j - \alpha_1$ and $\lambda_k - \alpha_1$ are linearly dependent, i.e. $\lambda_j - \alpha_1 = c(\lambda_k - \alpha_1)$, $c \in \mathbb{R}$. From the definition of k and $\lambda_j \neq \lambda_k$ we deduce that $-1 \leq c < 1$.
Thus

$$\begin{aligned}\lambda_j - \alpha_2 &= \lambda_j - \alpha_1 + \alpha_1 - \alpha_2 \\ &= \lambda_j - \alpha_1 + \frac{1}{n}(\lambda_k - \alpha_1) \\ &= (c + \frac{1}{n})(\lambda_k - \alpha_1).\end{aligned}$$

Hence in any case we have $|\lambda_j - \alpha_2| < \frac{n+1}{n}|\lambda_k - \alpha_1|$.
Thus, we may take $k = k(\alpha_2)$, and eventually $k = k(\alpha_i)$ for all i. To prove the latter, we proceed by induction to

$$\alpha_{i-1} - \alpha_i = \frac{\lambda_k - \alpha_1}{n^{i-1}}, \qquad (*)$$

and, therefore,

$$\begin{aligned}\lambda_k - \alpha_i &= (\lambda_k - \alpha_1) + (\alpha_1 - \alpha_2) + \cdots + (\alpha_{i-1} - \alpha_i) \\ &= (\lambda_k - \alpha_1)\left(1 + \frac{1}{n} + \cdots + \frac{1}{n^{i-1}}\right) \\ &= (\lambda_k - \alpha_1)\frac{1 - 1/n^i}{1 - 1/n}.\end{aligned}$$

Using this we obtain

$$\begin{aligned}|\lambda_j - \alpha_i| &\leq |\lambda_j - \alpha_{i-1}| + |\alpha_{i-1} - \alpha_i| \\ &< |\lambda_k - \alpha_1|\left(\frac{1 - 1/n^{i-1}}{1 - 1/n} + \frac{1}{n^{i-1}}\right) \\ &\leq |\lambda_k - \alpha_1|\frac{1 - 1/n^i}{1 - 1/n} = |\lambda_k - \alpha_i|.\end{aligned}$$

It follows that we may take $k(\alpha_i) = k(\alpha_{i-1})$, and hence, $k(\alpha_i)$ can be kept the same on all iterations.

Little effort is now needed to prove that

$$\alpha_i \to \alpha_0 \equiv \frac{1}{n-1} \sum_{j \neq k} \lambda_j.$$

Evidently, we may take $k(\alpha_0) = k$, too. Moreover, the relation (∗) implies that

$$\alpha_0 - \alpha_i = \frac{\alpha_1 - \lambda_k}{n^i} \left(1 + \frac{1}{n} + \frac{1}{n^2} + \ldots\right) = \frac{\alpha_1 - \lambda_k}{(n-1)\, n^{i-1}},$$

and therefore results in the estimate claimed for the convergence rate.

We still need to prove that the IR algorithm converges to the global minimum. In our case, the latter is computed explicitly. As is readily seen, the squared minimum is equal to

$$m^2 = \min_{\alpha,u,v} \|A - \alpha I - uv^*\|_F^2 = \min_\alpha \sum_{j \neq k(\alpha)} |\lambda_j - \alpha|^2 = \min_k \sum_{j \neq k} \left|\lambda_j - \frac{1}{n-1} \sum_{l \neq k} \lambda_l\right|^2.$$

Furthermore,

$$\begin{aligned}
m^2 &= \min_k \left(\sum_{j \neq k} \left|(\lambda_j - \alpha_1) - \frac{1}{n-1} \sum_{l \neq k}(\lambda_l - \alpha_1)\right|^2\right) \\
&= \min_k \left(\sum_{j \neq k} |\lambda_j - \alpha_1|^2 - \frac{1}{n-1} \left|\sum_{j \neq k}(\lambda_j - \alpha_1)\right|^2\right) \\
&= \min_k \left(\sum_{j \neq k} |\lambda_j - \alpha_1|^2 - \frac{1}{n-1} |\lambda_k - \alpha_1|^2\right) \\
&= \sum_{j=1}^n |\lambda_j - \alpha_1|^2 - \frac{n}{n-1} \max_k |\lambda_k - \alpha_1|^2 \\
&= \sum_{j=1}^n |\lambda_j - \alpha_1|^2 - \frac{n}{n-1} |\lambda_{k(\alpha_1)} - \alpha_1|^2 \\
&= \sum_{j \neq k(\alpha_1)} |\lambda_j - \alpha_0|^2
\end{aligned}$$

and since $k(\alpha_1) = k(\alpha_0)$, this completes the proof. □

4. Discussion

The idea behind the IR algorithm suggests a more or less general approach. For example, let $\mathcal{S}$ be the set of diagonal matrices. It should also be borne in mind that anything we can say in this case applies equally to circulant matrices in the role of $\mathcal{S}$. This follows from the well-known fact that any circulant C is of the form $C = FDF^*$ for some diagonal matrix D; here F is a unitary matrix which is the same for all circulants. Specifically, F is the so-called *Fourier matrix* (cf. [5, 7]).

Two extreme cases of the $D + R$ approximation are one with $R = 0$ and another with $D = 0$. Both are no problem. By analogy with the IR algorithm, we can proceed as follows.

The DR algorithm:

$$R_0 = 0;$$
$$D_i = \operatorname{diag}(A - R_{i-1}), \quad R_i = \operatorname{skeleton}_r(A - D_i), \quad i = 1, 2, \ldots.$$

It is easy to verify that $||A - D_i - R_i||_F$ decreases monotonously in i. However, the DR algorithm can produce a local minimum. We are still uncertain about when this occurs, about how the algorithm behaves and what can really be proved. One source of difficulties is the following.

The set of diagonal matrices D of order n is obviously closed, as is the set of matrices R with rank $R \leq r$. However, matrices of the form $D + R$ are not a closed set.

We produce a sequence of matrices $A_k = D_k + R_k$, with rank $R_k = 1$ and D_k a diagonal matrix, that converges, as $k \to \infty$, to a matrix A that enjoys the inequality rank $(A - D) \geq 2$ for any diagonal matrix D. The latter condition means that A cannot be split as $A = D + R$ with rank $R \leq 1$.

Let $n = 3$ and set

$$D_k = \begin{bmatrix} -k^2 & 0 & 0 \\ 0 & 0 & 0 \\ 0 & 0 & 0 \end{bmatrix}, \quad R_k = \begin{bmatrix} k \\ 1/k \\ 1/k \end{bmatrix} \begin{bmatrix} k & 1/k & 1/k \end{bmatrix}.$$

It is easy to check that

$$D_k + R_k \to A = \begin{bmatrix} 0 & 1 & 1 \\ 1 & 0 & 0 \\ 1 & 0 & 0 \end{bmatrix}.$$

Now, take up a diagonal matrix D and estimate the rank of $A - D$. The minor located in the columns 1 and 3 and rows 1 and 2 is

$$\det \begin{bmatrix} -d_1 & 1 \\ 1 & 0 \end{bmatrix} = -1 \quad \forall\, d_1 \quad \Rightarrow \quad \operatorname{rank}(A - D) \geq 2 \quad \forall\, D.$$

Note that, in contrast to the result obtained, matrices of the form $\alpha I + R$ with rank $R \leq 1$ constitute a closed set. This is probably why we are going to have more success for this particular case.

We stress that the convergence proof for the IR algorithm was obtained under the assumption that, for the skeleton$_1$ operations involved, we choose some suitable implementations among all possible (and, in any other respect, equivalent) ones. This means, unfortunately, that when using the SVD algorithm straightforwardly we are not always guaranteed that the IR algorithm does not stall.

Note also that the convergence is linear, and, which may be amusing, the convergence factor is equal to $1/n$, so the IR converges faster for larger n.

For the DR algorithm, we still have a convergence theorem that stems from the IR case.

Theorem 2. *If $A \in \mathbb{C}^{n \times n}$ is a circulant, then, in the case $r = 1$, the skeleton$_1$ operations can be implemented so that the DR algorithm converges to the global minimum linearly.*

We realise, of course, that we have only made first steps, and that what we know about the $D + R$ approximation is far less than we wish to know. This applies even more so to more complicated structured matrices. Still, the DR algorithm can be easily written for the latter cases: we performed numerous experiments with it, and a way was found to obtain a really robust procedure [3]. It yielded the data with less skeletons as is shown in Section 2. This procedure [3] could be regarded as a (nontrivial) extension of the DR, yet it appeared after revising the setting of the approximation problem in question. The work on this procedure is still in progress and will be reported in a forthcoming paper.

References

[1] G. H. Golub, C. F. Van Loan: *Matrix Computations*, The Johns Hopkins University Press, Baltimore and London, 1989.

[2] S. Rjasanow: Effective algorithms with circulant-block matrices, *Linear Algebra Appl.* **202** (1994), 55–69.

[3] S. Rjasanow, E. E. Tyrtyshnikov: *The ATS algorithm*, manuscript, 1998.

[4] G. W. Stewart, J. Sun: *Matrix Perturbation Theory*, Academic Press, San Diego, 1990.

[5] E. E. Tyrtyshnikov: Optimal and superoptimal circulant preconditioners, *SIAM J. Matrix Anal. Appl.* **13** (2) (1992), 459–473.

[6] E. E. Tyrtyshnikov: Mosaic-skeleton approximations. *Calcolo* **33** (1996), 47–58.

[7] E. E. Tyrtyshnikov: *A Brief Introduction to Numerical Analysis*, Birkhäuser, Boston, 1997.

M. Bebendorf and S. Rjasanow: FB 9 – Mathematik, Universität des Saarlandes
Postfach 151150, 66041 Saarbrücken, Germany
Email: rjasanow@num.uni-sb.de

E. E. Tyrtyshnikov: Inst. of Numerical Mathematics, Russian Academy of Sciences, Gubkina 8, Moscow 117333, Russia
Email: tee@inm.ras.ru

M. BEN TAHAR, C. GRANAT and T. HA-DUONG

Variational integral formulation in the problem of elastic scattering by a buried obstacle

1. Introduction

Elastic wave propagation and scattering in solids has been under investigation in a wide range of engineering applications such as seismology, geophysics or nondestructive testing. In these applications, the solid medium is often semi-infinite and bounded by an infinite surface. Since the domain is infinite, the method of calculation is a boundary integral equation (BIE) technique. It is based on the potential theory.

The purpose of this paper is to analyze a two–dimensional half-plane elastodynamics problem of scattering by a buried obstacle. Since the half-plane boundary is of infinite length, the "classical" Boundary Integral Equations Method can not be applied directly. We here use an adapted Green's function taking into consideration the boundary condition on the infinite surface. Therefore, there is no need to model the infinite surface. This type of Green's function in the frequency domain has been constructed by Kobayashi in 1983 [1]. It can also be found in [2]. The classical paper of Lapwood [3] should also be mentioned. The use of this adapted Green's function, despite the reduction in discretization, increases the computational effort due to the high complexity of the involved functions.

In the present work, instead of using a collocation method as Nishimura did [2], a variational approach is developed.

The geometry and the governing equations of the problem are presented in Section 2. First of all, the Green's function adapted to the half–plane problem is incorporated in the BIE. Then, in Section 4, a variational formulation is associated to the BIE. The difficulties coming up within the formulation (hypersingularity of the BIE, computation of the kernel that contains the additional term of the adapted Green's function) are discussed. We will see that thanks to a regularization process [4] which removes the hypersingularity of the BIE, the variational formulation can be explicited on a tractable form ready to be implemented. This leads to a symmetrical linear matrix system, after discretization by boundary finite elements. This linear system can then be solved by conventional numerical algorithms.

2. Geometry and governing equations

The configuration of the problem is depicted in Figure 1. We deal with the propagation of a scattered wave displacement $\vec{u}^{sc}$ emitted by an incident plane wave $\vec{u}^{inc}$ in an isotropic and homogeneous elastic half-plane Ω bounded by an infinite traction free surface Γ. Thus, the characteristics of Ω are its density ρ and Lamé parameters λ, μ.

The scattering obstacle of arbitrary shape is bounded by a surface S on which the traction is known. The incident wave has a harmonic time dependence $(e^{-i\omega t})$. The state of plane strain is assumed for the analysis of this problem.

This work was supported by the "Conseil Regional de Picardie" under the regional group of modeling (project N 94-8).

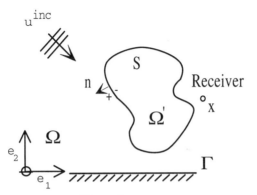

Figure 1:
Geometry of the problem

The scattered displacement field

The scattered displacement field $\vec{u}^{sc}$ satisfies:
the Navier equations

$$\nabla \cdot \sigma(\vec{u}^{sc}(x)) + \rho\omega^2 \vec{u}^{sc}(x) = \vec{0}, \qquad \forall x \in \Omega, \qquad (1)$$

with the position vector $x = (x_1, x_2)$;
the boundary conditions on the surface of the obstacle

$$\sigma(\vec{u}^{sc}(x))\vec{n} = \vec{t_0}(x), \qquad \forall x \in S, \qquad (2)$$

where $\vec{n}$ is the unit normal vector drawn out of the barrier to the elastic medium and $\vec{t_0}$ is the traction on S;
the boundary condition on the infinite surface

$$\sigma(\vec{u}^{sc}(x))\vec{e_2} = \vec{0}, \qquad \forall x \in \Gamma, \qquad (3)$$

and the radiation condition conditions at infinity

$$\lim_{R \to \infty} \sqrt{R} \left| \frac{\partial \vec{u}^{sc}_\alpha(R)}{\partial R} - ik_\alpha \vec{u}^{sc}_\alpha(R) \right| = 0 \quad \text{with} \quad R = |x| \text{ in } \Omega, \qquad (4)$$

where the subscript α is assigned to either a P–wave or a SV–wave according to the decomposition in Lamé potentials $\vec{u}^{sc} = \vec{u}^{sc}_p + \vec{u}^{sc}_{sv}$,
the regularity condition

$$\lim_{R \to \infty} |\vec{u}^{sc}_\alpha(x)| = 0 \quad \text{with} \quad R = |x|. \qquad (5)$$

These conditions guarantee that only an outward flow of energy is allowed at infinity. This eliminates reflections from infinity, as one expects physically.

Let us remark, that radiations in the theory of elasticity (isotropic case) have been first formulated by V.D. Kupradze [8].

In order to determine $\vec{u}^{sc}$ an indirect integral representation based on a double layer potential is used and consequently, a fundamental solution has to be established.

3. Expression of the half–plane fundamental solution

The major difficulty in the present formulation is the infinite feature of the traction free surface Γ. In order to avoid the discretization of this surface in the boundary element method (BEM), we use a special half-plane Green's matrix [2]. This Green's matrix satisfies:

$$\nabla \cdot \sigma(G(x,y)) + \rho\omega^2 G(x,y) = -\delta(x-y)I, \quad \forall (x,y) \in \Omega \times \Omega, \quad (6)$$
$$\sigma(G(x,y))\vec{e_2} = \vec{0}I, \quad \forall x \in \Gamma, \quad (7)$$

and the conditions at infinity.

To calculate G, the method of images can be used. If G_0 is the infinite 2D Green's function satisfying the radiation conditions (4 - 5), one can represent G as

$$G(x,y) = G_0(x,y) + G_0(x,y\prime) + P(x,y) \quad (8)$$

where $y\prime$ is the image under reflection of y with respect to Γ.

G_0 is also called the Helmholtz's solution in plane strain [5]

$$G_0(x,y) = \frac{i}{4\mu}\left[H_0^1(k_s r)I + \frac{1}{k_s^2}\vec{\nabla}\otimes\vec{\nabla}(H_0^1(k_s r) - H_0^1(k_p r))\right] \quad (9)$$

where $r = |x - y|$, k_p and k_s represent the wave numbers of the pressure wave and the shear wave respectively, and μ is a Lamé elastic constant.

Contrarily to the case of the free surface acoustic half–plane, the presence of an additional term $P(x,y)$ in (8) is imposed by the tangential derivatives in (7). One meets a similar situation when calculating the Green's function for an acoustic half–plane with impedance condition on Γ [6]: $\frac{\partial p(x)}{\partial n} - ik\beta p(x) = 0$ (where p denotes the pressure field in this acoustic problem).

It is clear that P satisfies the following equations:

$$\nabla \cdot \sigma(P(x,y)) + \rho\omega^2 P(x,y) = 0, \quad \forall (x,y) \in \Omega \times \Omega, \quad (10)$$
$$\sigma(P(x,y))\vec{e_2} = -\sigma(G_0(x,y) + G_0(x,y\prime))\vec{e_2}, \quad \forall x \in \Gamma, \quad (11)$$

and the radiation conditions at infinity.

Then, the expression of $P(x,y)$ is obtained by an inverse partial Fourier transform with respect to x_1. It results a tensor of four contour integrals in the complex plane. One can find the following expression in Nishimura's thesis [2]:

$$P(x,y) = FT^{-1}_{\xi \to (x_1-y_1)} \frac{1}{\mu k_s^2 F(\xi)} \begin{bmatrix} A & B \\ C & D \end{bmatrix}, \quad (12)$$

where $F(\xi) = (2\xi^2 - k_s^2)^2 - 4\xi^2 R_p(\xi) R_s(\xi)$ is the Rayleigh function,

$A = -\frac{i\xi^2}{R_p}(2R_p R_s e^{-x_2 R_s} - (2\xi^2 - k_s^2)e^{-x_2 R_p})(2R_p R_s e^{-y_2 R_s} - (2\xi^2 - k_s^2)e^{-y_2 R_p})$,
$B = i\xi(2\xi^2 e^{-x_2 R_p} - (2\xi^2 - k_s^2)e^{-x_2 R_s})(2R_p R_s e^{-y_2 R_p} - (2\xi^2 - k_s^2)e^{-y_2 R_s})$,
$C = -i\xi(2\xi^2 e^{-x_2 R_s} - (2\xi^2 - k_s^2)e^{-x_2 R_p})(2R_p R_s e^{-y_2 R_s} - (2\xi^2 - k_s^2)e^{-y_2 R_p})$,
$D = -\frac{\xi^2}{R_s}(2R_p R_s e^{-x_2 R_p} - (2\xi^2 - k_s^2)e^{-x_2 R_s})(2R_p R_s e^{-y_2 R_p} - (2\xi^2 - k_s^2)e^{-y_2 R_s})$.

The points x and y have the following components: $x = (x_1, x_2)$, $y = (y_1, y_2)$. Here $R_\alpha(\xi)$ ($\alpha = p$ or s) is the square root of $\xi^2 - k_\alpha^2$ with positive real part.

Despite the complexity of the functions involved in the correction term P, computation of P is conventional in principle; it has been presented in [2]. To clarify, we will detail the computation of a typical term of P in Section 5.

4. Integral representation and variational formulation

To find an integral representation of the scattered displacement, we use an indirect method based on a double layer potential. That corresponds to introduce the following interior problem:

$$\nabla \cdot \sigma(\vec{u}^{sc}(x)) + \rho\omega^2 \vec{u}^{sc}(x) = \vec{0}, \quad \forall x \in \Omega', \quad (13)$$
$$\sigma(\vec{u}^{sc-}(x))\vec{n} = \vec{t_0}(x), \quad \forall x \in S. \quad (14)$$

Since a half-plane Green's function satisfies the free boundary condition on Γ and the conditions at infinity is employed, $\vec{u}^{sc}$ admits the following integral representation:

$$\int_S (\sigma_y(G(x,y))\vec{n}_y)^T \vec{\varphi}(y) dS(y) = \begin{cases} \vec{u}^{sc}(x) & \forall x \in \Omega, \\ \frac{1}{2}(\vec{u}^{sc+}(x) + \vec{u}^{sc-}(x)) & \forall x \in S, \end{cases} \quad (15)$$

where $\vec{\varphi}$ represents the jump of $\vec{u}^{sc}$ through the surface S. The latter integral (15) has to be understood as a Cauchy principal value if $x \in S$.

The integral representation shows that the scattered displacement at any point in the elastic domain Ω or on the obstacle surface S can be evaluated from the value of the jump of $\vec{u}^{sc}$ through the boundary only.

The next step consists in determining the unknown density $\vec{\varphi}$.

Applying the boundary conditions (2) and (14) on the surface S to the integral representation of the scattered displacement, one gets:

$$\sigma_x\left(\int_S (\sigma_y(G(x,y))\vec{n}_y)^T \vec{\varphi}(y) dS(y)\right) \vec{n}_x = \vec{t_0}(x), \quad \forall x \in S. \quad (16)$$

At this step, a system of integral equations (16) with one unknown vector function $\vec{\varphi}$ is obtained. In general, however, (13), (14) might have for specific ω a nontrivial kernel; and also (16) might not be equivalent to (13), (14) at so-called irregular frequencies. These cases are avoided in what follows.

In (16) the kernel related to the full space Green's function is hypersingular. Then, in the next part, we will associate a variational formulation with the integral equation system (16) and use a regularization process [4].

The associated variational formulation

The associated variational formulation has the following expression:

$$\underbrace{\int_S \int_S \vec{\psi}(x) \left(\sigma_x (\sigma_y G(x,y) \vec{n}_y)^T \vec{\varphi}(y) \right) \vec{n}_x dS(y) dS(x)}_{A(\vec{\psi}, \vec{\varphi})} = \underbrace{\int_S \vec{\psi}(x) \vec{t_0}(x) dS(x)}_{L(\vec{\psi})}, \quad (17)$$

where $A(\vec{\psi}, \vec{\varphi})$ is a bilinear form, $L(\vec{\psi})$ is a linear form and $\vec{\psi}$ represents a regular test function defined on S. Through the next paragraphs we will show that $A(\vec{\psi}, \vec{\varphi})$ is symmetrical and can be written with only weakly singular kernels.

Symmetry of the bilinear form $A(\vec{\psi}, \vec{\varphi})$

Considering the principle of virtual work or applying the first Green's formula to the exterior problem Ω as well as to the interior problem Ω', one obtains:

$$\int_S \vec{v}^+(x) \cdot \sigma(\vec{u}^{sc+}(x)) \vec{n} \, dS(x) = \int_\Omega \left(\rho \omega^2 \vec{v}(x) \cdot \vec{u}^{sc}(x) - \sigma(\vec{u}^{sc}(x)) : \varepsilon(\vec{v}(x)) \right) dV(x)$$

$$\int_S \vec{v}^-(x) \cdot \sigma(\vec{u}^{sc-}(x)) \vec{n} \, dS(x) = \int_{\Omega'} \left(-\rho \omega^2 \vec{v}(x) \cdot \vec{u}^{sc}(x) + \sigma(\vec{u}^{sc}(x)) : \varepsilon(\vec{v}(x)) \right) dV(x)$$

with $\vec{v}(x)$ an arbitrary regular function in $\Omega \cup \Omega'$.
Since $\sigma(\vec{u}^{sc+}(x)) \vec{n} = \sigma(\vec{u}^{sc-}(x)) \vec{n} \equiv \vec{t_0}(x) \quad \forall x \in S$, one can easily find:

$$\int_S \left(\vec{v}^+(x) - \vec{v}^-(x) \right) \cdot \vec{t_0}(x) dS(x)$$
$$= \int_{\Omega \cup \Omega'} \left(\rho \omega^2 \vec{v}(x) \cdot \vec{u}^{sc}(x) - \sigma(\vec{u}^{sc}(x)) : \varepsilon(\vec{v}(x)) \right) dV(x).$$

We can denote $\vec{v}^+(x) - \vec{v}^-(x) = \vec{\psi}(x)$, thus, knowing (17)

$$\int_S \left(\vec{v}^+(x) - \vec{v}^-(x) \right) \cdot \vec{t_0}(x) dS(x) = L(\vec{\psi}).$$

As an isotropic elastic material has been considered for Ω and Ω',

$$\sigma(\vec{u}^{sc}(x)) : \varepsilon(\vec{v}(x)) = \sigma(\vec{v}(x)) : \varepsilon(\vec{u}^{sc}(x)),$$

then

$$\int_{\Omega \cup \Omega'} \left(\rho \omega^2 \vec{v}(x) \cdot \vec{u}^{sc}(x) - \sigma(\vec{u}^{sc}(x)) : \varepsilon(\vec{v}(x)) \right) dV(x)$$

is symmetrical with respect to $\vec{u}^{sc}$ and $\vec{v}(x)$; consequently $A(\vec{\psi}, \vec{\varphi})$ is a symmetrical bilinear form in terms of the jump functions $\vec{\varphi}$ and $\vec{\psi}$.

Therefore, the solution of (16) is given by the stationarity of the functional

$$F(\vec{\varphi}) = \frac{1}{2}A(\vec{\varphi},\vec{\varphi}) - L(\vec{\varphi}).\tag{18}$$

Regularization of the double layer integral

Taking into account the splitting (8), $A(\vec{\psi},\vec{\varphi})$ can be expressed as the sum of three bilinear forms:

$$A(\vec{\psi},\vec{\varphi}) = b_0(\vec{\psi},\vec{\varphi}) + b_0'(\vec{\psi},\vec{\varphi}) + b_P(\vec{\psi},\vec{\varphi})\tag{19}$$

where b_0 and b_0' are respectively related to the Helmholtz solution G_0 (9) and its image. b_P is related to the correction term P (12).

Since the kernel $\sigma(G_0(x,y))\vec{n}$ is hypersingular, b_0 is transformed through a regularization process [4]. This process rewrites b_0 as the sum of two parts containing weakly singular kernels:

$$\underbrace{\int_S\int_S \vec{\psi}(x) \cdot \left(\sigma_x\left(\sigma_y G_0(x,y)\vec{n}_y\right)^T \vec{\varphi}(y)\right) \vec{n}_x dS(y) dS(x)}_{b_0(\vec{\psi},\vec{\varphi})}$$

$$= \underbrace{\int_S\int_S \frac{\partial \vec{\psi}(x)}{\partial S} \cdot F(x,y) \frac{\partial \vec{\varphi}(y)}{\partial S} dS(x) dS(y)}_{b_1(\vec{\psi},\vec{\varphi})}$$

$$+ \underbrace{\int_S\int_S \vec{\psi}(x) \cdot \mathbf{N}_x \rho\omega^2 R(x,y) \cdot \mathbf{N}_y^T \vec{\varphi}(y) dS(x) dS(y)}_{b_2(\vec{\psi},\vec{\varphi})},$$

where F and R are weakly singular kernels expressed as some second derivatives of the kernel N:

$$N(r) = \frac{i}{4}\left(H_0^1(k_s r) - H_0^1(k_p r)\right), \quad r = |x-y|,$$

$$F(x,y) = \frac{4\mu^2}{\rho\omega^2}\begin{bmatrix} \partial_{22}^2 N(r) & -\partial_{12}^2 N(r) \\ -\partial_{12}^2 N(r) & \partial_{11}^2 N(r) \end{bmatrix},$$

$$R(x,y) = \frac{1}{\lambda+\mu}\left(\frac{\mu}{\rho\omega^2}\left(\mathbf{D}(r) - \nabla^2 N(r)\mathbf{A}\right) - N(r)\mathbf{C}\right)$$

with

$$\mathbf{D}(r) = \begin{bmatrix} 4(\lambda+\mu)\partial_{11}^2 N(r) & 2\lambda(\partial_{11}^2+\partial_{22}^2)N(r) & 2(\lambda+\mu)\partial_{12}^2 N(r) \\ 2\lambda(\partial_{11}^2+\partial_{22}^2)N(r) & 4(\lambda+\mu)\partial_{22}^2 N(r) & 2(\lambda+\mu)\partial_{12}^2 N(r) \\ 2(\lambda+\mu)\partial_{12}^2 N(r) & 2(\lambda+\mu)\partial_{12}^2 N(r) & \mu(\partial_{11}^2+\partial_{22}^2)N(r) \end{bmatrix},$$

$$\mathbf{A} = \begin{bmatrix} 5\lambda+6\mu & 3\lambda & 0 \\ 3\lambda & 5\lambda+6\mu & 0 \\ 0 & 0 & \lambda+3\mu \end{bmatrix},$$

$$\mathbf{C} = \begin{bmatrix} \lambda+2\mu & \lambda & 0 \\ \lambda & \lambda+2\mu & 0 \\ 0 & 0 & \mu \end{bmatrix} \text{ is the matrix of the elastic constants, and}$$

$$\mathbf{N}_x = \begin{bmatrix} n_1(x) & 0 & n_2(x) \\ 0 & n_2(x) & n_1(x) \end{bmatrix} \text{ is the tensor of the normal components at a point } x.$$

Remark 1.
This regularization process consists in building an expression of the bilinear form in terms of locally integrable kernels. To circumvent the hypersingularity, tangential derivatives of the unknown and test functions have been introduced. This idea was also followed by [7] for the Helmholtz equation.

Remark 2.
We describe the asymptotic behavior of N and its second derivatives when r tends to zero:
$$N(r) \simeq -\frac{1}{2\pi}\log\frac{k_s}{k_p}, \quad \partial_{ij}^2 N(r) \simeq Const - \frac{1}{4\pi}\left(k_p^2 \log\frac{k_p r}{2} - k_s^2 \log\frac{k_s r}{2}\right)\delta_{ij},$$
where $Const$ is independent of r.

$G_0(x, y')$ occurred to be singular when the surface of the buried obstacle S comes into contact with Γ. This particular case will not be developed in this paper.

According to the general expression of its terms in (24), $P(x,y)$ is not singular.

Thereby, all integrals in (17) are well-defined and their numerical computation is conventional.

Discretization by a finite element method

Taking into account the former regularization process, no integral in the variational formulation is singular. Thus, a discretization of the functional F with straight linear isoparametric finite elements for the boundary S can be performed. One gets after assembly:

$$F(\vec{\varphi}) = \frac{1}{2}\{\varphi\}\cdot D\{\varphi\} - \{\varphi\}\cdot\{t_0\}, \tag{20}$$

where $\{\varphi\}$ represents the vector of the nodal values of $\vec{\varphi}$.

The stationarity of the discretized functional (20) leads to the following linear symmetrical matrix system:

$$D\{\varphi\} = \{t_0\}. \qquad (21)$$

On the other hand, one can prove that the bilinear form $A(\vec{\psi}, \vec{\varphi})$ in (18) corresponds to a pseudo-differential operator of order 1, satisfying a Gårding inequality. That permits the numerical analysis of the discretization process by the finite element method as indicated for example in [9].

Finally, the application of the integral representation (15) gives the scattered displacement at any receiver point in the elastic medium Ω.

5. Numerical method to compute each term of the correction term

Expression of the bilinear form b_P in terms of Fourier Transform.

At this step of the study, the process followed for the computation of the bilinear form b_P related to the correction term is presented. It has not been developed above because we wished to clarify its principle.

According to (19)

$$b_P(\vec{\psi}, \vec{\varphi}) = \int_S \int_S \vec{\psi}(x) \cdot \left(\sigma_x \left(\sigma_y P(x,y) \vec{n}_y\right)^T \vec{\varphi}(y)\right) \cdot \vec{n}_x dS(y) dS(x)$$

$$= \int_S \int_S \vec{\psi}(x) \cdot \mathbf{N}_x \mathbf{C} E_P(x,y) \mathbf{C} \mathbf{N}_y^T \vec{\varphi}(y) dS(x) dS(y)$$

where

$$E_P(x,y) = \begin{bmatrix} P_{11,1\bar{1}} & P_{12,2\bar{1}} & P_{11,2\bar{1}} + P_{12,1\bar{1}} \\ P_{21,1\bar{2}} & P_{22,2\bar{2}} & P_{21,2\bar{2}} + P_{22,1\bar{2}} \\ P_{11,1\bar{2}} + P_{21,1\bar{1}} & P_{12,2\bar{2}} + P_{22,2\bar{1}} & P_{11,2\bar{2}} + P_{22,1\bar{2}} + P_{11,2\bar{1}} + P_{22,1\bar{1}} \end{bmatrix} \qquad (22)$$

with $P_{ij,n\bar{m}} = \frac{\partial^2 P_{ij}(x,y)}{\partial x_m \partial y_n}$ a term of the correction matrix $P(x,y) = \begin{bmatrix} P_{11}(x,y) & P_{12}(x,y) \\ P_{21}(x,y) & P_{22}(x,y) \end{bmatrix}$.

Since P is also expressed as an inverse Fourier Transform (12), P_{ij} can be given by the general expression:

$$P_{ij}(x,y) = FT^{-1}_{\xi \to (x_1-y_1)} \frac{M}{\mu k_s^2 F(\xi)} \quad \text{with} \quad \begin{cases} M \equiv A & \text{if } i = j = 1, \\ M \equiv D & \text{if } i = j = 2, \\ M \equiv B & \text{if } i = 1, j = 2, \\ M \equiv C & \text{if } i = 2, j = 1, \end{cases}$$

Therefore, the second derivatives which constitute the matrix E_P can be computed through the four following typical inverse Fourier Transforms:

$$P_{ij,1\bar{1}} = FT^{-1}_{\xi \to (x_1-y_1)} \frac{\xi^2}{\mu k_s^2 F(\xi)} M, \qquad P_{ij,2\bar{2}} = FT^{-1}_{\xi \to (x_1-y_1)} \frac{1}{\mu k_s^2 F(\xi)} \frac{\partial^2 M}{\partial x_2 \partial y_2},$$

$$P_{ij,1\bar{2}} = FT^{-1}_{\xi \to (x_1-y_1)} \frac{-i\xi}{\mu k_s^2 F(\xi)} \frac{\partial M}{\partial x_2}, \qquad P_{ij,2\bar{1}} = FT^{-1}_{\xi \to (x_1-y_1)} \frac{i\xi}{\mu k_s^2 F(\xi)} \frac{\partial M}{\partial y_2}. \qquad (23)$$

In order to determine any term of E_P, the inverse Fourier Transform of a term of P will be established in the first place. Afterwards, it will be sufficient to apply this numerical method of inversion to (23).

Computation of a typical term of P

In this subsection, a numerical method of inverting the Fourier Transform (12) is presented. It is based on a method of contour integration similar to that used by [6].

Since each term of the (2×2)–matrix $P(x, y)$ consists of 4 similar terms of the following typical form

$$FT^{-1}_{\xi \to (x_1-y_1)} \frac{2}{\mu k_s^2} \frac{\xi^2 R_s(\xi)(2\xi^2 - k_s^2)}{F(\xi)} e^{-x_2 R_s(\xi) - y_2 R_p(\xi)}, \tag{24}$$

the method of inversion will only be presented for this typical term.

In fact, the integral representation of (24)

$$I = \frac{1}{\pi \mu k_s^2} \int_{-\infty}^{+\infty} \frac{\xi^2 R_s (2\xi^2 - k_s^2)}{F(\xi)} e^{-x_2 R_s - y_2 R_p} e^{i\xi(x_1 - y_1)} d\xi$$

is not adapted to numerical integrations because of the branch point singularities at $\xi = \pm k_p$, $\xi = \pm k_s$, and the oscillatory nature of the integrand. To avoid these troubles, a method of contour integration is now employed.

Original path of integration

The branch cuts are determined by the radicals $R_\alpha(\xi)$ (where $\alpha = p$ or s). As ξ goes from $-\infty$ to $+\infty$ the original path of integration is shown in Figure 2.

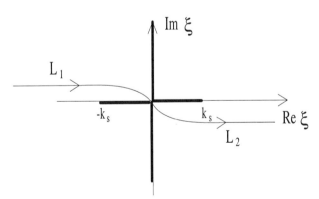

Figure 2: Original path of integration and branch cuts.

Condition for the exponents $e^{-x_2 R_s}$, $e^{-y_2 R_p}$ not to explode

The arguments of these exponents have to have asymptotically negative real parts: $\operatorname{Re} R_\alpha \geq 0$ since x_2 and y_2 are positive. This condition is ensured by our choice of the square root R_α.

Change of variable in the complex plane ξ

Asymptotically as $|\xi| \to \infty$, on L_2 $\operatorname{Re} R_\alpha \sim \xi$ and on L_1 $\operatorname{Re} R_\alpha \sim -\xi$. Therefore, we consider the following change of variable when $|\xi| \to \infty$ in such a way that the exponent $e^{-x_2 R_s - y_2 R_p} e^{i\xi(x_1-y_1)}$ will not increase exponentially:
$$\xi = (-(x_2+y_2) + ix)\frac{s}{d^2} \quad \text{on } L_1$$
$$\xi = ((x_2+y_2) + ix)\frac{s}{d^2} \quad \text{on } L_2,$$
with $s \in \mathbf{R}_+$, $d^2 = (x_2+y_2)^2 + x^2$, and $x = x_1 - y_1$.

New path of integration due to the previous change of variable

The previous change of variable transforms the original path of integration to the new one shown in Figure 3 and Figure 4.

Actually L_1 is transformed into C_1 the slope of which is $\frac{-x}{x_2+y_2}$ and L_2 is transformed into C_4 the slope of which is $\frac{x}{x_2+y_2}$. Then two configurations of the new contour have to be considered.

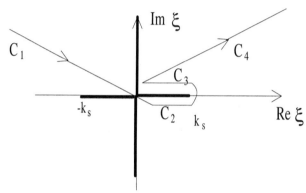

Figure 3: New path of integration for $x = x_1 - y_1 > 0$.

Computation of the inverse Fourier Transform on C_1 and C_4

We denote I_1 (respectively I_4) the integral of the inverse Fourier transform on C_1 (respectively C_4). As on C_1 (respectively C_4) s varies from $+\infty$ to 0 (respectively from 0 to $+\infty$) independently of the configuration of the contour ($x > 0$ or $x < 0$), one gets:

$$I_1 + I_4 = \frac{1}{\pi\mu k_s^2} \int_0^{+\infty} \left(\frac{(x_2+y_2) - ix\, \xi_2^2 \sqrt{\xi_2^2 - k_s^2}(2\xi_2^2 - k_s^2)}{d^2 \quad F(\xi_2)} e^{-x_2\sqrt{\xi_2^2-k_s^2} - y_2\sqrt{\xi_2^2-k_p^2}} e^{i\xi_2 x} \right.$$
$$\left. + \frac{(x_2+y_2) + ix\, \xi_1^2 \sqrt{\xi_1^2 - k_s^2}(2\xi_1^2 - k_s^2)}{d^2 \quad F(\xi_1)} e^{-x_2\sqrt{\xi_1^2-k_s^2} - y_2\sqrt{\xi_1^2-k_p^2}} e^{i\xi_1 x} \right) ds, \quad (25)$$

where we denote $\xi_1 = ((x_2+y_2) + ix)\frac{s}{d^2}$ and $\xi_2 = -\overline{\xi_1}$.

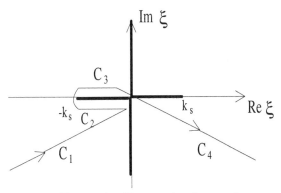

Figure 4: New path of integration for $x < 0$.

Since the square roots in $I_1 + I_4$ have to be taken in the sense of complex square roots with positive real parts, we can directly verify that the exponential functions in (25) are with negative real parts. Therefore they converge quickly and (25) is suitable for numerical calculation using Gauss-Laguerre integrations.

Computation of the inverse Fourier Transform on C_2 and C_3

We can split C_2 (respectively C_3) into two parts C_{21} and C_{22} (respectively C_{31} and C_{32}) where the radicals get different values.

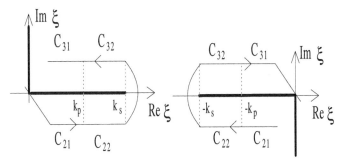

Figure 5: Parts of the path C_2 and C_3 for $x > 0$ and $x < 0$.

As the value of the square roots $R_\alpha(\xi)$ (where $\alpha = p$ or s) are chosen such as their real part are positive, one can obtain:

in the configuration $x > 0$

on C_{21} $R_\alpha = -i\sqrt{k_\alpha^2 - \xi^2}$,   on C_{22} $R_p = \sqrt{\xi^2 - k_p^2}$ $R_s = -i\sqrt{k_s^2 - \xi^2}$,

on C_{31} $R_\alpha = i\sqrt{k_\alpha^2 - \xi^2}$,   on C_{32} $R_p = \sqrt{\xi^2 - k_p^2}$ $R_s = i\sqrt{k_s^2 - \xi^2}$,

in the configuration $x < 0$

on C_{21} $R_\alpha = i\sqrt{k_{p,s}^2 - \xi^2}$,   on C_{22} $R_p = \sqrt{\xi^2 - k_p^2}$ $R_s = i\sqrt{k_s^2 - \xi^2}$,

on C_{31} $R_\alpha = -i\sqrt{k_{p,s}^2 - \xi^2}$,   on C_{32} $R_p = \sqrt{\xi^2 - k_p^2}$ $R_s = -i\sqrt{k_s^2 - \xi^2}$.

Therefore, if we denote by I_2 (respectively I_3) the integral of the inverse Fourier transform on C_2 (respectively C_3):

$$I_2 + I_3 = \frac{-2i}{\pi \mu k_s^2} \int_{k_1}^{k_2} \frac{\xi^2 \sqrt{k_s^2 - \xi^2}(2\xi^2 - k_s^2)\cos(x_2\sqrt{k_s^2 - \xi^2} + y_2\sqrt{k_p^2 - \xi^2})}{(2\xi^2 - k_s^2)^2 + 4\xi^2\sqrt{k_p^2 - \xi^2}\sqrt{k_s^2 - \xi^2}} e^{i\xi x} d\xi$$

$$- \frac{2i}{\pi \mu k_s^2} \int_{k_3}^{k_4} \left(\xi^2 \sqrt{k_s^2 - \xi^2}(2\xi^2 - k_s^2) e^{-y_2\sqrt{\xi^2 - k_p^2}} e^{i\xi x} \times \right.$$

$$\left. \times \operatorname{Re}\left(\frac{e^{-ix_2\sqrt{k_s^2 - \xi^2}}}{(2\xi^2 - k_s^2)^2 + 4i\xi^2\sqrt{\xi^2 - k_p^2}\sqrt{k_s^2 - \xi^2}} \right) \right) d\xi, \qquad (26)$$

where $\begin{cases} \text{in the configuration } x > 0, & k_1 = 0, \quad k_2 = k_p, \quad k_3 = k_p, \quad k_4 = k_s, \\ \text{in the configuration } x < 0, & k_1 = -k_p, \quad k_2 = 0, \quad k_3 = -k_s, \quad k_4 = -k_p. \end{cases}$

At this step, these integrals can be evaluated by Gauss-Legendre quadratures.

Calculus of the residue

As the integrand (24) is singular at the Rayleigh poles $\xi = \pm k_R$, the residue contribution to (24) has to be considered. We denote it I_R.

Applying the theorem of residues and as $R_\alpha = \sqrt{k_R^2 - k_\alpha^2}$ when $\xi = \pm k_R$, one gets:

$$I_R = \frac{2i}{\mu k_s^2} \operatorname{sign}(x) k_R^2 \sqrt{k_R^2 - k_s^2}(2k_R^2 - k_s^2) e^{-x_2\sqrt{k_R^2 - k_s^2} - y_2\sqrt{k_R^2 - k_p^2}} \left. \frac{e^{i\xi x}}{F'(\xi)} \right|_{\xi = \operatorname{sign}(x) k_R}, \qquad (27)$$

where $F'(\xi) = \frac{F(\xi)}{(\xi - k_R)}$ if $x > 0$ and $F'(\xi) = \frac{F(\xi)}{(\xi + k_R)}$ if $x < 0$.

We have thus discussed a numerical method of computing any term of the correction term P. Finally, applying this numerical method to (23) allows the computation of the matrix E_p (22) associated to the bilinear form b_p.

6. Conclusion

In this paper, the solution of an half-plane elastodynamics problem of scattering by a buried obstacle has been presented. It occurred firstly in the special Green's function which incorporates the boundary condition at the free surface, and secondly in the use of a variational approach.

Although no infinite free surface requires to be discretized, this special Green's function generates an increase of the CPU cost.

The variational approach leads after discretization to a symmetrical matrix system. Therefore, substantial computer time and memory are saved during the resolution of the system. On the other hand a price has to be paid for the creation of double spatial integrals. But, since the variational approach is more stable for a given number of boundary finite elements than a collocation approach, its CPU cost is widely counterbalanced. Moreover, this variational approach has the advantage to provide a well-suited framework for future coupling with the interior domain.

References

[1] S. Kobayashi: Some problems of the boundary equation method in elastodynamics, *Boundary Elements*. C.A. Brebbia, T. Futagami and M. Tanaka, Springer-Verlag, Berlin (1983), 353-362.

[2] N. Nishimura: Applications of the boundary integral equation methods to solid mechanics. Ph D Thesis, Kyoto University, 1988.

[3] E.R Lapwood: The disturbance due to a line source in a semi-infinite elastic medium. *Phil. Trans. R. Soc. Lond. A* **242**, (1949), 63-100.

[4] E. Becache, T. Ha Duong: A space-time variational formulation for the boundary integral equation in a 2D elastic crack problem. *Mathematical Modelling and Numerical Analysis* **28** (1994), 141-176.

[5] M. Bonnet: Equations integrales et elements finis de frontiere. *CNRS Editions Eyrolles*, 1995.

[6] D.C Hothersall, S.N. Chandler-Wilde, M.N. Hajmirzae: Efficiency of single noise barriers. *J. Sound. Vib.* **146**(2) (1991), 303-322.

[7] M.A. Hamdi, J.M Ville: Sound radiation from ducts theory and experiment. *J. Sound. Vib.* **107**(2) (1986), 231-246.

[8] V.D. Kupradze: *Potential Methods in The Theory of Elasticity*, Israel Program for Scientific Translations, Jerusalem 1965.

[9] W.L. Wendland: Boundary Element Methods and their asymptotic convergence. *CISM Courses Lectures* (1983), 135-216.

M. Ben Tahar and C. Granat: DAVI/LG2mS, UPRESA CNRS 6066, Universite Technologique de Compiegne , France .
Email: mabrouk.ben-tahar@utc.fr and cristel.granat@utc.fr

T. Ha-Duong: DMA/GI , Universite Technologique de Compiegne, France.
Email: tuong.ha-duong@utc.fr

M. BOCHNIAK and A.–M. SÄNDIG

Sensitivity Analysis for Elastic Fields in Non Smooth Domains

1. Introduction

The goal of shape and structure optimization in mechanics is to determine the shape of a mechanical system which is optimal with respect to objective and constraint functionals. For example, if one wants to avoid plastification the values of the von Mises yield functional should be small enough, or if one wants to avoid crack growth, the energy release rate (or the stress intensity factors) should not exceed their critical values.

The influence of the shape or the structure of the domain on the stress behaviour has been studied by many authors [6, 16] and the corresponding sensitivity analysis is well developed for problems in smooth domains. In this paper we focus on the sensitivity analysis for elastic fields in two-dimensional non smooth domains. It is well known [12] that in this case material discontinuities and geometrical peculiarities like corners, cracks and notches lead to stress singularities. The singular behaviour can be described by means of asymptotic expansions with respect to the distance to the geometrical and structural singularities. If we denote by (r, ω) the local polar coordinates with origin in an interface corner point, then the asymptotics of the two–dimensional displacement or harmonic fields u_1, u_2 in the subdomains Ω_1, Ω_2 read as

$$\begin{pmatrix} u_1 \\ u_2 \end{pmatrix}(r,\omega) \sim \sum_j c_j r^{\alpha_j} \begin{pmatrix} \varphi_{1,j}(\log r, \omega) \\ \varphi_{2,j}(\log r, \omega) \end{pmatrix}. \tag{1}$$

Here we study the sensitivity of both elastic fields and generalized stress intensity factors c_j with respect to shape perturbation. To this end we apply the method of interior variations [5, 6, 16], i.e. we introduce a fixed reference configuration and consider a class of small perturbations of the reference domain. The state equations as well as all quantities which are defined over the actual configuration are transformed to the fixed reference configuration. Thus the investigation of shape sensitivity can be reduced to the investigation of a regular perturbed boundary value problem for the transformed elastic fields. We expand the transformed quantities asymptotically with respect to the perturbation parameter and justify the asymptotics with the help of a–priori estimates in weighted Sobolev spaces. In this way we obtain existence and regularity results for the material derivatives of the displacement fields and of the weight functions and derive explicit formulas for the sensitivity of the stress intensity and notch factors. Numerical experiments using boundary element techniques illustrate the results. For a rigorous sensitivity analysis of stress intensity factors for a wider class of problems (three–dimensional problems, influence of material parameters) we refer to [1].

This work was supported by the German Research Foundation in the framework of the Sonderforschungsbereich 404.

2. Formulation of the problem

2.1. Two-dimensional boundary transmission problems

Let $\Omega \subset \mathbb{R}^2$ be a bounded domain, whose boundary consists of two smooth open manifolds Γ^D, Γ^N and a set S of isolated points where stress singularities can occur, i.e. corner points and points where the boundary conditions change. We assume that $\Gamma^D \neq \emptyset$ and that the domain Ω is locally diffeomorph in a neighbourhood of each point from S to a wedge. Furthermore, we allow that $\overline{\Omega} = \overline{\Omega}_1 \cup \overline{\Omega}_2$ is a coupled structure consisting of two bounded domains such that $\partial\Omega_1 \cap \partial\Omega_2 \neq \emptyset$ (Fig. 1). In this case we denote by $\Gamma \subset \partial\Omega_1 \cap \partial\Omega_2$ the interface and by $\Gamma_i = \partial\Omega_i \setminus \overline{\Gamma}$ the outer boundary pieces.

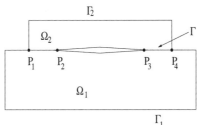

Figure 1: 2D coupled structure

The domains Ω_i are occupied by different isotropic linear elastic materials. The linearized elasticity equations for the displacement vectors u_i in Ω_i together with the Dirichlet conditions on Γ^D, the Neumann conditions on Γ^N and the standard interface conditions on Γ read

$$\begin{aligned}
-\mu_i \Delta u_i - (\lambda_i + \mu_i)\nabla \mathrm{div}\, u_i &= f_i & &\text{in } \Omega_i, \\
u_i &= 0 & &\text{on } \Gamma_i^D, \\
\sigma_i(u_i)\vec{n}_i &= h_i & &\text{on } \Gamma_i^N, \\
u_1 - u_2 &= 0 & &\text{on } \Gamma, \\
\sigma_1(u_1)\vec{n}_1 + \sigma_2(u_2)\vec{n}_2 &= t_2 & &\text{on } \Gamma.
\end{aligned} \qquad (2)$$

Here, $\sigma_i(u_i)$ is the stress tensor and $\vec{n}_i$ is the exterior normal vector on $\partial\Omega_i$. In the following we assume for simplicity that $\lambda_i + \mu_i > 0$ and $\mu_i > 0$, $i = 1, 2$. These conditions guarantee that the transmission problem (2) is elliptic [15].
The behaviour of the two-dimensional displacement field $u = (u_1, u_2)$ in a neighbourhood of a corner point $P \in S \cap \overline{\Gamma}$ (see Fig.1 where $P \in \{P_1, ..., P_4\}$) can be described by an asymptotic expansion, provided the right-hand sides of (2) are sufficiently smooth and satisfy some compatibility conditions in P [14]:

$$\begin{pmatrix} u_1 \\ u_2 \end{pmatrix}(r,\omega) = \sum_{j=1}^{N} \sum_{k=1}^{m_g(\alpha_j)} \sum_{l=0}^{\kappa_{k,j}-1} c_{j,k,l} \begin{pmatrix} v_{1,j}^{l,k} \\ v_{2,j}^{l,k} \end{pmatrix}(r,\omega) + \begin{pmatrix} v_1 \\ v_2 \end{pmatrix}(r,\omega), \qquad (3)$$

where

$$\begin{pmatrix} v_{1,j}^{l,k} \\ v_{2,j}^{l,k} \end{pmatrix}(r,\omega) = r^{\alpha_j} \sum_{s=0}^{l} \frac{1}{s!} (\log r)^s \begin{pmatrix} \varphi_{1,j}^{l-s,k} \\ \varphi_{2,j}^{l-s,k} \end{pmatrix}(\omega), \qquad (4)$$

(r,ω) are polar coordinates with origin in P and $v = (v_1, v_2) \sim O(r)$. The functions $(v_{1,j}^{l,k}, v_{2,j}^{l,k})$ are solutions of the problem (2) with vanishing right-hand sides in the double-wedge $C_P = \{(r,\omega) : 0 < r < \infty, \omega \in (\omega_0, \omega_1) \cup (\omega_1, \omega_2)\}$, which is locally diffeomorph to $\Omega = \Omega_1 \cup \Omega_2$ in a neighbourhood of P. Writing the boundary transmission problem in polar coordinates (r,ω) and applying the Mellin transform with respect to r we proceed to a generalized eigenvalue problem $\mathcal{A}_P(\alpha)u = 0$, whose eigensolutions coincide with the functions $(\varphi_{1,j}^{0,k}, \varphi_{2,j}^{0,k})$. The complex exponents α_j are eigenvalues of the operator pencil $\mathcal{A}_P$ and the functions $\{(\varphi_{1,j}^{l,k}, \varphi_{2,j}^{l,k}) : 1 \le k \le m_g(\alpha_j), 0 \le l \le \kappa_{k,j} - 1\}$ form a system of Jordan–chains of $\mathcal{A}_P$. Furthermore, N is the number of eigenvalues in the strip $0 < \mathrm{Re}\,\alpha < 1$, $m_g(\alpha_j)$ denotes the number of Jordan–chains to the eigenvalue α_j and $\kappa_{k,j}$ is the length of the k-th Jordan–chain. In the neighbourhood of a corner point $P \notin S \cap \bar{\Gamma}$ with $P \in \partial\Omega_i$ the field u_i has an asymptotics analogously to (3).

2.2. Generalized stress intensity factors

The coefficients $c_{j,k,l}$ in (3) are called generalized stress intensity factors (abbr. GSIF's). They are a measure for the intensity of stress concentrations near singular points and play an important role in classical fracture criteria of the Irwin–Griffith-type and in delamination criteria for composite laminates [7].
For a fixed corner point $P \in S$ and every eigenvalue α_j of the operator pencil $\mathcal{A}_P$ we introduce the non–energetic functions

$$\begin{pmatrix} w_{1,j}^{l,k} \\ w_{2,j}^{l,k} \end{pmatrix}(r,\omega) = r^{-\overline{\alpha}_j} \sum_{s=0}^{l} \frac{1}{s!} (\log r)^s \begin{pmatrix} \psi_{1,j}^{l-s,k} \\ \psi_{2,j}^{l-s,k} \end{pmatrix}(\omega), \tag{5}$$

where $\{(\psi_{1,j}^{l,k}, \psi_{2,j}^{l,k}) : 1 \le k \le m_g(\alpha_j), 0 \le l \le \kappa_{k,j} - 1\}$ is a canonical system of Jordan–chains of the operator pencil $\mathcal{A}_P$ corresponding to the eigenvalue $-\overline{\alpha}_j$. Thus the functions $(w_{1,j}^{l,k}, w_{2,j}^{l,k})$ are solutions of the homogeneous problem (2) in the double–wedge C_P. Furthermore, we assume that the systems $\{(\psi_{1,j}^{l,k}, \psi_{2,j}^{l,k})\}$ and $(v_{1,j}^{l,k}, v_{2,j}^{l,k})$, $1 \le k \le m_g(\alpha_j), 0 \le l \le \kappa_{k,j} - 1$ satisfy a certain biorthonormality condition [9, 2].

The coefficients $c_{j,k,l}$ in (3) can be computed by inserting the solution (u_1, u_2) of the problem (2) and the function $(w_{1,j}^{l,k}, w_{2,j}^{l,k})$ into Green's formulae in $\Omega_i \setminus B_\delta(P)$, substituting the transmission conditions and taking the limit as $\delta \to 0$. Here, $B_\delta(P)$ is a circle with center in P and radius δ. Assuming that the right-hand sides of the two–dimensional problem (2) are regular enough we get the following

Theorem 1. *[9, 2] Let the functions $(v_{1,j}^{l,k}, v_{2,j}^{l,k})$ and $(w_{1,j}^{l,k}, w_{2,j}^{l,k})$ be defined by (4) and (5). Then the coefficients $c_{j,k,l}$ in (3) are given by*

$$c_{j,k,l} = \sum_{i=1}^{2} \left(\int_{\Omega_i} L_i(\eta u_i) \cdot \overline{w_{i,j}^{l,k}} dx + \int_{\Gamma_i} \left(\sigma_i(\eta u_i)\vec{n}_i \cdot \overline{w_{i,j}^{l,k}} - \eta u_i \cdot \overline{\sigma_i(w_{i,j}^{l,k})\vec{n}_i} \right) ds_x \right)$$
$$+ \int_{\Gamma} (\sigma_1(\eta u_1)\vec{n}_1 + \sigma_2(\eta u_2)\vec{n}_2) \cdot \overline{w_{1,j}^{l,k}} ds_x \tag{6}$$

with a cut-off function η whose support lies in the neighbourhood of the point P.

Theorem 2. *[9, 2] Let $(W_{1,j}^{l,k}, W_{2,j}^{l,k})$ be a solution of the two-dimensional transmission problem (2) with vanishing right-hand sides, which admits the decomposition*

$$\begin{pmatrix} W_{1,j}^{l,k} \\ W_{2,j}^{l,k} \end{pmatrix} = \begin{pmatrix} w_{1,j}^{l,k} \\ w_{2,j}^{l,k} \end{pmatrix} + \begin{pmatrix} \tilde{w}_{1,j}^{l,k} \\ \tilde{w}_{2,j}^{l,k} \end{pmatrix} \tag{7}$$

with $\tilde{w}_{i,j}^{l,k} \in [H^1(\Omega_i)]^2$, $i = 1, 2$. Then the following formula holds

$$c_{j,k,l} = \sum_{i=1}^{2} \left(\int_{\Omega_i} f_i \cdot \overline{W_{i,j}^{l,k}} dx + \int_{\Gamma_i^N} h_i \cdot \overline{W_{i,j}^{l,k}} ds_x \right) + \int_{\Gamma} t_2 \cdot \overline{W_{1,j}^{l,k}} ds_x. \tag{8}$$

2.3. Shape sensitivity

In order to describe the shape sensitivity, i.e., the influence of the shape of the domain on the displacement fields and the GSIFs we describe small perturbations of a fixed configuration $(\Omega_i, \Gamma_i^D, \Gamma_i^N, \Gamma)$ as its image under some C^2–diffeomorphism $\Phi_\varepsilon = I + \varepsilon \Phi$ with $\varepsilon \in [0, \varepsilon_0]$, i.e.

$$\begin{aligned}
\Omega_{i,\varepsilon} &= \Phi_\varepsilon(\Omega_i) = (I + \varepsilon\Phi)(\Omega_i), & \Gamma_\varepsilon &= \Phi_\varepsilon(\Gamma) = (I + \varepsilon\Phi)(\Gamma) \\
\Gamma_{i,\varepsilon}^D &= \Phi_\varepsilon(\Gamma_i^D) = (I + \varepsilon\Phi)(\Gamma_i^D), & \Gamma_{i,\varepsilon}^N &= \Phi_\varepsilon(\Gamma_i^N) = (I + \varepsilon\Phi)(\Gamma_i^N).
\end{aligned}$$

Furthermore, we assume that the mapping Φ is constant in the vicinity $\mathcal{U}(P)$ of every singular point $P \in S$. Examples of such perturbations are: smooth perturbation of the smooth boundary or interface pieces described by a local Hadamard parameterization such that a neighbourhood of the points from S is unchanged; local translation of points from S without change of the opening angles [1].

We can formulate analogously to (2) a boundary transmission problem in the perturbed domains $(\Omega_{1,\varepsilon}, \Omega_{2,\varepsilon})$

$$\begin{aligned}
L_i u_{i,\varepsilon} := -\mu_i \Delta u_{i,\varepsilon} - (\lambda_i + \mu_i)\nabla \mathrm{div} u_{i,\varepsilon} &= f_{i,\varepsilon} & \text{in } \Omega_{i,\varepsilon} \\
u_{i,\varepsilon} &= 0 & \text{on } \Gamma_{i,\varepsilon}^D, \\
\sigma_i(u_{i,\varepsilon})\vec{n}_{i,\varepsilon} &= h_{i,\varepsilon} & \text{on } \Gamma_{i,\varepsilon}^N, \\
u_{1,\varepsilon} - u_{2,\varepsilon} &= 0 & \text{on } \Gamma_\varepsilon, \\
\sigma_1(u_{1,\varepsilon})\vec{n}_{1,\varepsilon} + \sigma_2(u_{2,\varepsilon})\vec{n}_{2,\varepsilon} &= t_{2,\varepsilon} & \text{on } \Gamma_\varepsilon.
\end{aligned} \tag{9}$$

The generalized stress intensity factors in the asymptotic expansion (3) of the solution $(u_{1,\varepsilon}, u_{2,\varepsilon})$ of the perturbed problem (9) will be denoted by $c_{j,k,l}(\Phi_\varepsilon)$.

We assume for the right-hand sides of (9)

$$f_{i,\varepsilon} \circ \Phi_\varepsilon \in C^2([0,\varepsilon_0], L_2(\Omega_i)), \quad h_{i,\varepsilon} \circ \Phi_\varepsilon \in C^2([0,\varepsilon_0], H^{1/2}(\Gamma_i)), \tag{10}$$

$$t_{2,\varepsilon} \circ \Phi_\varepsilon \in C^2([0,\varepsilon_0], H^{1/2}(\Gamma)). \tag{11}$$

The change of variables $x_\varepsilon = \Phi_\varepsilon(x)$ leads to an elliptic transmission problem for the functions $(u_{1,\varepsilon} \circ \Phi_{1,\varepsilon}, u_{2,\varepsilon} \circ \Phi_{2,\varepsilon})$ in the domains Ω_1, Ω_2

$$\begin{aligned}
L_i^\varepsilon(\partial_x)(u_{i,\varepsilon} \circ \Phi_\varepsilon) &= f_{i,\varepsilon} \circ \Phi_\varepsilon & \text{in } \Omega_i, \\
u_{i,\varepsilon} \circ \Phi_\varepsilon &= 0 & \text{on } \Gamma_i^D, \\
\sigma_i^\varepsilon(\partial_x)(u_{i,\varepsilon} \circ \Phi_\varepsilon)(\vec{n}_{i,\varepsilon} \circ \Phi_\varepsilon) &= h_{i,\varepsilon} \circ \Phi_\varepsilon & \text{on } \Gamma_i^N, \\
u_{1,\varepsilon} \circ \Phi_\varepsilon - u_{2,\varepsilon} \circ \Phi_\varepsilon &= 0 & \text{on } \Gamma, \\
\sigma_1^\varepsilon(\partial_x)(u_{1,\varepsilon} \circ \Phi_\varepsilon)(\vec{n}_{1,\varepsilon} \circ \Phi_\varepsilon) + \sigma_2^\varepsilon(\partial_x)(u_{2,\varepsilon} \circ \Phi_\varepsilon)(\vec{n}_{2,\varepsilon} \circ \Phi_\varepsilon) &= t_{2,\varepsilon} \circ \Phi_\varepsilon & \text{on } \Gamma.
\end{aligned} \tag{12}$$

Now we consider the formal asymptotic expansion of the perturbed displacement fields and of the perturbed GSIFs

$$(u_{i,\varepsilon} \circ \Phi_\varepsilon)(x) = u_{i,0}(x) + \varepsilon \dot{u}_i(x) + O(\varepsilon^2), \tag{13}$$

$$c_{j,k,l}(\Phi_\varepsilon) = c_{j,k,l}(I) + \varepsilon \delta c_{j,k,l}(I, \Phi) + O(\varepsilon^2). \tag{14}$$

Our Problem is : *Justify the formal expansions (13) and (14) and compute the material derivative*

$$\dot{u}_i = \frac{d}{d\varepsilon}(u_{i,\varepsilon} \circ \Phi_\varepsilon)\bigg|_{\varepsilon=0} = \lim_{\varepsilon \to 0} \frac{u_{i,\varepsilon} \circ \Phi_\varepsilon - u_{i,0}}{\varepsilon}$$

and the variation

$$\delta c_{j,k,l}(I, \Phi) = \frac{d}{d\varepsilon} c_{j,k,l}(I + \varepsilon \Phi)\bigg|_{\varepsilon=0} = \lim_{\varepsilon \to 0} \frac{c_{j,k,l}(I + \varepsilon \Phi) - c_{j,k,l}(I)}{\varepsilon}.$$

3. Main results

3.1. Justification of the asymptotic expansion of the displacement fields

The operators $L_i^\varepsilon(\partial_x)$ and $\sigma_i^\varepsilon(\partial_x)$ have variable coefficients which depend smoothly on the perturbation parameter ε. Expanding the coefficients in Taylor series with respect to ε we obtain the decompositions [1]

$$L_i^\varepsilon(\partial_x) = L_i(\partial_x) + \varepsilon L_i^1(x, \partial_x) + \varepsilon^2 L_{i,R}(\varepsilon, \partial_x), \tag{15}$$

$$\sigma_i^\varepsilon(\partial_x) = \sigma_i(\partial_x) + \varepsilon \sigma_i^1(x, \partial_x) + \varepsilon^2 \sigma_{i,R}(\varepsilon, \partial_x). \tag{16}$$

Similarly, we have

$$(\vec{n}_{i,\varepsilon} \circ \Phi_\varepsilon)(x) = \vec{n}_{i,0}(x) + \varepsilon \dot{\vec{n}}_i(x) + O(\varepsilon^2), \tag{17}$$

where $\dot{\vec{n}}_i = \langle \vec{n}_{i,0}, D\Phi^\top \vec{n}_{i,0} \rangle \vec{n}_{i,0} - D\Phi^\top \vec{n}_{i,0}$ [6].

We derive a boundary transmission problem for the material derivatives $\dot{u}_i$ inserting into (9) the formal ansatz (13), the Taylor expansions (15, 16, 17) and the Taylor expansions of the transformed right-hand sides $f_{i,\varepsilon} \circ \Phi_\varepsilon, h_{i,\varepsilon} \circ \Phi_\varepsilon, t_{2,\varepsilon} \circ \Phi_\varepsilon$. Comparing the terms of the same order in ε we obtain a transmission problem for the functions $(u_{1,0}, u_{2,0})$ defined on the reference configuration

$$\begin{aligned}
L_i(\partial_x) u_{i,0} &= f_{i,0} && \text{in } \Omega_i, \\
u_{i,0} &= 0 && \text{on } \Gamma_i^D, \\
\sigma_i(\partial_x)(u_{i,0})\vec{n}_{i,0} &= h_{i,0} && \text{on } \Gamma_i^N, \\
u_{1,0} - u_{2,0} &= 0 && \text{on } \Gamma, \\
\sigma_1(\partial_x)(u_{1,0})\vec{n}_{1,0} + \sigma_2(\partial_x)(u_{2,0})\vec{n}_{2,0} &= t_{2,0} && \text{on } \Gamma
\end{aligned} \tag{18}$$

and a transmission problem for the material derivative $(\dot{u}_1, \dot{u}_2)$

$$\begin{aligned}
L_i(\partial_x)\dot{u}_i &= \dot{f}_i - L_i^1(\partial_x)u_{i,0} && \text{in } \Omega_i, \\
\dot{u}_i &= 0 && \text{on } \Gamma_i^D, \\
\sigma_i(\partial_x)(\dot{u}_i)\vec{n}_{i,0} &= \dot{h}_i - \sigma_i^1(\partial_x)(u_{i,0})\vec{n}_{i,0} - \sigma_i(\partial_x)(u_{i,0})\dot{\vec{n}}_i && \text{on } \Gamma_i^N, \\
\dot{u}_1 - \dot{u}_2 &= 0 && \text{on } \Gamma, \\
\sigma_1(\partial_x)(\dot{u}_1)\vec{n}_{1,0} & && \\
+\sigma_2(\partial_x)(\dot{u}_2)\vec{n}_{2,0} &= \dot{t}_2 - \sigma_1^1(\partial_x)(u_{1,0})\vec{n}_{1,0} - \sigma_1(\partial_x)(u_{1,0})\dot{\vec{n}}_1 && \\
& \quad -\sigma_2^1(\partial_x)(u_{2,0})\vec{n}_{2,0} - \sigma_2(\partial_x)(u_{2,0})\dot{\vec{n}}_2 && \text{on } \Gamma
\end{aligned} \quad (19)$$

We discuss the existence and the regularity of solutions of the problems (18) and (19) in weighted Sobolev spaces, which take care for the resulting non-regularities of the solutions near singular points.

The weighted Sobolev spaces $V_\beta^d(\Omega_i)$, $d = 0, 1, 2, \ldots, \beta \in \mathbb{R}$, are defined as the closure of $C_0^\infty(\overline{\Omega} \setminus S)$ with respect to the norm

$$\|u; V_\beta^d(\Omega)\|^2 = \sum_{|\gamma| \le d} \|\tilde{r}(x)^{\beta-d+|\gamma|} |D_x^\gamma u|; L_2(\Omega)\|^2, \quad (20)$$

where $\tilde{r}(x) = \text{dist}(x, S)$. For an open part $T \subset \partial\Omega_i \setminus S$ we define the trace spaces $V_\beta^{d-1/2}(T)$ in the usual way [12].

Theorem 3. [1] *Let a_0 be the greatest real number for which the strip $0 < \text{Re}\alpha \le a_0$ is free of eigenvalues of the operator pencil $\mathcal{A}_P$ for every corner point $P \in S$ and let $0 < \beta < 1 - a_0 < 1$. Then there is a positive real constant d such that the following estimate is valid*

$$\sum_{i=1}^{2} \|u_{i,\varepsilon} \circ \Phi_\varepsilon - u_{i,0} - \varepsilon\dot{u}_i; V_\beta^2(\Omega_i)\| \le d\varepsilon^2. \quad (21)$$

Sketch of the proof : From the formal ansatz (13), the decompositions (15,16) and the problems (12,18,19) it follows that the functions $v_{i,\varepsilon} = u_{i,\varepsilon} \circ \Phi_\varepsilon - u_{i,0} - \varepsilon\dot{u}_i$, $i = 1, 2$ are a solution of the transmission problem

$$\begin{aligned}
L_i^\varepsilon(\partial_x)v_{i,\varepsilon} &= f_{i,\varepsilon} \circ \Phi_\varepsilon - f_{i,0} - \varepsilon\dot{f}_i + f_{i,R} && \text{in } \Omega_i \\
v_{i,\varepsilon} &= 0 && \text{on } \Gamma_i^D, \\
\sigma_i^\varepsilon(\partial_x)(v_{i,\varepsilon})(\vec{n}_{i,\varepsilon} \circ \Phi_\varepsilon) &= h_{i,\varepsilon} \circ \Phi_\varepsilon - h_{i,0} - \varepsilon\dot{h}_i + h_{i,R} && \text{on } \Gamma_i^N, \\
v_{1,\varepsilon} - v_{2,\varepsilon} &= 0 && \text{on } \Gamma, \\
\sigma_1^\varepsilon(\partial_x)(v_{1,\varepsilon})(\vec{n}_{1,\varepsilon} \circ \Phi_\varepsilon) & && \\
+\sigma_2^\varepsilon(\partial_x)(v_{2,\varepsilon})(\vec{n}_{2,\varepsilon} \circ \Phi_\varepsilon) &= t_{2,\varepsilon} \circ \Phi_\varepsilon - t_{2,0} - \varepsilon\dot{t}_2 + t_{2,R} && \text{on } \Gamma,
\end{aligned} \quad (22)$$

where $f_{i,R} \in L_2(\Omega_i)$, $h_{i,R} \in H^{1/2}(\Gamma_i^N)$, $t_{2,R} \in H^{1/2}(\Gamma)$ and $f_{i,R} = O(\varepsilon^2), f_{i,R} = O(\varepsilon^2), t_{2,R} = O(\varepsilon^2)$ as $\varepsilon \to 0$. Since the transmission problem (22) is elliptic we can estimate the norm of $v_{i,\varepsilon}$ by the norms of the right-hand sides of (22) [8, 14].

3.2. Formula for the sensitivity of the GSIFs, one material

For the sake of simplicity we give here a detailed derivation of the formula for the computation of $\delta c_{j,k,l}(I,\Phi)$ only for the case of homogeneous elastic fields. A generalization to the case of elastic transmission problems is straightforward.
Let us consider the two–dimensional mixed boundary value problem

$$\begin{aligned} L(\partial_{x_\varepsilon})u_\varepsilon &= f_\varepsilon \quad \text{in } \Omega_\varepsilon, \\ u_\varepsilon &= 0 \quad \text{on } \Gamma_\varepsilon^D, \\ \sigma(\partial_{x_\varepsilon})(u_\varepsilon)\vec{n}_\varepsilon &= h_\varepsilon \quad \text{on } \Gamma_\varepsilon^N. \end{aligned} \qquad (23)$$

We assume that for $\varepsilon \in [0,\varepsilon_0]$ the number of Jordan chains $m_g(\alpha_j(\varepsilon))$ of $\mathcal{A}_{P_\varepsilon}$ to the eigenvalue $-\overline{\alpha_j(\varepsilon)}$ and their lengths $\kappa_{k,j}$ are constant. The weak solution u_ε of the problem (23) has in the neighbourhood of a corner point $P_\varepsilon = \Phi_\varepsilon(P_0) \in S_\varepsilon$ of the boundary $\partial\Omega_\varepsilon$ the asymptotics (3) with coefficients $c_{j,k,l}(\Phi_\varepsilon)$ depending on the diffeomorphism Φ_ε which can be calculated by an analogon of formula (6)

$$c_{j,k,l}(\Phi_\varepsilon) = \int_{\Omega_\varepsilon} L(\partial_{x_\varepsilon})(\eta u_\varepsilon) \cdot \overline{w_{j,\varepsilon}^{l,k}} dx_\varepsilon + \int_{\partial\Omega_\varepsilon} \sigma(\partial_{x_\varepsilon})(\eta u_\varepsilon)\vec{n}_\varepsilon \cdot \overline{w_{j,\varepsilon}^{l,k}} ds_{x_\varepsilon}, \qquad (24)$$

where the singular functions $w_{j,\varepsilon}^{l,k}$ are formed analogously to (5).
The change of variables $x_\varepsilon = \Phi_\varepsilon(x)$ leads to

$$c_{j,k,l}(\Phi_\varepsilon) = \int_\Omega L^\varepsilon(\partial_x)(\eta u_\varepsilon \circ \Phi_\varepsilon) \cdot \overline{(w_{j,\varepsilon}^{l,k} \circ \Phi_\varepsilon)} \det(D\Phi_\varepsilon) dx \qquad (25)$$

$$+ \int_{\partial\Omega} \sigma^\varepsilon(\partial_x)(\eta u_\varepsilon \circ \Phi_\varepsilon)(\vec{n}_\varepsilon \circ \Phi_\varepsilon) \cdot \overline{(w_{j,\varepsilon}^{l,k} \circ \Phi_\varepsilon)} \det(D\Phi_\varepsilon)\|D\Phi_\varepsilon^{-T}\vec{n}_0\| ds_x.$$

Since Φ is constant in $\mathcal{U}(P_0)$ we have

$$\begin{aligned} \det(D\Phi_\varepsilon)|_{\mathcal{U}(P_0)} &= 1, & D\Phi_\varepsilon^{-T}|_{\mathcal{U}(P_0)} &= I, \\ \vec{n}_\varepsilon \circ \Phi_\varepsilon|_{\mathcal{U}(P_0)} &= \vec{n}_0, & w_{j,\varepsilon}^{l,k} \circ \Phi_\varepsilon|_{\mathcal{U}(P_0)} &= w_{j,0}^{l,k}, \\ L^\varepsilon(\partial_x)|_{\mathcal{U}(P_0)} &= L(\partial_x), & \sigma^\varepsilon(\partial_x)|_{\mathcal{U}(P_0)} &= \sigma(\partial_x). \end{aligned} \qquad (26)$$

We choose the cut–off function η in (24) in such a way that $\operatorname{supp} \eta \subset \mathcal{U}(P_0)$ and apply (26). In this way we obtain

$$c_{j,k,l}(\Phi_\varepsilon) = \int_\Omega L(\partial_x)(\eta u_\varepsilon \circ \Phi_\varepsilon) \cdot \overline{w_{j,0}^{l,k}} dx + \int_{\partial\Omega} \sigma(\partial_x)(\eta u_\varepsilon \circ \Phi_\varepsilon)\vec{n}_0 \cdot \overline{w_{j,0}^{l,k}} ds_x. \qquad (27)$$

Inserting the expansion (13) into (27) we obtain the following

Theorem 4. *Under above assumptions we have*

$$\delta c_{j,k,l}(I,\Phi) = \int_\Omega L(\partial_x)(\eta \dot{u}) \cdot \overline{w_{j,0}^{l,k}} dx + \int_{\partial\Omega} \sigma(\partial_x)(\eta \dot{u})\vec{n}_0 \cdot \overline{w_{j,0}^{l,k}} ds_x, \qquad (28)$$

i.e. $\delta c_{j,k,l}(I,\Phi)$ *coincides with the stress intensity factor* $c_{j,k,l}(\dot{u})$.

Sketch of the proof: From the formulae (27,28) follows that $c_{j,k,l}(\Phi_\varepsilon) - c_{j,k,l}(I) - \varepsilon \delta c_{j,k,l}(I, \Phi)$ coincides with the stress intensity factor $c_{j,k,l}$ in the asymptotics (3) of the function $v_\varepsilon = u_\varepsilon \circ \Phi_\varepsilon - u_0 - \varepsilon \dot{u}$. Since v_ε is a solution of an elliptic transmission problem (22), the stress intensity factors in the asymptotics of $v_{i,\varepsilon}$ can be estimated by the norms of the right-hand sides of (22) [12, 1]. As in Theorem 3 we obtain therefore the estimate

$$|c_{j,k,l}(\Phi_\varepsilon) - c_{j,k,l}(I) - \varepsilon \delta c_{j,k,l}(I, \Phi)| \leq d\varepsilon^2 \tag{29}$$

with a positive real constant d.

4. Example: straight crack increment
4.1. Sensitivity formulas

Let us apply the results of the last section to a simple two–dimensional straight crack increment problem in an isotropic homogeneous material. We consider the mixed boundary value problem (23) in a domain Ω_ε with a straight crack and the crack tip in $P_\varepsilon = (\varepsilon, 0)$. The straight crack increment can be described by a diffeomorphism $\Phi_\varepsilon(x) = x + \varepsilon \eta(x) \Phi(x)$, where

$$\Phi\begin{pmatrix}x_1\\x_2\end{pmatrix} = \begin{pmatrix}1\\0\end{pmatrix} \tag{30}$$

and $\eta \in C_0^\infty(\mathbb{R}^2)$ is a cut–off function with support in some neighbourhood $\mathcal{U}(P_0)$ of the unperturbed crack tip P_0. If homogeneous Neumann conditions are given on the crack lips then the asymptotic expansion (3) of the displacement vector u_ε in the neighbourhood of the perturbed crack tip P_ε has the form (see e.g. [12, 13])

$$u_\varepsilon(r_\varepsilon, \omega_\varepsilon) = \sum_{j=1}^{2} K_j(\Phi_\varepsilon) r_\varepsilon^{1/2} \varphi_j(\omega_\varepsilon) + u_{R,\varepsilon}, \quad u_{R,\varepsilon} \in H^2(\Omega_\varepsilon) \tag{31}$$

with

$$\varphi_1(\omega) = \frac{1}{4\mu\sqrt{2\pi}} \begin{pmatrix} -\cos(3\omega/2) + (2\kappa - 1)\cos(\omega/2) \\ \sin(3\omega/2) - (2\kappa + 1)\sin(\omega/2) \end{pmatrix}, \tag{32}$$

$$\varphi_2(\omega) = \frac{1}{4\mu\sqrt{2\pi}} \begin{pmatrix} 3\sin(3\omega/2) - (2\kappa - 1)\sin(\omega/2) \\ 3\cos(3\omega/2) - (2\kappa + 1)\cos(\omega/2) \end{pmatrix}, \tag{33}$$

where $\kappa = (\lambda + 3\mu)(\lambda + \mu)^{-1}$. Here, $(r_\varepsilon, \omega_\varepsilon)$ are polar coordinates with origin in P_ε and the angular variable ω_ε is oriented in such a way that the crack lips correspond to the angles $\pi, -\pi$. Furthermore we have written the singular functions $\varphi_j, j = 1, 2$, in polar components $\varphi_j = (\varphi_j^r, \varphi_j^\omega)^\top$. The coefficients $K_1(\Phi_\varepsilon)$ and $K_2(\Phi_\varepsilon)$ are called stress intensity factors of mode I and mode II, respectively, and are given by

$$K_j(\Phi_\varepsilon) = \int_{\Omega_\varepsilon} f_\varepsilon \cdot \zeta_{j,\varepsilon} dx_\varepsilon + \int_{\Gamma_\varepsilon^N} h_\varepsilon \cdot \zeta_{j,\varepsilon} ds_{x_\varepsilon}, \tag{34}$$

where the weight functions $\zeta_{j,\varepsilon} \in L_2(\Omega_\varepsilon)$, $j = 1, 2$, satisfy the homogeneous problem (23) in Ω_ε and admit the asymptotics

$$\zeta_{j,\varepsilon}(r_\varepsilon, \omega_\varepsilon) = r_\varepsilon^{-1/2}\psi_j(\omega_\varepsilon) + O(r_\varepsilon^{1/2}). \tag{35}$$

with

$$\psi_1(\omega) = \frac{1}{(1+\kappa)\sqrt{8\pi}} \begin{pmatrix} -3\cos(\omega/2) + (2\kappa+1)\cos(3\omega/2) \\ 3\sin(\omega/2) - (2\kappa-1)\sin(3\omega/2) \end{pmatrix}, \tag{36}$$

$$\psi_2(\omega) = \frac{1}{(1+\kappa)\sqrt{8\pi}} \begin{pmatrix} \sin(\omega/2) - (2\kappa+1)\sin(3\omega/2) \\ \cos(\omega/2) - (2\kappa-1)\cos(3\omega/2) \end{pmatrix}. \tag{37}$$

According to Theorem 4 we have now

$$\delta K_j(I, \Phi) = K_j(\dot{u}). \tag{38}$$

The numerical realization of this formula requires the calculation of the material derivative $\dot{u}$ by solving of a boundary value problem with non vanishing body forces of a complicated form (see (19)). We refer to [3] where boundary element methods are used for the computation of $\dot{u}$.

For problems with vanishing body forces f one can obtain a further representation for $\delta K_j(I, \Phi)$ which is well suited for a numerical realization with boundary element methods. To this end we replace $\dot{u}$ in (38) by

$$\dot{u} = u' + (\Phi \cdot \nabla)u_0 \quad \text{where} \quad u' = \left.\frac{du_\varepsilon}{d\varepsilon}\right|_{\varepsilon=0} \tag{39}$$

is the shape derivative of u_ε. We use the fact that the stress intensity factors K_j in the asymptotics of ∇u_0 can be expressed with the help of higher order stress intensity factors in the asymptotics of u_0. Suppose that the right-hand sides of (23) are smooth enough so that the decomposition

$$u_\varepsilon(r_\varepsilon, \omega_\varepsilon) = \sum_{j=1}^{2} K_j(\Phi_\varepsilon) r_\varepsilon^{1/2} \varphi_j(\omega_\varepsilon) + \sum_{j=3}^{4} K_j(\Phi_\varepsilon) r_\varepsilon^{3/2} \varphi_j(\omega_\varepsilon) + u_{R,\varepsilon}, \quad u_{R,\varepsilon} \in H^3(\Omega_\varepsilon) \tag{40}$$

is valid. Here is

$$\varphi_3(\omega) = \frac{1}{12\mu\sqrt{2\pi}} \begin{pmatrix} \cos(5\omega/2) + (2\kappa-3)\cos(\omega/2) \\ -\sin(5\omega/2) + (2\kappa+3)\sin(\omega/2) \end{pmatrix}, \tag{41}$$

$$\varphi_4(\omega) = \frac{1}{12\mu\sqrt{2\pi}} \begin{pmatrix} 5\sin(5\omega/2) + (2\kappa-3)\sin(\omega/2) \\ 5\cos(5\omega/2) - (2\kappa+3)\cos(\omega/2) \end{pmatrix}. \tag{42}$$

Since $\Phi \cdot \nabla = \partial_{x_1}$ in $\mathcal{U}(P_0)$ we obtain for $j = 1, 2$

$$\delta K_j(I, \Phi) = K_j(\dot{u}) = K_j(u') + K_j(\partial_{x_1} u_0). \tag{43}$$

Furthermore, (see [13])

$$\partial_{x_1}(r^{1/2}\varphi_j(\omega)) = -\alpha r^{-1/2}\psi_j(\omega), \quad j = 1, 2, \qquad (44)$$

$$\partial_{x_1}(r^{3/2}\varphi_j(\omega)) = \frac{1}{2}r^{1/2}\varphi_{j-2}(\omega), \quad j = 3, 4 \qquad (45)$$

with $\alpha = (1+\kappa)/(4\mu)$ and thus

$$\partial_{x_1}u_0(r_0, \omega_0) = -\alpha \sum_{j=1}^{2} K_j(u_0) r_0^{-1/2} \psi_j(\omega_0)$$

$$+ \frac{1}{2} \sum_{j=3}^{4} K_j(u_0) r_0^{1/2} \varphi_{j-2}(\omega_0) + \partial_{x_1} u_{R,0}, \quad u_{R,0} \in H^3(\Omega_0). \qquad (46)$$

Therefore

$$\delta K_j(I, \Phi) = K_j(u') + \frac{1}{2} K_{j+2}(u_0), \quad j = 1, 2. \qquad (47)$$

It can be shown (see [12, Chapter 7]) that $u' \in L_2(\Omega_0)$ is a non-trivial solution of the homogeneous problem (23) in the non-perturbed domain Ω_0. According to the results of Maz'ya/Plamenevsky [9] the shape derivative u' is a linear combination of the weight functions $\zeta_{i,0} \in L_2(\Omega_0)$, $i = 1, 2$. On the other hand, we know from (39) that the main parts of the asymptotics of the functions u' and $-(\Phi \cdot \nabla)u_0$ coincide. From (46) follows

$$u' = \alpha \sum_{i=1}^{2} K_i(u_0) \zeta_i. \qquad (48)$$

Inserting (48) into (47) we obtain

$$\delta K_j(I, \Phi) = \alpha \sum_{i=1}^{2} K_i(u_0) K_j(\zeta_i) + \frac{1}{2} K_{j+2}(u_0), \quad j = 1, 2. \qquad (49)$$

These kinds of representations were obtained for the Gâteaux derivative $\delta K_1(I, \Phi)$ by S.A Nazarov and others in [11, 10, 4] under the assumption that $K_2(u_0) = K_4(u_0) = 0$.

4.2. Numerical results

We investigate the dependence of the mode I stress intensity factor $K_1(u_\varepsilon)$ on the length of the crack for the mixed boundary value problem in Fig. 2 using formula (49). Note that in this case $K_2(u_\varepsilon) = 0$ because of the symmetry of boundary forces. The stress intensity factors $K_j(u_0)$, $j = 1, 3$ are computed via formula (34) for the Poisson ratio $\nu = 0.3$. This requires the numerical computation of the unknown boundary data of the regular parts

$$\tilde{\zeta}_{i,0} := \zeta_{i,0} - r_0^{-1/2} \psi_i(\omega_0), \quad i = 1, 3, \qquad (50)$$

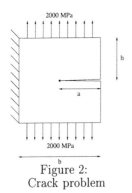

Figure 2:
Crack problem

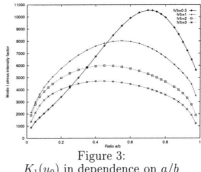

Figure 3:
$K_1(u_0)$ in dependence on a/b

of the weight functions $\zeta_{i,0}$. This is done by symmetric boundary element methods (see [2, 17] for a detailed description).

The coefficient $K_1(\zeta_{1,0})$ in the asymptotics of the regular part of $\zeta_{1,0}$ is calculated in the same way. The boundary is discretized by a uniform mesh with 190 elements. We approximate the boundary displacements by piecewise linear and the boundary stresses by piecewise constant splines. In Fig. 3 the stress intensity factor K_1 in dependence on the ratio a/b is plotted for different ratios h/b and Fig. 4 shows the corresponding numerical approximations of $\delta K_1(I, \Phi)$. Finally we compare in Fig. 5 the values obtained by using formula (49) with the finite difference approximations calculated from the values in Fig. 3.

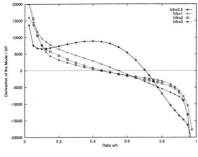

Figure 4: $\delta K_1(I, \Phi)$
in dependence on a/b

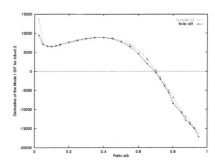

Figure 5: Finite difference
approximations

References

[1] M. Bochniak, A.-M. Sändig: Sensitivity analysis of stress intensity and notch factors in elastic structures. Preprint 97/54, SFB 404, Univ. Stuttgart, 1997.

[2] M. Bochniak, A.-M. Sändig: Computation of generalized stress intensity factors for bonded elastic structures. To appear in *Math. Model. Numer. Anal.*

[3] M. Bonnet: Regularized BIE formulations for first- and second-order shape sensitivity of elastic fields. *Computers and Structures* **56**, (1995) 799–811.

[4] C. Eck, S.A. Nazarov, W.L. Wendland: Asymptotic analysis for the mixed boundary value problem associated to a contact problem. Preprint 97–9, Mathematisches Institut A, Univ. Stuttgart, 1997.

[5] P.R. Garabedian, M. Schiffer: Convexity of Domain Functionals. *J. d'Anal. Math.* **2**, (1952/53) 281–368.

[6] E.J. Haug, K.K. Choi, V. Komkov: *Design Sensitivity Analysis of Structural Systems*. Academic Press Inc., Orlando, 1986.

[7] C. Hwu, C.J. Kao, L.E. Chang: Delamination fracture criteria for composite laminates. *J. Composite Materials* **29**, (1995) 1962–1987.

[8] V.A. Kondrat'ev: Boundary problems for elliptic equations in domains with conical or angular points. *Trans. Moscow Math. Soc.* **16**, (1967) 209–292.

[9] V.G. Maz'ya, B.A. Plamenevsky: Coefficients in the asymptotics of the solution of elliptic boundary value problems. *Math. Nachr.* **76**, (1977) 29–60.

[10] A.B. Movchan, S.A. Nazarov, O.R. Polyakova: Growth of stress intensity factors by increment of a curvilinear crack. *Mech. Tv. Tela* **1** (1992) 84–92.

[11] S. Nazarov: Derivation of the variational inequality for small increase of mode–one crack. *Mech. Solids* **24**, (1989) 145–152.

[12] S.A. Nazarov, B.A. Plamenevsky: *Elliptic Problems in Domains with Piecewise Smooth Boundaries*. Walter de Gruyter, Berlin, 1994.

[13] S.A. Nazarov, O.R. Polyakova: Rupture criteria, asymptotic conditions at crack tips, and selfadjoint extensions of the Lamé operator. *Trans. Moscow Math. Soc.* **57** (1996) 13-66.

[14] S. Nicaise, A.–M. Sändig: Transmission problems for the Laplace and elasticity operators: Regularity and boundary integral formulation. To appear in *Math. Model. Meth. Appl. Sci.*

[15] A.–M. Sändig: The Shapiro–Lopatinskij condition for boundary value and transmission problems for the Lamé system. Preprint 98/14, SFB 404, Univ. Stuttgart, 1998.

[16] J. Sokolowski, J.P. Zolesio: *Introduction to Shape Optimization*. Springer Verlag, Berlin, 1992.

[17] O. Steinbach: On the realization of boundary element methods for mixed boundary value problems. Preprint 98/07, SFB 404, Univ. Stuttgart, 1998.

Mathematical Institute A, University of Stuttgart,
Pfaffenwaldring 57, D–70569 Stuttgart, Germany.
Email: bochniak@mathematik.uni-stuttgart.de,
         anna@mathematik.uni-stuttgart.de

M. BONNET

A Formulation for Crack Shape Sensitivity Analysis Based on Galerkin BIE, Domain Differentiation and Adjoint Variable

1. Introduction and motivation

The consideration of sensitivity analysis of integral functionals with respect to shape parameters arises in many situations where a geometrical domain plays a primary role; shape optimization and inverse problems are the most obvious, as well as possibly the most important, of such instances. In addition to numerical differentition, shape sensitivities can be evaluated using either a direct differentiation or the adjoint variable approach [10, 19], the present paper being focused on the latter. Besides, consideration of shape changes in otherwise linear problems makes it very attractive to use boundary integral equation (BIE) formulations, which constitute the minimal modelling as far as the geometrical support of unknown field variables is concerned.

In the BIE context, the direct differentiation approach rests upon an application of the material differentiation formula for surface integrals to the governing integral equations, in either singular form [1, 12] or regularized form [4, 11, 15] (the material differentiation formula is shown in [5] to remain valid for strongly singular or hypersingular surface integrals); the direct differentiation approach is thus in particular applicable in the presence of cracks. Sensitivity evaluation then rests upon solving as many new boundary-value problems as the number of shape parameters present. The fact that they all involve the original governing operator reduces the computational effort to building new right-hand sides and solving linear systems by backsubstitution.

The adjoint variable approach is even more attractive: it needs to solve only one new boundary-value problem (the so-called adjoint problem) per integral functional present (often only one), whatever the number of shape parameters. In connexion with BIE formulations alone, the adjoint variable approach has been successfully applied to many shape sensitivity problems [3, 9, 10, 13]. It relies critically upon the possibility of formulating the final, analytical expression of the shape sensitivity of a given integral functional as a *boundary* integral that involves the boundary traces of the primary and adjoint states. However, this step raises difficulties when cracks (elasticity) or screens (acoustics) are present: non-integrable quantities associated with crack front singularity of field variables arise in the process.

This paper purports to show that this difficulty is avoided by formulating, in the Lagrangian, the direct problem constraint as a symmetric Galerkin BIE. Explicit boundary-only expressions of sensitivities are obtained for shape perturbations of either voids or cracks, in linear acoustics and elastodynamics.

The shape sensitivity problem. Consider a domain B of $\mathbb{R}^m$ ($m = 2$ or 3) with external boundary S, containing an internal defect in the form of either a void V of boundary Γ (Fig. 1a) or a crack with crack surface Γ (Fig. 1b). Let Ω denote the actual body (i.e.containing the defect): $\Omega = B \setminus V$ or $\Omega = B \setminus \Gamma$ and put $A = \partial\Omega = S \cup \Gamma$.

For ease of exposition, the main developments concentrate on scalar wave propagation in the frequency-domain, whereby some complex-valued field variable u (e.g.

the acoustic pressure) satisfies the Helmholtz equation $(\Delta + k^2)u = 0$, where $k = \omega/c$ is the wave number (ω: angular frequency, c: wave velocity). However, the treatment of shape sensitivity to follow is applicable to many other linear direct problems; its extension to steady-state elastodynamics is addressed in Sec 6.

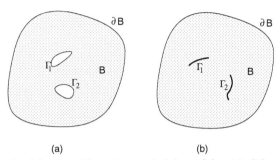

(a) (b)

Figure 1: A body with an internal defect: (a) void, (b) crack.

The shape and position of the boundary Γ characterizing the defect are unknown. Suppose that a given flux $\tilde{q}$ is imposed on B, while the defect surface is flux-free. The boundary value problem for given defect reads:

$$(\Delta + k^2)u = 0 \quad (\text{in } \Omega) \qquad q(u) = \tilde{q} \quad (\text{on } S) \qquad q(u) = 0 \quad (\text{on } \Gamma) \qquad (1)$$

where q denotes the normal derivative of u: $q = \partial u / \partial n$. In the case of a crack, the variable u is allowed to jump across Γ; $[\![u]\!] \equiv u^+ - u^- \neq 0$.

Let us thus introduce the generic objective function $\mathcal{J}$:

$$\mathcal{J}(\Gamma) = \int_S \varphi(u_\Gamma, \boldsymbol{x}) \, dS + \int_\Gamma \psi(\boldsymbol{x}) \, dS \qquad (2)$$

where u_Γ denotes the solution of problem (1) for a given Γ. For instance, in the problem of determining the shape and position of the defect using experimental data for a physical quantity governed by problem (1), one may use $2\varphi(u_\Gamma, \boldsymbol{x}) = \chi(S_m) |\hat{u} - u_\Gamma|^2$ where $\hat{u}(\boldsymbol{x})$ is a measurement of u for $\boldsymbol{x} \in S_m \subseteq S$ and $\psi(\boldsymbol{x})$ serves to define a regularization term. Other kinds of sensitivity problems with different motivations (e.g. optimization) can be considered as well.

Evaluating the gradient of $\mathcal{J}$ with respect to perturbations of Γ allows to use standard minimization algorithms, e.g. quasi-Newton. In view of the fact that $\mathcal{J}$ defined by Eq. (2) depends only on boundary quantities and the problem (1) does not involve sources distributed over Ω, it is natural to base the evaluation of $\mathcal{J}$ and its gradient upon a boundary element treatment.

2. Shape sensitivity analysis

The shape of the body Ω is now assumed to evolve with the values of a (small) time-like parameter p, according to a continuum kinematics-type Lagrangian description (with the "initial" configuration Ω_0 conventionally associated with $p = 0$):

$$\boldsymbol{x} \in \Omega_0 \to \boldsymbol{x}^p = \boldsymbol{\Phi}(\boldsymbol{x}, p) \in \Omega_p \qquad \text{where} \quad (\forall \boldsymbol{x} \in \Omega_0), \, \boldsymbol{\Phi}(\boldsymbol{x}, 0) = \boldsymbol{x} \qquad (3)$$

The *geometrical transformation* $\mathbf{\Phi}(\boldsymbol{x}, p)$ must possess a strictly positive Jacobian.

Differentiation of field variables and integrals in a domain perturbation is a well-documented subject [17, 19]; a few basic definitions and facts are recalled now, using notations introduced in [4]. The *initial transformation velocity* field $\boldsymbol{\theta}(\boldsymbol{x})$, defined by

$$\boldsymbol{\theta}(\boldsymbol{x}) = \boldsymbol{\Phi}_{,p}(\boldsymbol{x}, 0) \tag{4}$$

is the "initial" velocity of the "material" point which coincides with the geometrical point $\boldsymbol{x}$ at "time" $p = 0$. The "Lagrangian" derivative at $p = 0$ of a field quantity $f(\boldsymbol{y}, p)$ in a geometrical transformation, denoted by $\overset{\star}{f}$, is defined by:

$$\overset{\star}{f} = \lim_{p \searrow 0} p^{-1}[f(\boldsymbol{x}^p, p) - f(\boldsymbol{x}, 0)] = f_{,p} + \boldsymbol{\nabla} f \cdot \boldsymbol{\theta} \tag{5}$$

The Lagrangian derivative of the gradient of a field quantity is then given by:

$$(\boldsymbol{\nabla} f)^* = \boldsymbol{\nabla} \overset{\star}{f} - \boldsymbol{\nabla} f \cdot \boldsymbol{\nabla} \boldsymbol{\theta} \tag{6}$$

where the symbol "$\cdot$" indicates the dot product of two vectors or tensors (e.g. $\boldsymbol{\nabla} f . \boldsymbol{\nabla}\boldsymbol{\theta} = f_{,i}\theta_{i,j}\boldsymbol{e}_j$). Besides, the material derivatives of the unit normal $\boldsymbol{n}$ to a "moving" surface S_p and the area differential element dS are given by:

$$\overset{\star}{dS} = \operatorname{div}_S \boldsymbol{\theta} \, dS = D_i \theta_i \quad \overset{\star}{\boldsymbol{n}} = -\boldsymbol{n} \cdot \boldsymbol{\nabla}_S \boldsymbol{\theta} = -n_j D_i \theta_j \boldsymbol{e}_i \tag{7}$$

in terms of the surface gradient: $\boldsymbol{\nabla}_S f = \boldsymbol{\nabla} f - (\boldsymbol{\nabla} f \cdot \boldsymbol{n})\boldsymbol{n} = (f_{,i} - n_i f_{,n})\boldsymbol{e}_i \equiv (D_i f)\boldsymbol{e}_i$ and the surface divergence $\operatorname{div}_S \boldsymbol{u} = \operatorname{div} \boldsymbol{u} - (\boldsymbol{\nabla} \boldsymbol{u}.\boldsymbol{n}) \cdot \boldsymbol{n} = D_i u_i$. The material derivatives of domain and surface integrals are expressed by:

$$\frac{d}{dp}\int_\Omega f \, d\Omega = \int_\Omega (\overset{\star}{f} + f \operatorname{div} \boldsymbol{\theta}) \, d\Omega \quad \Omega\text{: any domain} \tag{8}$$

$$\frac{d}{dp}\int_S f \, dS = \int_S (\overset{\star}{f} + f \operatorname{div}_S \boldsymbol{\theta}) \, dS \quad S\text{: any surface} \tag{9}$$

From formula (9), the Lagrangian derivative of a double surface integral is given by:

$$\frac{d}{dp}\int_\Gamma \int_\Gamma f(\boldsymbol{x},\boldsymbol{y}) \, dS_x dS_y = \int_\Gamma \int_\Gamma \{\overset{\star}{f}(\boldsymbol{x},\boldsymbol{y}) + f[\operatorname{div}_S \boldsymbol{\theta}(\boldsymbol{x}) + \operatorname{div}_S \boldsymbol{\theta}(\boldsymbol{y})]\} \, dS_x dS_y \tag{10}$$

Remark 1. The external boundary S and its neighbourhood are here assumed to be unaffected by the shape transformation, so $\boldsymbol{\theta} = \boldsymbol{0}$ and $\boldsymbol{\nabla}\boldsymbol{\theta} = \boldsymbol{0}$ on S. However, if emerging cracks are considered, $\boldsymbol{\theta}$ and $\boldsymbol{\nabla}\boldsymbol{\theta}$ do not vanish on some neighbourhood of the emerging edge (3-D problems) or point (2-D problems).

Remark 2. The pseudo-time p can be understood either as an abstract parameter, in the fashion of the calculus of variations (arbitrary shape perturbations being accommodated using $\boldsymbol{\theta}$) or a geometrical parameter having a definite practical meaning (a radius, a major semiaxis, a centroid coordinate ...). As far as first-order derivatives are concerned, the analysis to follow considers for convenience a single shape parameter p but is in fact applicable to the case of finitely many shape parameters.

3. Adjoint problem for domain integral formulation

In this section, the classical way of defining an adjoint state, which rests upon a (weak) domain formulation of the direct problem (1), is recalled. The resulting sensitivity expression involves a domain integral which, when the defect is a void, can be converted into a boundary integral (section 3.2). However, when the defect is a crack, the conversion into boundary integrals will be seen to break down (section 3.3).

3.1. Definition of an adjoint state. Introduce the following Lagrangian:

$$\mathcal{L}(u,v,\Gamma) = \mathcal{J}(\Gamma) + \int_\Omega (\nabla u \cdot \nabla \bar{v} - k^2 u \bar{v})\, d\Omega - \int_S \tilde{q}\bar{v}\, dS \qquad (v \in H^1(\Omega)) \qquad (11)$$

in which the weak formulation of the direct wave problem (1) appears as an equality constraint term, the test function v acting as the Lagrange multiplier. Taking into account Eqs. (5)–(9), the total material derivative of the Lagrangian with respect to a variation of the domain can be expressed as:

$$\overset{\star}{\mathcal{L}}(u,v,\Gamma) = \int_S \varphi_{,u}\,\overset{\star}{u}\, dS + \int_\Gamma (\overset{\star}{\psi} \cdot \boldsymbol{\theta} + \psi \mathrm{div}_S \boldsymbol{\theta})\, dS + \int_\Omega (\nabla \overset{\star}{u} \cdot \nabla \bar{v} - k^2 \overset{\star}{u}\, \bar{v})\, d\Omega$$
$$- \int_\Omega \{[\nabla u \cdot \nabla \bar{v} - k^2 u\,\bar{v}]\mathrm{div}\boldsymbol{\theta} - [\nabla u \otimes \nabla \bar{v} + \nabla \bar{v} \otimes \nabla u] \cdot \nabla \boldsymbol{\theta}\}\, d\Omega \qquad (12)$$

For cracks, the partial "time" derivative $\nabla u_{,p}$ has generally a $d^{-3/2}$ singularity along the crack edge $\partial \Gamma$, while $\nabla(\overset{\star}{u})$ and ∇u have the same $d^{-1/2}$ singularity, where d is the distance to $\partial \Gamma$. For this reason, the Lagrangian derivative $\overset{\star}{u}$ has been introduced instead of the partial derivative $u_{,p}$. The derivations made in this section are therefore valid for both void and crack problems. Note that the terms containing $\overset{\star}{v}$ have been omitted in Eq. (12), because they merely repeat the equality constraint.

In equation (12), the test function v is now chosen so that the terms which contain $\overset{\star}{u}$ combine to zero for any $\overset{\star}{u}$, i.e. solves the *adjoint problem*:

$$\int_\Omega (\nabla \bar{v} \cdot \nabla \overset{\star}{u} - k^2 \bar{v}\, \overset{\star}{u})\, d\Omega + \int_S \varphi_{,u}\,\overset{\star}{u}\, dS = 0 \qquad (\forall\, \overset{\star}{u} \in H^1(\Omega)) \qquad (13)$$

whose strong formulation is:

$$(\Delta + k^2)v = 0 \quad (\text{in } \Omega) \qquad q(v) = -\bar{\varphi}_{,u} \quad (\text{on } S) \qquad q(v) = 0 \quad (\text{on } \Gamma) \qquad (14)$$

Finally, $\overset{\star}{\mathcal{J}}$ is expressed from Eq. (12) in terms of the direct and adjoint states (u_Γ, v_Γ):

$$\overset{\star}{\mathcal{J}}(\Gamma) = \overset{\star}{\mathcal{L}}(u_\Gamma, \bar{v}_\Gamma, \Gamma) = \int_\Gamma (\overset{\star}{\psi} \cdot \boldsymbol{\theta} + \psi \mathrm{div}_S \boldsymbol{\theta})\, dS \qquad (15)$$
$$+ \int_\Omega \{[\nabla u_\Gamma \cdot \nabla \bar{v}_\Gamma - k^2 u_\Gamma \bar{v}_\Gamma]\mathrm{div}\boldsymbol{\theta} - (\nabla u_\Gamma \cdot \nabla \boldsymbol{\theta}) \cdot \nabla \bar{v}_\Gamma - (\nabla \bar{v}_\Gamma \cdot \nabla \boldsymbol{\theta}) \cdot \nabla u_\Gamma\}\, d\Omega$$

3.2. Conversion into boundary integrals (void problem). The formula (16) for the sensitivity of $\mathcal{J}$, expressed by a domain integral, is not suitable for BEM-based computations. However, for void defects, it can be recast in boundary-only form. In order to do so, one first notices that u_Γ and v_Γ verify:

$$[\nabla u \otimes \nabla \bar{v} + \nabla \bar{v} \otimes \nabla u].\nabla \boldsymbol{\theta}$$
$$= \mathrm{div}[(\nabla u_\Gamma \cdot \boldsymbol{\theta})\nabla \bar{v}_\Gamma + (\nabla \bar{v}_\Gamma \cdot \boldsymbol{\theta})\nabla u_\Gamma] + k^2 \nabla(u_\Gamma \bar{v}_\Gamma) \cdot \boldsymbol{\theta} - \nabla(\nabla u_\Gamma \cdot \nabla \bar{v}_\Gamma) \cdot \boldsymbol{\theta} \qquad (16)$$

Then, using this identity and after some additional manipulation, one obtains:

$$(\nabla u_\Gamma \cdot \nabla \bar{v}_\Gamma - k^2 u_\Gamma \bar{v}_\Gamma) \text{div}\boldsymbol{\theta} - (\nabla u_\Gamma \cdot \nabla \boldsymbol{\theta}) \cdot \nabla \bar{v}_\Gamma - (\nabla \bar{v}_\Gamma \cdot \nabla \boldsymbol{\theta}) \cdot \nabla u_\Gamma$$
$$= \text{div}[(\nabla u_\Gamma \cdot \nabla \bar{v}_\Gamma - k^2 u_\Gamma \bar{v}_\Gamma)\boldsymbol{\theta} - (\nabla u_\Gamma \cdot \boldsymbol{\theta})\nabla \bar{v}_\Gamma - (\nabla \bar{v}_\Gamma \cdot \boldsymbol{\theta})\nabla u_\Gamma] \quad (17)$$

which allows us to recast Eq. (16) into a boundary integral by means of the divergence formula. One obtains (using the fact that S is unperturbed, i.e. $\boldsymbol{\theta} = \mathbf{0}$ on S):

$$\overset{\star}{\mathcal{J}}(\Gamma) = \int_\Gamma [\nabla u_\Gamma \cdot \nabla \bar{v}_\Gamma - k^2 u_\Gamma \bar{v}_\Gamma] \theta_n \, dS$$
$$- \int_\Gamma \{(\nabla u_\Gamma \cdot \boldsymbol{\theta})q(\bar{v}_\Gamma) + (\nabla \bar{v}_\Gamma \cdot \boldsymbol{\theta})q(u_\Gamma)\} \, dS + \int_\Gamma (\overset{\star}{\psi} \cdot \boldsymbol{\theta} + \psi \text{div}_S \boldsymbol{\theta}) \, dS \quad (18)$$

(where $\theta_n \equiv \boldsymbol{\theta}.\mathbf{n}$). Further, splitting gradients into tangential and normal parts and using the boundary conditions $q(u) = q(v) = 0$ on the void surface Γ, Eq. (18) becomes:

$$\overset{\star}{\mathcal{J}}(\Gamma) = \int_\Gamma [\nabla_S u_\Gamma \cdot \nabla_S \bar{v}_\Gamma - k^2 u_\Gamma \bar{v}_\Gamma] \theta_n \, dS + \int_\Gamma (\overset{\star}{\psi} \cdot \boldsymbol{\theta} + \psi \text{div}_S \boldsymbol{\theta}) \, dS \quad (19)$$

Formula (19) allows the computation of the derivatives of any objective functional $\mathcal{J}$ of the type (2) with respect to shape parameters. In particular, since $\nabla_S u_\Gamma, \nabla_S v_\Gamma$ depend only on the boundary traces of u_Γ and v_Γ, the sensitivity (19) is computable directly from the BEM solution of the primary and adjoint problems.

3.3. Inapplicability to crack problems. Consider the case where the unknown defect is a crack, i.e. the limiting case of a void bounded by two surfaces Γ^+ and Γ^- identical and of opposite orientations (Fig. 2).

Figure 2: A crack bounded by two almost identical surfaces Γ^+ and Γ^-.

It is tempting to still apply Eq. (19) to compute sensitivities with respect to crack perturbations. However, Eq. (19) is not applicable to crack defects. For instance, considering a domain shape transformation such that $\theta_n = 0$ on the crack surface Γ (i.e. allowing crack perturbations along the tangent plane at the crack front, i.e. crack extensions), Eq. (19) gives $d\mathcal{J}/dp = 0$, which is certainty not true in general. In contrast, when Γ is the piecewise smooth boundary of a void, $\theta_n = 0$ means that the void is unperturbed. This apparent paradox may be explained as follows: identity (16), which plays a key role in establishing the boundary-only expression (19), involves the field quantity $\nabla(\nabla u \cdot \nabla v)$. For cracks, u and v behave like $d^{1/2}$ in the vicinity of the crack front (d: distance to $\partial \Gamma$). Then, $\nabla(\nabla u \cdot \nabla v)$ behaves like d^{-2} and the domain integral $\int_\Omega \nabla(\nabla u \cdot \nabla v) \cdot \boldsymbol{\theta} \, d\Omega$ is divergent.

Hence, the adjoint state defined by Eqs. (13) or (14) leads to boundary-only sensitivity expressions for functionals of the form (2) only when the defect is a void. To retain the boundary-only character of sensitivity expressions when the defect is a crack, a suitable and different definition of the adjoint problem approach is called for. In what follows, this objective is shown to be fulfilled by enforcing the direct problem constraint as a symmetric Galerkin boundary integral equation (SGBIE).

4. Symmetric Galerkin BIE formulation

The direct problem (1) admits, if the defect is a void, the following regularized SGBIE formulation, obtained from a weighted residual statement of the (initially hypersingular) flux boundary integral equation:

$$\mathcal{B}_{AA}(u,w) = \mathcal{F}(w) \qquad \forall w \in H^{1/2}(A) \tag{20}$$

where

$$\mathcal{B}_{XY}(u,w) = \int_X \int_Y G(\boldsymbol{x}-\boldsymbol{y})\{R_i u(\boldsymbol{y}) R_i \bar{w}(\boldsymbol{x}) - k^2 n_i(\boldsymbol{y}) n_i(\boldsymbol{x}) u(\boldsymbol{y}) \bar{w}(\boldsymbol{x})\} \, \mathrm{d}S_y \, \mathrm{d}S_x \tag{21}$$

$$\mathcal{F}(w) = \int_S \int_S [H(\boldsymbol{y},\boldsymbol{x})\bar{w}(\boldsymbol{x}) - H^0(\boldsymbol{y},\boldsymbol{x})\bar{w}(\boldsymbol{y})]\tilde{q}(\boldsymbol{y}) \, \mathrm{d}S_y \, \mathrm{d}S_x$$

$$+ \int_\Gamma \int_S H(\boldsymbol{y},\boldsymbol{x})\bar{w}(\boldsymbol{x})\tilde{q}(\boldsymbol{y}) \, \mathrm{d}S_y \, \mathrm{d}S_x - \int_S \tilde{q}(\boldsymbol{y})\bar{w}(\boldsymbol{y}) \, \mathrm{d}S_y \tag{22}$$

In these expressions, $R_i u$ denotes the i-th component of the *surface curl* of the scalar function u: $R_i u = e_{ijk} n_j u_{,k}$, which is a tangential differential operator [14] associated to a variant of the Stokes' formula:

$$\int_C R_i f \, \mathrm{d}S = 0 \qquad C: \text{ any piecewise smooth } closed \text{ surface} \tag{23}$$

$G(\boldsymbol{x}-\boldsymbol{y}) = e^{ikr}/(4\pi r)$ is the infinite-space fundamental solution of the Helmholtz equation ($r = |\boldsymbol{y} - \boldsymbol{x}|$), and $H(\boldsymbol{y},\boldsymbol{x}) = G_{,j}(\boldsymbol{x}-\boldsymbol{y})n_j(\boldsymbol{x})$, $H^0(\boldsymbol{y},\boldsymbol{x}) = -(x_j - y_j)n_j(\boldsymbol{x})/(4\pi r^3)$. The expression (21) of $\mathcal{B}(u,w)$ stems in fact from the following identity verified by G:

$$-n_i(\boldsymbol{y})n_j(\boldsymbol{x})G_{,ij}(\boldsymbol{x}-\boldsymbol{y}) = k^2 G(\boldsymbol{x}-\boldsymbol{y}) - R_i^y R_i^x G(\boldsymbol{x}-\boldsymbol{y}) \tag{24}$$

When the defect is a crack, the corresponding SGBIE formulation is obtained by taking the limiting form of Eqs. (20–21–22) when the cavity becomes infinitely thin. For example, one can consider a cavity of boundary $\Gamma^+ \cup \Gamma^-$, where Γ^+, Γ^- (the crack faces) are identical open surfaces with opposite orientations. Let $\Gamma \equiv \Gamma^-$ and $\boldsymbol{n} \equiv \boldsymbol{n}^-$. Using test functions w which are continuous over $\Gamma^+ \cup \Gamma^-$ and vanish on Γ^+ (in particular, such w vanish on the edge $\partial\Gamma$), the SGBIE formulation turns out to be:

$$\mathcal{B}_{AS}(u,w) + \mathcal{B}_{A\Gamma}(\phi,w)\mathcal{F}(w) \tag{25}$$

where $\phi(\boldsymbol{y}) = [\![u]\!](\boldsymbol{y}) \equiv u^+(\boldsymbol{y}) - u^-(\boldsymbol{y})$ is the discontinuity of u across the crack.

When there is no external boundary (void or crack embedded in an infinite medium, known incident wave u^I), the SGBIE formulation reads (u: total field):

$$\mathcal{B}_{\Gamma\Gamma}(u,w) = \int_\Gamma q^I \bar{w} \, \mathrm{d}S \quad \text{(void)} \qquad \mathcal{B}_{\Gamma\Gamma}(\phi,w) = -\int_\Gamma q^I \bar{w} \, \mathrm{d}S \quad \text{(crack)} \qquad \forall w \in H^{1/2}$$

5. Adjoint problem formulation for crack defects

In order to define a suitable adjoint problem, the constraint corresponding to the direct wave problem is now introduced in its SGBIE form (20) in a Lagrangian $\mathcal{L}_B$:

$$\mathcal{L}_B(u, w, \Gamma) = \mathcal{J}(\Gamma) + \mathcal{B}(u, w) - \mathcal{F}(w) \tag{26}$$

where $\mathcal{B}_{AA}(u, w)$ (noted $\mathcal{B}(u, w)$ for brevity) and $\mathcal{F}(w)$ are given by Eqs. (21) and (22), respectively. Since the latter are respectively bilinear and linear, the material derivative of $\mathcal{L}_B$ is expected to take the form:

$$\overset{\star}{\mathcal{L}}_B(u, w, \Gamma) = \int_S \varphi_{,u} \overset{\star}{u} \, dS + \int_\Gamma (\overset{\star}{\psi} \cdot \boldsymbol{\theta} + \psi \mathrm{div}_S \boldsymbol{\theta}) \, dS$$
$$+ \mathcal{B}(\overset{\star}{u}, w) + \mathcal{B}^1(u, w; \boldsymbol{\theta}) - \mathcal{F}^1(w; \boldsymbol{\theta}) \tag{27}$$

having, again, omitted the terms containing $\overset{\star}{w}$ for the same reasons as in Sec. 3.1. The new terms $\mathcal{B}^1(u, w; \boldsymbol{\theta})$ and $\mathcal{F}^1(w; \boldsymbol{\theta})$, respectively bilinear in (u, w) and linear in w, are both linear in $\boldsymbol{\theta}$; they result from application of the differentiation formulas (9) and (10) to Eqs. (21) and (22) and will be detailed later.

Again, the adjoint state w_Γ is defined by imposing that all terms containing $\overset{\star}{u}$ vanish in Eq. (27). Therefore, w_Γ is the solution to the adjoint SGBIE formulation:

$$\mathcal{B}(w, \overset{\star}{u}) = -\int_S \varphi_{,u} \overset{\star}{u} \, dS \qquad \forall \overset{\star}{u} \in H^{1/2}(A) \tag{28}$$

where of course the symmetric character of the bilinear form $\mathcal{B}(u, w)$ has been used.

Remark 3. The adjoint problem Eq. (28) is the SGBIE formulation of problem (13), but in *indirect* form. Therefore, the solution w_Γ to Eq. (28) is a double-layer density, and is *not* equal to the boundary trace of the solution v_Γ to Eq. (13), except in the special case of cracks in infinite media.

The material derivative of the functional $\mathcal{J}$ is then expressed as:

$$\overset{\star}{\mathcal{J}}(\Gamma) = \overset{\star}{\mathcal{L}}_B(u_\Gamma, w_\Gamma, \Gamma) = \int_\Gamma (\overset{\star}{\psi} \cdot \boldsymbol{\theta} + \psi \mathrm{div}_S \boldsymbol{\theta}) \, dS + \mathcal{B}^1(u_\Gamma, w_\Gamma; \boldsymbol{\theta}) - \mathcal{F}^1(w_\Gamma; \boldsymbol{\theta}) \tag{29}$$

The remaining task is to establish explicit expressions for the terms $\mathcal{B}^1(u, w; \boldsymbol{\theta})$ and $\mathcal{F}^1(w; \boldsymbol{\theta})$. This is achieved in quite straightforward fashion, by applying the differentiation formulas (9) and (10) to the simple and double integrals involved in Eqs. (21) and (22) which define $\mathcal{B}(u, w)$ and $\mathcal{F}(w)$. It is important to note at this point that all double integrations in Eqs. (21) and (22) consist in a weakly singular inner integral followed by a nonsingular outer integral.

In order to perform this calculation, let us first collect some useful formulas. First, for any kernel function of the form $K(\boldsymbol{y} - \boldsymbol{x})$, one has:

$$\overset{\star}{K}(\boldsymbol{y} - \boldsymbol{x}) = [\theta_i(\boldsymbol{x}) - \theta_i(\boldsymbol{y})] K_{,i}(\boldsymbol{y} - \boldsymbol{x}) \tag{30}$$

Then, from Eqs. (7), it is easy to show that

$$(n_i \, dS)^* = (n_i D_j \theta_j - n_j D_i \theta_j) \, dS = e_{ijk} R_k \theta_j \, dS \tag{31}$$

In addition, using Eqs. (6), (7) and (31), one has:

$$(R_i u \, dS)^* = e_{ijk}(e_{jab} R_b \theta_a u_{,k} - n_j u_a \theta_{a,k}) \, dS = (R_k \theta_i - R_i \theta_k) u_{,k} - R_i \theta_a u_{,a}$$
$$= -R_a \theta_i u_{,a} = R_a u D_i \theta_a \tag{32}$$

The last equality results from invoking again the definition of the surface curl and its consequence $n_a R_a u = 0$. Using Eqs. (10), (30), (31) and (32), it is then a simple matter to show (again assuming that $\overset{*}{w}= 0$) that:

$$\overset{*}{\mathcal{B}}(u,w) = \mathcal{B}(\overset{*}{u},w) + \mathcal{B}^1(u,w;\boldsymbol{\theta}) \tag{33}$$

where $\mathcal{B}^1(u,w;\boldsymbol{\theta})$ is given by:

$$\mathcal{B}^1(u,w;\boldsymbol{\theta}) = \int_A \int_A \{R_i u(\boldsymbol{y}) R_j \bar{w}(\boldsymbol{x}) B^1_{ij}(\boldsymbol{y},\boldsymbol{x};\boldsymbol{\theta}) - k^2 u(\boldsymbol{y}) \bar{w}(\boldsymbol{x}) A^1(\boldsymbol{y},\boldsymbol{x};\boldsymbol{\theta})\} \, dS_y \, dS_x \tag{34}$$

with the new kernel functions:

$$B^1_{ij}(\boldsymbol{y},\boldsymbol{x};\boldsymbol{\theta}) = [\theta_k(\boldsymbol{x}) - \theta_k(\boldsymbol{y})] G_{,k}(\boldsymbol{x}-\boldsymbol{y}) \delta_{ij} + [\theta_{i,j}(\boldsymbol{x}) + \theta_{i,j}(\boldsymbol{y})] G(\boldsymbol{x}-\boldsymbol{y}) \tag{35}$$

$$A^1(\boldsymbol{y},\boldsymbol{x};\boldsymbol{\theta}) = n_i(\boldsymbol{y}) n_i(\boldsymbol{x})[\theta_j(\boldsymbol{x}) - \theta_j(\boldsymbol{y})] G_{,j}(\boldsymbol{x}-\boldsymbol{y})$$
$$+ e_{ijk}[n_i(\boldsymbol{y}) R_k \theta_j(\boldsymbol{x}) + n_i(\boldsymbol{x}) R_k \theta_j(\boldsymbol{y})] G(\boldsymbol{x}-\boldsymbol{y}) \tag{36}$$

Under the conditions of the present study, $\boldsymbol{\theta} = \mathbf{0}$ on S should be enforced in Eq. (34). Next, one has, from the definition of $H(\boldsymbol{y},\boldsymbol{x})$ and using again Eqs. (30) and (31):

$$(H(\boldsymbol{y},\boldsymbol{x}) \, dS_x)^* = (G_{,i}(\boldsymbol{x}-\boldsymbol{y}) n_i(\boldsymbol{x}) \, dS_x)^*$$
$$= (\theta_i(\boldsymbol{x}) G_{,ij}(\boldsymbol{x}-\boldsymbol{y}) n_i(\boldsymbol{x}) + G_{,i}(\boldsymbol{x}-\boldsymbol{y}) e_{ijk} R_k \theta_j(\boldsymbol{x})) \, dS_x$$

where $\boldsymbol{x}$ and $\boldsymbol{y}$ are assumed to lie on Γ and S, respectively, so that $\boldsymbol{\theta}(\boldsymbol{y}) = \mathbf{0}$. In addition, one has (since here $\boldsymbol{x} \neq \boldsymbol{y}$):

$$G_{,ij}(\boldsymbol{x}-\boldsymbol{y}) n_i(\boldsymbol{x}) = e_{ijk} R^x_k G_{,i}(\boldsymbol{x}-\boldsymbol{y}) + n_j(\boldsymbol{x}) G_{,ii}(\boldsymbol{x}-\boldsymbol{y})$$
$$= e_{ijk} R^x_k G_{,i}(\boldsymbol{x}-\boldsymbol{y}) - k^2 n_j(\boldsymbol{x}) G(\boldsymbol{x}-\boldsymbol{y})$$

and hence, combining these relations:

$$(H(\boldsymbol{y},\boldsymbol{x}) \, dS_x)^* = e_{ijk} R^x_k (G_{,i}(\boldsymbol{x}-\boldsymbol{y}) \theta_i(\boldsymbol{x})) - k^2 \theta_n(\boldsymbol{x}) G(\boldsymbol{x}-\boldsymbol{y}) \tag{37}$$

One is now in a position to calculate the material derivative of $\mathcal{F}(w)$. The differentiation of Eq. (22) is performed under the assumption that the surface S is left unperturbed by the domain transformation $\boldsymbol{\theta} = \mathbf{0}$ and hence affects only *single* integrals. By virtue of the identity (37) and using the Stokes identity (23), one finds:

$$\overset{\star}{\mathcal{F}}(w) = \mathcal{F}^1(w;\boldsymbol{\theta})$$
$$= \int_\Gamma \int_S \{e_{ijk}R_k^x\big(G_{,i}(\boldsymbol{x}-\boldsymbol{y})\theta_i(\boldsymbol{x})\big) - k^2\theta_n(\boldsymbol{x})G(\boldsymbol{x}-\boldsymbol{y})\}\bar{w}(\boldsymbol{x})\tilde{q}(\boldsymbol{y})\,\mathrm{d}S_y\,\mathrm{d}S_x$$
$$= -\int_\Gamma \int_S \{G_{,i}(\boldsymbol{x}-\boldsymbol{y})\theta_i(\boldsymbol{x})e_{ijk}R_k\bar{w}(\boldsymbol{x}) + k^2\theta_n(\boldsymbol{x})G(\boldsymbol{x}-\boldsymbol{y})\bar{w}(\boldsymbol{x})\}\tilde{q}(\boldsymbol{y})\,\mathrm{d}S_y\,\mathrm{d}S_x \quad (38)$$

Equations (33) to (36) and (38) provide the sought-for explicit expressions of $\mathcal{B}^1(u,w;\boldsymbol{\theta})$ and $\mathcal{F}^1(w;\boldsymbol{\theta})$. The sensitivity formula (29) is then completely determined.

Remark 4. Equations (33) to (36) and (38), and hence the sensitivity formula (29), are applicable for both voids and cracks. The derivation of $\mathcal{B}^1(u,w;\boldsymbol{\theta})$, presented for Eq. (21) (void), is essentially repeated when applied to Eq. (25) (crack).

Remark 5. When dealing with void defects, two alternative sensitivity expressions, Eqs. (19) and (29), are thus available.

Remark 6. The inner integrals in Eqs. (33) to (36) and (38) are weakly singular if $\boldsymbol{\theta} \in C^{0,1}(\Gamma)$, while the outer ones are regular. Moreover, due to the double surface integration, the integrals in Eqs. (25), (29) are convergent even in the presence of a $d^{-1/2}$ crack front singularity of $R_i\phi_\Gamma$ and $R_i\bar{w}_\Gamma$.

Remark 7. The approach developed in Sections 5. and 6. for the first-order material derivative can be carried out one step further for the evaluation of second-order derivatives. Indeed this has been done in [6] to formulate and compute first- and second-order material derivatives of the elastic potential energy at equilibrium, as part of a numerical scheme to study the stability of quasi-static crack extension. Recall, however, that the second-order material derivative (assuming a finite number of geometrical parameters) and domain derivative do not coincide in general [18].

6. Frequency-domain elastodynamics

This section presents the generalization of the previous formulations to linear isotropic elastodynamics. The displacement vector $\boldsymbol{u}$, strain tensor $\boldsymbol{\varepsilon}$ and stress tensor $\boldsymbol{\sigma}$ are related through the equilibrium, constitutive and compatibility field equations:

$$\mathrm{div}\boldsymbol{\sigma} + \rho\omega^2\boldsymbol{u} = \boldsymbol{0} \quad \boldsymbol{\sigma} = \boldsymbol{C}{:}\boldsymbol{\varepsilon} \quad \boldsymbol{\varepsilon} = \frac{1}{2}(\nabla\boldsymbol{u} + \nabla^T\boldsymbol{u}) \quad \text{in } \Omega \quad (39)$$

where $\boldsymbol{C}$ is the fourth-order Hooke tensor of elastic moduli, whose components possess the symmetry properties $C_{ijk\ell} = C_{jik\ell} = C_{ij\ell k} = C_{k\ell ij}$. For isotropic elasticity, one has $C_{ijk\ell} = \mu(\kappa\delta_{ij}\delta_{k\ell} + \delta_{j\ell}\delta_{ik} + \delta_{i\ell}\delta_{jk})$ (μ: shear Lamé modulus, ν: Poisson ratio, $\kappa = 2\nu/(1-2\nu)$). Equations (39) are supplemented by boundary conditions:

$$\boldsymbol{t}(\boldsymbol{u}) = \tilde{\boldsymbol{t}} \quad (\text{on } S) \qquad \boldsymbol{t}(\boldsymbol{u}) = \boldsymbol{0} \quad (\text{on } \Gamma) \quad (40)$$

where $\boldsymbol{t}(\boldsymbol{u}) = \boldsymbol{\sigma}(\boldsymbol{u}).\boldsymbol{n}$ is the elastic traction vector.

6.1. Adjoint formulation for the void defect. Considering objective functions similar to Eq. (2), the Lagrangian incorporating the problem (39–40) in weak form is:

$$\mathcal{L}(\boldsymbol{u},\boldsymbol{v},\Gamma) = \mathcal{J}(\Gamma) + \int_\Omega (\boldsymbol{\sigma}(\boldsymbol{u}){:}\boldsymbol{\varepsilon}(\bar{\boldsymbol{v}}) - \rho\omega^2 \boldsymbol{u}\cdot\bar{\boldsymbol{v}})\,\mathrm{d}\Omega - \int_S \tilde{\boldsymbol{t}}\cdot\bar{\boldsymbol{v}}\,\mathrm{d}S \quad (\boldsymbol{v}\in [H^1(\Omega)]^3) \tag{41}$$

Its material derivative is given by:

$$\overset{\star}{\mathcal{L}}(\boldsymbol{u},\boldsymbol{v},\Gamma) = \int_S \varphi_{,u}\cdot\overset{\star}{\boldsymbol{u}}\,\mathrm{d}S + \int_\Omega (\boldsymbol{\sigma}(\overset{\star}{\boldsymbol{u}}){:}\boldsymbol{\varepsilon}(\bar{\boldsymbol{v}}) - \rho\omega^2\overset{\star}{\boldsymbol{u}}\cdot\bar{\boldsymbol{v}})\,\mathrm{d}\Omega + \int_\Gamma (\overset{\star}{\psi}\cdot\boldsymbol{\theta} + \psi\mathrm{div}_S\boldsymbol{\theta})\,\mathrm{d}S$$
$$+ \int_\Omega \{[\boldsymbol{\sigma}(\boldsymbol{u}){:}\boldsymbol{\varepsilon}(\bar{\boldsymbol{v}}) - \rho\omega^2\boldsymbol{u}\cdot\bar{\boldsymbol{v}}]\mathrm{div}\boldsymbol{\theta} - \boldsymbol{\sigma}(\boldsymbol{u}){:}(\nabla\bar{\boldsymbol{v}}.\nabla\boldsymbol{\theta}) - \boldsymbol{\sigma}(\bar{\boldsymbol{v}}){:}(\nabla\boldsymbol{u}\cdot\nabla\boldsymbol{\theta})\}\,\mathrm{d}\Omega \tag{42}$$

(note that the terms containing $\overset{\star}{\boldsymbol{v}}$ have been, again, dropped). The adjoint field $\boldsymbol{v}$ is thus found to solve the field equations (39) together with the boundary conditions:

$$\boldsymbol{t}(\boldsymbol{v}) = -\bar{\varphi}_{,u} \quad (\text{on } S) \qquad \boldsymbol{t}(\boldsymbol{v}) = 0 \quad (\text{on } \Gamma) \tag{43}$$

Besides, the solutions $\boldsymbol{u}_\Gamma$ and $\boldsymbol{v}_\Gamma$ to the direct and adjoint problems (and, indeed, any pair $(\boldsymbol{u},\boldsymbol{v})$ solving the field equations (39) as well) are found to verify:

$$[\boldsymbol{\sigma}(\boldsymbol{u}){:}\boldsymbol{\varepsilon}(\bar{\boldsymbol{v}}) - \rho\omega^2\boldsymbol{u}\cdot\bar{\boldsymbol{v}}]\mathrm{div}\boldsymbol{\theta} - \boldsymbol{\sigma}(\boldsymbol{u}){:}(\nabla\bar{\boldsymbol{v}}\cdot\nabla\boldsymbol{\theta}) - \boldsymbol{\sigma}(\bar{\boldsymbol{v}}){:}(\nabla\boldsymbol{u}\cdot\nabla\boldsymbol{\theta})$$
$$= \mathrm{div}\big([\boldsymbol{\sigma}(\boldsymbol{u}){:}\boldsymbol{\varepsilon}(\bar{\boldsymbol{v}}) - \rho\omega^2\boldsymbol{u}\cdot\bar{\boldsymbol{v}}]\boldsymbol{\theta} - \boldsymbol{\sigma}(\boldsymbol{u})(\nabla\bar{\boldsymbol{v}}.\boldsymbol{\theta}) - \boldsymbol{\sigma}(\bar{\boldsymbol{v}})(\nabla\boldsymbol{u}\cdot\boldsymbol{\theta})\big)$$

so that, *for void defects*, the sensitivity of $\mathcal{J}$ is expressed in boundary-only form as:

$$\overset{\star}{\mathcal{J}} = \int_\Gamma [\boldsymbol{\sigma}(\boldsymbol{u}){:}\boldsymbol{\varepsilon}(\bar{\boldsymbol{v}}) - \rho\omega^2\boldsymbol{u}\cdot\bar{\boldsymbol{v}}]\theta_n\,\mathrm{d}S + \int_\Gamma (\overset{\star}{\psi}\cdot\boldsymbol{\theta} + \psi\mathrm{div}_S\boldsymbol{\theta})\,\mathrm{d}S \tag{44}$$

having taken into account the homogeneous boundary conditions on Γ. Further, the strain energy density $\boldsymbol{\sigma}(\boldsymbol{u}){:}\boldsymbol{\varepsilon}(\bar{\boldsymbol{v}})$ can be expressed in terms of the tangential derivatives $\nabla_S\boldsymbol{u}, \nabla_S\bar{\boldsymbol{v}}$ and of the tractions $\boldsymbol{t}(\boldsymbol{u}), \boldsymbol{t}(\bar{\boldsymbol{v}})$. Since the latter vanish on Γ, one obtains:

$$\boldsymbol{\sigma}(\boldsymbol{u}){:}\boldsymbol{\varepsilon}(\bar{\boldsymbol{v}}) = \mu\big\{\frac{2\nu}{1-\nu}\mathrm{div}_S\boldsymbol{u}\,\mathrm{div}_S\bar{\boldsymbol{v}} + (\nabla_S\boldsymbol{u}+\nabla_S^T\boldsymbol{u}){:}\nabla_S\bar{\boldsymbol{v}} - (\boldsymbol{n}\cdot\nabla_S\boldsymbol{u})\cdot(\boldsymbol{n}\cdot\nabla_S\bar{\boldsymbol{v}})\big\} \tag{45}$$

6.2. Adjoint formulation for the crack defect. The direct problem (39) admits, if the defect is a void, the following regularized SGBIE formulation, obtained from a weighted residual statement of the traction boundary integral equation:

$$\mathcal{B}_{AA}(\boldsymbol{u},\boldsymbol{w}) = \mathcal{F}(\boldsymbol{w}) \qquad \forall \boldsymbol{w}\in [H^{1/2}(A)]^3 \tag{46}$$

where

$$\mathcal{B}_{XY}(\boldsymbol{u},\boldsymbol{v}) = \int_X \int_Y \big\{ R_q u_i(\boldsymbol{y}) R_s \bar{w}_k(\boldsymbol{x}) B_{ikqs}(\boldsymbol{x}-\boldsymbol{y}) $$
$$+ k_T^2 n_j(\boldsymbol{y}) n_\ell(\boldsymbol{x}) u_i(\boldsymbol{y}) \bar{w}_k(\boldsymbol{x}) A_{ijk\ell}(\boldsymbol{x}-\boldsymbol{y}) \big\} \, \mathrm{d}S_y \, \mathrm{d}S_x \quad (47)$$

$$\mathcal{F}(\boldsymbol{w}) = \int_S \int_S [T_i^k(\boldsymbol{y},\boldsymbol{x}) \bar{w}_k(\boldsymbol{x}) - T_i^{(0)k}(\boldsymbol{y},\boldsymbol{x}) \bar{w}_k(\boldsymbol{y})] \tilde{t}_a(\boldsymbol{y}) \, \mathrm{d}S_y \, \mathrm{d}S_x $$
$$+ \int_\Gamma \int_S T_i^k(\boldsymbol{y},\boldsymbol{x}) \bar{w}_k(\boldsymbol{x}) \bar{w}_k(\boldsymbol{x}) \tilde{t}_a(\boldsymbol{y}) \, \mathrm{d}S_y \, \mathrm{d}S_x - \int_S \tilde{t}_a(\boldsymbol{y}) \bar{w}_k(\boldsymbol{y}) \, \mathrm{d}S_y \quad (48)$$

In Eq. (48), $T_i^k(\boldsymbol{y},\boldsymbol{x})$ (resp. $T_i^{(0)k}(\boldsymbol{y},\boldsymbol{x})$) denotes the traction vector associated with the dynamical (resp. static) infinite-space fundamental solution $U_i^k(\boldsymbol{x}-\boldsymbol{y})$ (resp. $U_i^{(0)k}(\boldsymbol{x}-\boldsymbol{y})$) of (39):

$$U_i^k = 2(1-\nu)[\Delta F + k_L^2 F]\delta_{ik} - F_{,ik} \qquad U_i^{(0)k} = 2(1-\nu)\Delta F^0 \delta_{ik} - F_{,ik}^0$$
$$\Sigma_{ij}^k = \lambda \delta_{ij} U_{i,i}^k + \mu(U_{i,j}^k + U_{j,i}^k) \qquad \Sigma_{ij}^{(0)k} = \lambda \delta_{ij} U_{i,i}^{(0)k} + \mu(U_{i,j}^{(0)k} + U_{j,i}^{(0)k})$$
$$T_i^k = \Sigma_{ij}^k n_j \qquad T_i^{(0)k} = \Sigma_{ij}^{(0)k} n_j$$

with

$$F(\boldsymbol{x}-\boldsymbol{y}) = \frac{1}{4\pi\mu k_T^2}(e^{ik_L r} - e^{ik_T r})\frac{1}{r} \quad F^0(\boldsymbol{x}-\boldsymbol{y}) = \frac{1}{16\pi\mu(1-\nu)}r \quad (49)$$

with $k_L^2 = \rho\omega^2/(\lambda+2\mu)$, $k_T^2 = \rho\omega^2/\mu$. The latter functions verify:

$$\Delta^2 F = \frac{1}{4\pi\mu k_T^2}(k_L^4 e^{ik_L r} - k_T^4 e^{ik_T r})\frac{1}{r} \qquad \Delta^2 F^0 = 0 \quad (\boldsymbol{x} \neq \boldsymbol{y}) \quad (50)$$

Besides, Eq. (47) stems from the following identity (given in [16] and generalized to anisotropic elasticity in [2]):

$$-C_{ijab}C_{k\ell cd}U_{a,bd}^c(\boldsymbol{x}-\boldsymbol{y})n_j(\boldsymbol{x})n_\ell(\boldsymbol{y}) = R_q^x R_s^y B_{ikqs}(\boldsymbol{x}-\boldsymbol{y}) + k_T^2 A_{ijk\ell}(\boldsymbol{x}-\boldsymbol{y})n_j(\boldsymbol{x})n_\ell(\boldsymbol{y})$$

where the kernel functions $B_{ikqs}(\boldsymbol{x}-\boldsymbol{y})$ and $A_{ijk\ell}(\boldsymbol{x}-\boldsymbol{y})$ are given by:

$$B_{ikqs}(\boldsymbol{x}-\boldsymbol{y}) = -e_{iep}e_{kgr}\mu^2[4\nu\delta_{pq}\delta_{rs} + 2(1-\nu)(\delta_{pr}\delta_{qs} + \delta_{ps}\delta_{qr})]F_{,eg} \quad (51)$$
$$A_{ijk\ell}(\boldsymbol{x}-\boldsymbol{y}) = (1-2\nu)k_T^2 C_{ijab}C_{k\ell cd}\delta_{ac}F_{,bd}$$
$$+ [2(1-\nu)(\delta_{ik}\delta_{j\ell} + \delta_{jk}\delta_{i\ell}) + 4\nu\delta_{ij}\delta_{k\ell}]\Delta^2 F \quad (52)$$

and are both weakly singular in view of Eqs. (49a) and (50).

Introducing the Lagrangian $\mathcal{L}_B(\boldsymbol{u},\boldsymbol{w},\Gamma) = \mathcal{J}(\Gamma) + \mathcal{B}(\boldsymbol{u},\boldsymbol{w}) - \mathcal{F}(\boldsymbol{w})$, the analysis of Sec. 5. is essentially repeated: the adjoint state $\boldsymbol{w}_\Gamma$ is the solution to the adjoint SGBIE formulation:

$$\mathcal{B}(\boldsymbol{w},\overset{\star}{\boldsymbol{u}}) = -\int_S \bar{\varphi}_{,u} \cdot \overset{\star}{\boldsymbol{u}} \, \mathrm{d}S \qquad \forall \overset{\star}{\boldsymbol{u}} \in [H^{1/2}(A)]^3 \quad (53)$$

The material derivative of the functional $\mathcal{J}$ is then expressed as:

$$\overset{\star}{\mathcal{J}}(\Gamma) = \int_{\Gamma} (\overset{\star}{\psi} \cdot \boldsymbol{\theta} + \psi \mathrm{div}_S \boldsymbol{\theta}) \, \mathrm{d}S + \mathcal{B}^1(\boldsymbol{u}_\Gamma, \boldsymbol{w}_\Gamma; \boldsymbol{\theta}) - \mathcal{F}^1(\boldsymbol{w}_\Gamma; \boldsymbol{\theta}) \tag{54}$$

A calculation similar to that performed in Sec. 5. then gives:

$$\mathcal{B}^1(\boldsymbol{u}, \boldsymbol{w}) = \int_A \int_A \Big\{ R_q u_i(\boldsymbol{y}) R_s \bar{w}_k(\boldsymbol{x}) B^1_{ikqs}(\boldsymbol{y}, \boldsymbol{x}; \boldsymbol{\theta})$$
$$+ k_T^2 n_j(\boldsymbol{y}) n_\ell(\boldsymbol{x}) u_i(\boldsymbol{y}) \bar{w}_k(\boldsymbol{x}) A^1_{ijk\ell}(\boldsymbol{y}, \boldsymbol{x}; \boldsymbol{\theta}) \Big\} \, \mathrm{d}S_y \, \mathrm{d}S_x \tag{55}$$

with

$$B^1_{ikqs}(\boldsymbol{y}, \boldsymbol{x}; \boldsymbol{\theta}) = \theta_{q,m}(\boldsymbol{x}) B_{ikms}(\boldsymbol{x}-\boldsymbol{y}) + \theta_{s,m}(\boldsymbol{y}) B_{ikqm}(\boldsymbol{x}-\boldsymbol{y})$$
$$+ [\theta_m(\boldsymbol{x}) - \theta_m(\boldsymbol{y})] B_{ikqs,m}(\boldsymbol{x}-\boldsymbol{y}) \tag{56}$$
$$A^1_{ijk\ell}(\boldsymbol{y}, \boldsymbol{x}; \boldsymbol{\theta}) = [\theta_m(\boldsymbol{x}) - \theta_m(\boldsymbol{y})] B_{ikqs,m}(\boldsymbol{x}-\boldsymbol{y}) + [D_m \theta_m(\boldsymbol{x}) + D_m \theta_m(\boldsymbol{y})] A_{ijk\ell}(\boldsymbol{x}-\boldsymbol{y})$$
$$- D_m \theta_j(\boldsymbol{x}) A_{imk\ell}(\boldsymbol{x}-\boldsymbol{y}) - D_m \theta_\ell(\boldsymbol{y}) A_{ijkm}(\boldsymbol{x}-\boldsymbol{y}) \tag{57}$$

and

$$\mathcal{F}^1(\boldsymbol{w}; \boldsymbol{\theta}) = -\int_\Gamma \int_S \Big\{ e_{jma} \theta_m(\boldsymbol{x}) \Sigma^k_{ij}(\boldsymbol{x}-\boldsymbol{y}) R^x_a \bar{w}_k(\boldsymbol{y})$$
$$+ \mu k_T^2 \theta_n(\boldsymbol{x}) U^k_i(\boldsymbol{x}-\boldsymbol{y}) \bar{w}_k(\boldsymbol{x}) \Big\} \tilde{t}_a(\boldsymbol{y}) \, \mathrm{d}S_y \, \mathrm{d}S_x \tag{58}$$

which completes our result for the elastodynamic case.

References

[1] Barone, M. R., Yang, R. J.: A Boundary Element Approach for Recovery of Shape Sensitivities in Three-dimensional Elastic Solids. *Comp. Meth. in Appl. Mech. Engng.*, **74** (1989), 69–82.

[2] Becache, E, Nedelec, J-C, Nishimura, N: Regularization in 3D for anisotropic elastodynamic crack and obstacle problems. *J. Elast.*, **31** (1993), 25–46.

[3] Bonnet, M.: BIE and material differentiation applied to the formulation of obstacle inverse problems. *Engng. Anal. with Bound. Elem.*, **15** (1995), 121–136.

[4] Bonnet, M.: Regularized BIE formulations for first- and second-order shape sensitivity of elastic fields. *Computers and Structures*, **56** (1995), 799–811. (Invited Paper, Special Issue, S. Saigal, Guest Editor).

[5] Bonnet, M.: Differentiability of strongly singular and hypersingular boundary integral formulations with respect to boundary perturbations. *Comp. Mech.*, **19** (1997), 240–246.

[6] Bonnet, M.: Stability of crack fronts under Griffith criterion: a computational approach using integral equations and domain derivatives of potential energy. *Comp. Meth. in Appl. Mech. Engng.* **173** (1999), 337–364.

[7] Burczynski, T.: Application of BEM in sensitivity analysis and optimization. *Comp. Mech.*, **13** (1993), 29–44.

[8] Burczyński, T.: Recent advances in boundary element approach to design sensitivity analysis – a survey. In T. Hisada M. Kleiber (ed.), *Design Sensitivity Analysis*. Atlanta Technology Publications (1993), 1–25.

[9] Burczynski, T., Fedelinski, P.: Boundary elements in shape design sensitivity analysis and optimal design of vibrating structures. *Engng. Anal. with Bound. Elem.*, **9** 1 (1992), 195–20.

[10] Choi, J. O., Kwak, B. M.: Boundary Integral Equation Method for Shape Optimization of Elastic Structures. *Int. J. Num. Meth. in Eng.*, **26** (1988), 1579–1595.

[11] Matsumoto, T., Tanaka, M., Miyagawa, M., Ishii, N.: Optimum design of cooling lines in injection moulds by using boundary element design sensitivity analysis. *Finite Elements in Analysis and Design*, **14** (1993), 177–185.

[12] Mellings, S. C., Aliabadi, M. H.: Flaw identification using the boundary element method. *Int. J. Num. Meth. in Eng.*, **38** (1995), 399–419.

[13] Meric, R. A.: Differential and integral sensitivity formulations and shape optimization by BEM. *Engng. Anal. with Bound. Elem.*, **15** (1995), 181–188.

[14] Nedelec, J. C.: Integral equations with non integrable kernels. *Integral equations and operator theory*, **5** (1982), 562–572.

[15] Nishimura, N.: Application of boundary integral equation method to various crack determination problems. In: M.H. Aliabadi (ed.), *Dynamic Fracture Mechanics*, Chapter 7. Comp. Mech. Publ., Southampton (1995).

[16] Nishimura, N., Kobayashi, S.: A regularized boundary integral equation method for elastodynamic crack problems. *Comp. Mech.*, **4** (1989), 319–328.

[17] Petryk, H., Mroz, Z.: Time derivatives of integrals and functionals defined on varying volume and surface domains. *Arch. Mech.*, **38** (1986), 694–724.

[18] Simon, J.: Second variations for domain optimization problems. In: W. Schappacher F. Kappel, K. Kunisch (ed.), *Control Theory of Distributed Parameter Systems and Applications.*, *International Series of Numerical Mathematics*. Birkhäuser Verlag, Basel, **91** (1989), 361–378.

[19] Sokolowski, J., Zolesio, J. P.: Introduction to shape optimization. Shape sensitivity analysis, *Springer Series in Computational Mathematics*. Springer-Verlag **16** (1992).

Laboratoire de Mécanique des Solides (UMR CNRS 7649), Ecole Polytechnique, 91128 Palaiseau cedex, France.
E-mail: bonnet@lms.polytechnique.fr

D. CLOUTEAU, D. AUBRY, M. L. ELHABRE and E. SAVIN

Periodic and Stochastic BEM for Large Structures Embedded in an Elastic Half-Space

1. Introduction

Modeling wave propagation around very long structures such as bridges or tunnels is a major issue in the fields of either earthquake engineering or ground borne vibrations induced by car or railway traffic. Indeed despite a seemingly bidimensional or periodic geometry, a true three-dimensional analysis has to be carried out since the loads are fully three-dimensional. Unfortunately usual 3D models are not able to account for such large structures either from the theoretical or the numerical point of view. The development of a periodic approach able to account for 3D loadings is addressed in the paper. Moreover, for such large geometries an accurate knowledge of either the loads or the soil parameters cannot be usually achieved. Consequently the analysis is also carried out in a stochastic sense using the deterministic tools previously defined. The cases of random moving loads and random incident field is studied in detail.

Let us consider a very long structure modeled as an unbounded open set Ω_l with given elastic properties and which is embedded in an elastic half-space $\Omega_s = D - (D \cap \Omega_l)$, D being the full half-space. $\Omega_g = \Omega_s \cup \Omega_l$ will denote the global domain. Ω_l is supposed to be periodic : an elementary bounded cell $\tilde{\Omega}_l$ exists such that $\Omega_l = \cup_{n=-\infty}^{+\infty} \tilde{\Omega}_{ln}$, $\tilde{\Omega}_{ln}$ being the translation of $\tilde{\Omega}_l$ of length L in direction d. The interfaces between Ω_l and Ω_s will be denoted Σ. The part of the boundary of each domain Ω_β on which Neumann boundary conditions are applied will be denoted $\Gamma_{\sigma\beta}$. u_β will be the displacement field in each domain Ω_β, $\epsilon(u_\beta)$ and $\sigma_\beta(u_\beta)$ will be the strain and stress tensors associated to these fields and $t_\beta(u_\beta)$ the traction vector on the boundary using the outer normal convention (see figure 1). Moreover one will use the following notations, a and b being any two vectors of $\mathbb{R}^3$ and A and B being any two tensors of $\mathbb{R}^3 \times \mathbb{R}^3$, $(DivA)_j = \sum_i \partial_i A_{ji}$ is the divergence of the tensor, $a.b = \sum_i a_i b_i$ is the scalar product, $A : B = \sum_{ij} A_{ij} B_{ij}$ is the contraction of two tensors and $(a \otimes b)_{ij} = a_i b_j$ is the tensorial product. The loads are either forces f applied in Ω_l or incident fields u_i satisfying the Navier equation in D and the free-surface boundary conditions on ∂D.

As we assume a linear elastic behaviour for each domain the equations can be written in the frequency domain for a given circular frequency ω, and every field will depend implicitly on ω. To avoid the definition of proper radiation conditions either in the half-space or in the structure, one will assume that some damping occurs in the materials modeled as a small imaginary part added to the elastic constants being either constant for hysteretic damping or proportional to ω for a viscous one. Thanks to this hypothesis, one can work in the usual Sobolev spaces even for the unbounded domains.

2. The generalized periodicity

In this section we will assume that the loads are exactly known. In a first step following the framework of Floquet [15] we will show that the analysis can be performed on a set of independent fields defined on the reference cell (see [4] or [21] for more details). On this cell we apply the classical subdomain approach [2] [7] allowing us to use boundary elements in $\tilde{\Omega}_s$ and finite elements in $\tilde{\Omega}_l$. For these methods the periodic assumption is addressed in detail.

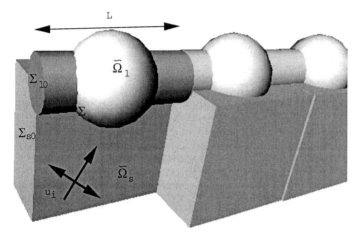

Figure 1: Model layout

2.1. The periodic decomposition

Before dealing with 3D periodic structures we first account for the one-dimensional case.

1D Floquet periodicity Let us first recall some classical results due to Floquet [15]:

Definition 1. *A complex valued function f defined on $\mathbb{R}$ is periodic of the second kind with period L and wave number κ if for any x in $\mathbb{R}$:*

$$f(x+L) = e^{-i\kappa L} f(x) \tag{1}$$

This means that one can build this function for any x once it is known on $]0, L[$. The following theorem shows that any function can be written as the superposition of a set of periodic functions of the second kind :

Theorem 1. *Given a function f defined on $\mathbb{R}$ and a period L, its Floquet-transform $\tilde{f}$ is defined on $]0, L[\times] - \pi/L, \pi/L[$ as follows :*

$$\tilde{f}(\tilde{x}, \kappa) = \sum_{n=-\infty}^{+\infty} f(\tilde{x} + nL) e^{in\kappa L} \tag{2}$$

$$= \sum_{n=-\infty}^{+\infty} \hat{f}(\kappa + 2n\pi/L) e^{-i(\kappa + 2n\pi/L)\tilde{x}} \tag{3}$$

Where $\hat{f}$ is the Fourier Transform of f. For any $x = \tilde{x} + nL$ and $k = \kappa + 2n\pi/L$, f and $\hat{f}$ may be recovered from its Floquet transform $\tilde{f}$ by :

$$f(x) = \frac{L}{2\pi}\int_{-\pi/L}^{\pi/L}\tilde{f}(\tilde{x},\kappa)e^{-in\kappa L}d\kappa \ , \ \hat{f}(k) = \frac{1}{L}\int_{0}^{L}\tilde{f}(\tilde{x},\kappa)e^{ik\tilde{x}}d\tilde{x} \qquad (4)$$

This property is of a great practical importance when dealing with periodic operators defined as follows :

Definition 2. *A linear differential operator A with domain $D(A)$ depending on x is periodic of period L if it satisfies for any x and for any u in $D(A)$:*

$$A(x+L)u = A(x)u$$

Then family of operators $\tilde{A}_\kappa$ defined formally as the restrictions of A on the reference cell $]0, L[$ with respect to x and applied on functions $\tilde{u}$ being the restriction of functions $u \in D(A)$ whose traces at the two ends of the reference cell satisfy the periodicity condition $u(L) = e^{-i\kappa L}u(0)$.

Indeed one can easily prove that as $\widetilde{(Au)}(\tilde{x},\kappa) = \tilde{A}_\kappa(\tilde{x})\tilde{u}(\tilde{x},\kappa)$ for any $u \in D(A)$ so that the following theorem holds :

Theorem 2. *Let (A, f) be a periodic operator and a function. If the following equations have unique solutions u in $D(A)$ and $\tilde{u}$ in $D(\tilde{A}_\kappa)$ for any $\kappa \in [-\pi/L, \pi/L[$ then $\check{u} = \tilde{u}$, the Floquet transform of u :*

$$Au = f, \quad \tilde{A}_\kappa\check{u} = \tilde{f}, \quad \check{u}(L) = e^{-i\kappa L}\check{u}(0) \qquad (5)$$

This means that instead of solving the first equation in (5) one can solve the second one on the generic cell for any κ such that $\tilde{f}$ does not vanish and then build solution u using the reconstruction formula (4). The key point in using this theorem is that each equation must have a unique solution. In the following this will be achieved as long as damping is accounted for.

3D domains and fields decomposition The aformentionned framework can be easily extended to three-dimensional domains having a periodicity L along direction d (x is now a vector in $\mathbb{R}^3$). The analysis can be restricted to the generic domains $\tilde{\Omega}_\beta = \Omega_\beta \cap S_o$ ($\beta \in \{l, s\}$) where $S_o = \{x \in \mathbb{R}^3$ such that $0 < x.d < L\}$. It is worth noticing that the boundary of $\tilde{\Omega}_\beta$ not only includes $\tilde{\Sigma}$ and $\tilde{\Gamma}_{\sigma\beta}$ the restrictions of Σ and $\Gamma_{\sigma\beta}$ on the generic cell but also $\Sigma_{\beta 0}$ and $\Sigma_{\beta L}$ the additional interfaces between previous and following cells. With the help of the Floquet transform of either the incident field $\tilde{u}_i(\tilde{x}, \kappa)$ or the force $\tilde{f}(\tilde{x})$ defined as follows :

$$\tilde{f}(\tilde{x},\kappa) = \sum_{n=-\infty}^{+\infty} f(\tilde{x}+nLd)e^{in\kappa L} \ , \ \tilde{u}_i(\tilde{x},\kappa) = \sum_{n=-\infty}^{+\infty} u_i(\tilde{x}+nLd)e^{in\kappa L}$$

one has to solve the generic problem for any κ, the global solution $u(x)$ being recovered using formula (4):

93

Problem 1. Find $(\tilde{u}_s(\tilde{x}, \kappa), \tilde{u}_l(\tilde{x}, \kappa))$ defined on $\tilde{\Omega}_s X \tilde{\Omega}_l$ satisfying :

$$Div\sigma_s(\tilde{u}_s - \tilde{u}_i) = \rho_s\omega^2(\tilde{u}_s - \tilde{u}_i) \quad in \quad \tilde{\Omega}_s \quad (6)$$

$$Div\sigma_l(\tilde{u}_l) = \rho_l\omega^2\tilde{u}_l \quad in \quad \tilde{\Omega}_l \quad (7)$$

$$\tilde{u}_s = \tilde{u}_l \ , \ t_s(\tilde{u}_s) + t_l(\tilde{u}_l) = 0 \quad on \quad \tilde{\Sigma} \quad (8)$$

$$t_\beta(\tilde{u}_\beta) = 0 \quad on \quad \tilde{\Gamma}_{\sigma\beta}, \beta \in \{s, l\} \quad (9)$$

$$\tilde{u}_\beta(x) = e^{-i\kappa L}\tilde{u}_\beta(x - L) \quad for \quad x \in \Sigma_{\beta L}, \beta \in \{s, l\} \quad (10)$$

2.2. The subdomain approach

In this section a numerical solution of problem (1) is built using the classical domain decomposition approach [2]. As $\tilde{\Omega}_l$ is bounded one can decompose the displacement field $\tilde{u}_l$ on a given basis $\{\phi_I(\kappa)\}_{I=1,N}$ that has to satisfy the periodic conditions (10). Moreover let where $\tilde{u}_{do} + \tilde{u}_i$ and $\tilde{u}_{dI}$ be fields defined in $\tilde{\Omega}_s$ satisfying the homogeneous Navier equation (6), the periodicity conditions (10), the homogeneous boundary conditions (9) and the following boundary conditions on $\tilde{\Sigma}$:

$$\tilde{u}_{dI} = \phi_I \ , \ \tilde{u}_{do} + \tilde{u}_i = 0 \quad on \quad \tilde{\Sigma} \quad (11)$$

Then one has the following decomposition either in $\tilde{\Omega}_l$ and in $\tilde{\Omega}_s$:

$$\tilde{u}_l(\tilde{x}) = \sum_{I=1}^N c_I\phi_I(\tilde{x}) \ , \ \tilde{u}_s(\tilde{x}) = \tilde{u}_i(\tilde{x}) + \tilde{u}_{do}(\tilde{x}) + \sum_{I=1}^N c_I\tilde{u}_{dI}(\tilde{x}) \quad (12)$$

At last, using a standard Galerkin approximation procedure in writing the equilibrium of $\tilde{\Omega}_l$ in a weak sense for any ϕ_J in the basis, one comes up with the following linear system :

$$\{K(\kappa) - \omega^2 M(\kappa) + K_s(\omega, \kappa)\} C(\omega, \kappa) = F(\kappa) + F_s(\omega, \kappa) \quad (13)$$

where :

$$K_{IJ} = \int_{\tilde{\Omega}_l} \sigma_l(\phi_I) : \epsilon(\bar{\phi}_J) dV \ , \ M_{IJ} = \int_{\tilde{\Omega}_l} \rho_l \phi_I \cdot \bar{\phi}_J dV \ , \ F_J = \int_{\tilde{\Omega}_l} \tilde{f} \cdot \bar{\phi}_J dV$$

$$K_{sIJ} = \int_{\tilde{\Sigma}} t_s(\tilde{u}_{dI}) \cdot \bar{\phi}_J dS \ , \ F_{sJ} = -\int_{\tilde{\Sigma}} (t_s(\tilde{u}_i) + t_s(\tilde{u}_{do})) \cdot \bar{\phi}_J dS \quad (14)$$

In order to solve this equation for any ω and κ one has first to compute the unknown traction fields $t_s(\tilde{u}_{dI})$ and $t_s(\tilde{u}_{do})$. The next subsection is devoted to this task using a boundary element technique. Another issue consists in building the basis $\phi_I(\kappa)$ using a standard Finite Element technique and will be presented in 2.4. The main point is thus that the classical domain decomposition approach is then extended to the case of periodic domain very easily.

2.3. Periodic Boundary elements

We propose here to compute the fields $\tilde{u}_{dI}$ and $\tilde{u}_{do}$, solutions of local boundary value problems of the following type :

Problem 2. *Find u in $\tilde{\Omega}_s$ such that :*

$$Div\sigma(u) = -\rho\omega^2 u \quad in \quad \tilde{\Omega}_s \tag{15}$$

$$u = u_o \quad on \quad \tilde{\Sigma} \tag{16}$$

$$t_s(u) = 0 \quad on \quad \tilde{\Gamma}_{\sigma s} \tag{17}$$

$$u(x) = e^{-i\kappa L}u(x-L) \quad for \quad x \in \Sigma_{sL} \tag{18}$$

using an Indirect Boundary Element Method [6] [11]. However as neither the left periodic interface Σ_{s0} nor the right one Σ_{sL} is bounded (see figure 1), a standard BEM cannot be directly used. Moreover, although it may be possible to account for periodic conditions in a BEM framework, this would require some heavy developments. To avoid these drawbacks, let us use the following periodic fundamental solution and integral operators [1][20]:

Definition 3. *U_s^G being the Green tensor of the elastic half-space D, let $\tilde{U}_s^G$ be the periodic Green Tensor and $\tilde{\mathcal{U}}_s^G$ the associated integral operator defined as follows :*

$$\tilde{U}_s^G(\tilde{x}, \tilde{y}) = \sum_{-\infty}^{+\infty} e^{in\kappa L} U_s^G(\tilde{x}, \tilde{y} + nLd) \tag{19}$$

$$\tilde{\mathcal{U}}_s^G(q)(\tilde{y}) = \int_{\tilde{\Sigma}} \tilde{U}_s^G(\tilde{x}, \tilde{y})\tilde{q}(\tilde{x})dS(\tilde{x}) \tag{20}$$

From these definitions one can easily remark that $\tilde{U}_s^G(\tilde{x}, \tilde{y})$ and $\tilde{\mathcal{U}}_s^G(\tilde{q})$ are periodic of the second kind with respect to $\tilde{y}$ and with a wave number equal to κ. Moreover one can remark that locally $\tilde{U}_s^G$ has the same singularities as U_s^G. As a consequence problem (2) is equivalent to the following Boundary integral equation where the integral is only on the bounded interface $\tilde{\Sigma}$ as periodic boundary conditions are automatically accounted for :

Problem 3. *Find $\tilde{q}$ on $\tilde{\Sigma}$ satisfying for any $\tilde{y} \in \tilde{\Sigma}$:*

$$\int_{\tilde{\Sigma}} \tilde{U}_s^G(\tilde{x}, \tilde{y})\tilde{q}(\tilde{x})dS(\tilde{x}) = u_o(\tilde{y}) \tag{21}$$

The tractions $t_s(u)$ needed in (14) are then given for any $\tilde{y} \in \tilde{\Sigma}$:

$$t_s(u)(\tilde{y}) = -1/2\tilde{q}(\tilde{y}) + \int_{\tilde{\Sigma}} t_s(\tilde{U}_s^G)(\tilde{x}, \tilde{y})\tilde{q}(\tilde{x})dS(\tilde{x}) \tag{22}$$

The numerical solution of this integral equation is handled using standard three-dimensional BEM. The only modification consists in computing the periodic Green tensors using formula (19). In this sum the singular terms arising for $n = 0$ are carried out using the 3D-BEM, the other terms being computed using standard procedures. This sum is truncated when convergence is reached [14].

2.4. The periodic structure

When dealing with Finite Elements, the computation of the matrices K and M defined in (14) is not straightforward as fields ϕ_I depend explicitly on κ (See [19] for periodic FEM). Using classical dynamic substructuring it is that K and M have an explicit dependence on κ by an extension of the Craig-Bampton [12] substructuring technique. It consists in the expansion of the displacement field of the structure $\tilde{\Omega}_l$ on dynamic eigenmodes ϕ_α with a fixed interface and on static modes ($\omega = 0$ in (7) generated by given displacements of the interface (practically unitary displacements of the nodes belonging to the interface). In the present case the interface consists of three parts $\tilde{\Sigma}$, $\tilde{\Sigma}_{l0}$ and $\tilde{\Sigma}_{lL}$. Let us first call ψ_β the static modes that vanish on $\tilde{\Sigma}_{l0}$ and $\tilde{\Sigma}_{lL}$. As a consequence they satisfy the periodic condition (10). Then, as long as the structure is periodic, one can find couples of the remaining static modes $(\varphi_{0\gamma}, \varphi_{L\gamma})$ satisfying:

$$\varphi_{0\gamma}(\tilde{x}) = \varphi_{L\gamma}(\tilde{x} + Ld) \quad \varphi_{0\gamma}(\tilde{x} + Ld) = \varphi_{L\gamma}(\tilde{x}) = 0 \quad \text{for } \tilde{x} \in \tilde{\Sigma}_{l0} \quad (23)$$

Then one can build new static modes $\tilde{\varphi}_\gamma$ combining these ones such that they also satisfy the periodic condition (10):

$$\tilde{\varphi}_\gamma(\tilde{x}) = e^{i\kappa L/2}\varphi_{0\gamma}(\tilde{x}) + e^{-i\kappa L/2}\varphi_{L\gamma}(\tilde{x}) \quad (24)$$

One is then able to compute easily the stiffness and mass matrices arising in (13) as a function of the FEM stiffness and mass matrices K^o and M^o, using a simple projection technique on this new basis. For example, denoting $\lambda = e^{-i\kappa L/2}$, the stiffness coefficient for two modes $\tilde{\varphi}_\gamma(\tilde{x})$ and $\tilde{\varphi}_{\gamma'}(\tilde{x})$ is then given by:

$$K_{\gamma\gamma'} = \lambda^{-2}\varphi_{0\gamma}^T K^o \varphi_{0\gamma'} + (\varphi_{0\gamma}^T K^o \varphi_{L\gamma'} + \varphi_{L\gamma}^T K^o \varphi_{0\gamma'}) + \lambda^2 \varphi_{L\gamma}^T K^o \varphi_{L\gamma'} \quad (25)$$

In this manner it is clearly seen that the stiffness and the mass matrices have an explicit expression with respect to λ and thus need not be computed each time.

2.5. The Fundamental solution of the composite domain

Provided with the results of this section one is now able to compute at least numerically either in the Floquet domain $(\tilde{x}, \kappa)$ or in the physical one the solution in each domain for any loads defined on the domains.

In particular one is able to compute the Green tensor $U_g^G(x, y)$ of the composite domain Ω_g made of the assembling of the individual ones giving the displacement field at any point $x \in \Omega_g$ created by any ponctual forces applied at any point $y \in \Omega_g$. In domains modeled using Finite Elements, the ponctual force becomes nodal forces when in domains modeled by Boundary Elements one can still account for ponctual forces as long as they are applied inside the domain. Indeed, in such domains one can define analytically an incident field u_i being nothing but the Green function of this local domain having a known singularity around y and then, computed numerically, the regular diffracted field u_d in the domain using the classical BEM techniques. This means that in either case one can effectively and accurately uses this fundamental solution in either integral representations or integral equations as long as the domain of integration does not include or cross an internal boundary of the composite domain. In this latter case some additional work has to be done. At last let us remark that this procedure has been implicitly used when using the Green functions of the elastic layered media in (21) as described in [3](see also [10] for BEM and Ray coupling).

3. Load variability

The aim of this section is to account for the variability on the load that arises when dealing with very large structures. We will first recall the usefulness of Karhunen-Loeve [22] expansions allowing us to deal with random variables instead of random fields. Then using the classical theory of linear filtering of Gaussian process [17], two particular cases of practical interest are investigated; the first one consists in a random moving source, and the second one in a random incident field.

Given stochastic inputs such as the preceding the response will be characterized by its first and second order moment. We assume here that the input is centered so that the output is centered too. The second moment is embodied by the auto-correlation function and it is the purpose of the following sections to sketch how it may be determined.

3.1. Karhunen-Loeve expansion

Let us consider a second order Gaussian centered random field $F(x)$ indexed on Ω, an open set of $\mathbb{R}^n$, with values in $\mathbb{R}^p$. It is therefore characterized by its auto-correlation tensor $c_F(x, x') = E[F(x) \otimes F^*(x')]$, $E[.]$ denoting the mathematical expectation and *, the conjugate transpose. One can then define the covariance operator $\mathcal{C}_F$ and its associated Karhunen-Loeve modes Φ_n and eigenvalues λ_n as follows ([22]):

Definition 4. *For any v defined on $L_2(\Omega)$ let $\mathcal{C}_F$ be the hermitian positive definite operator defined by :*

$$\mathcal{C}_F(v)(x') = \int_\Omega c_F(x, x').v(x)dV(x) \tag{26}$$

and let Φ_n be the normalized eigenfunctions of operator $\mathcal{C}_F$ and λ_n^2 the associated positive eigenvalues. Then any random field F can be decomposed using the Karhunen-Loeve modes Φ_n and its auto-correlation tensor $c_F(x, x')$ takes the following form :

$$\mathcal{C}_F(x, x') = \sum_{n=1}^{+\infty} \lambda_n^2 \Phi_n(x) \otimes \Phi_n^*(x') \tag{27}$$

For numerical purposes one can truncate the infinite sum in (27) to get an approximation which is controlled by the covariance norm.

3.2. Random loads

As far as Gaussian random loads F are concerned (either applied forces in the structures or incident fields in the soil) one can use the classical theory of linear filtering of Gaussian random fields to show that the response $u(y, t)$ at a given point in the domain is also Gaussian. In the following analysis stationary fields with respect to time will be accounted for, a more general development for non-stationary fields can be found in [8]. As a consequence, one can work directly in the frequency domain where the covariance C_u of the response is given by :

$$C_u(y, y') = \int_\Omega \int_\Omega H(y, x) \mathcal{C}_F(x, x') H^*(y', x') dV(x) dV(x') \tag{28}$$

where H is the deterministic transfer function. Using the Karhunen-Loeve expansion the covariance of the load takes a much simpler expression :

$$C_u(y,y') = \sum_{n=1}^{+\infty} \lambda_n^2 H_n(y) \otimes H_n^*(y) \quad , \quad H_n(y) = \int_\Omega H(y,x)\Phi_n(x)dV(x) \qquad (29)$$

where $\Phi_n(x)$ appears as a deterministic loading mode, H_n being the associated deterministic response. As a consequence a stochastic analysis consists on one hand of computing the covariance of the input and on the second hand of computing the transfer function as it is done in the next two examples where the modeling tools defined in the first section will be used.

3.3. Random moving sources

Vibrations induced in the ground by cars or trains are mainly due to vertical irregularities of either the wheels or the rolling area. Both of these are added in the following and denoted u_o. It is assumed that u_o can be modeled by a centered second-order Gaussian homogeneous random field of known spectral density $\hat{c}_{u_o}$ and that the force applied under one wheel is vertical and linearly dependent on u_o at the wheel location. Then the applied force $f(x_d,t)$ along the road or the rail at point x by a wheel moving at speed v along direction d, is :

$$f(x_d,t) = k_o \delta(x_d - vt) u_o(x_d) \qquad (30)$$

where $x_d = x.d$ is the coordinate of point x along direction d, $x^\perp = x - dx_d$ is the location vector in the plane normal to direction d, δ is the Dirac distribution and k_o the local stiffness. Taking the Fourier transform of $f(x_d,t)$ with respect to time one gets the following autocorrelation function in the frequency domain, denoting $k_v = \omega/v$:

$$\begin{aligned} C_f(x_d, x_d') &= \int_{-\infty}^{\infty} h_f(x_d,k) h_f^*(x_d',k) dk \\ h_f(x_d,k) &= k_o e^{ikx_d} \hat{c}_{u_o}^{1/2}(k_v - k) \end{aligned} \qquad (31)$$

Splitting the integral in (31) into an infinite series of integrals on $[(2n-1)\pi/L, (2n+1)\pi/L]$ and taking $x_d = \tilde{x}_d + n_x L$ and $x_d' = \tilde{x}_d' + n_x' L$ one gets :

$$\begin{aligned} C_f(x_d, x_d') &= \int_{-\pi/L}^{\pi/L} e^{i\kappa(n_x - n_x')L} \sum_{n=-\infty}^{\infty} \lambda_n^2 \Phi_n(\tilde{x}_d, \kappa) \otimes \Phi_n^*(\tilde{x}_d', \kappa) d\kappa \\ \Phi_n(\tilde{x}_d, \kappa) &= e^{2in\pi \tilde{x}_d/L} e^{i\kappa \tilde{x}_d} \quad , \quad \lambda_n(\kappa) = \hat{c}_{u_o}^{1/2}(k_v - \kappa - 2n\pi/L) \end{aligned} \qquad (32)$$

which is the Karhunen-Loeve expansion of $\tilde{c}_{u_o}(\tilde{x}_d, \tilde{x}_d', \kappa)$ on the generic cell for a given κ, $\Phi_n(\tilde{x}_d, \kappa)$ being the Karhunen-Loeve modes and λ_n their eigenvalues. The transfer function $H(y, x_d)$ in the present case is nothing but the Fundamental solution of the global domain including the soil and the structure for a vertical source located on the rail at position $x^\perp + x_d d$. Thus C_u reads :

$$C_u(y, y') = \int_{-\infty}^{\infty} \int_{-\infty}^{\infty} H(y, x_d) C_f(x_d, x_d') H^*(y', x_d') dx_d dx_d' \qquad (33)$$

Noticing that the effective computation of $H(y, x_d)$ is performed according to section 2 as :

$$H(y, x_d) = \int_{-\pi/L}^{\pi/L} \tilde{H}(\tilde{y}, \tilde{x}_d, \kappa) e^{i\kappa L(n_x - n_y)} d\kappa \qquad (34)$$

with $y = \tilde{y} + n_y L d$, $0 < \tilde{x}_d.d < L$ and $0 < \tilde{y}.d < L$, and splitting the integrals in (33) into pieces of length L, one can then finally get the auto-covariance of the response :

$$C_u(y, y') = \int_{-\pi/L}^{\pi/L} \sum_{n=-\infty}^{\infty} \lambda_n^2(\kappa) H_n(y, \kappa) \otimes H_n^*(y', \kappa) d\kappa$$

$$H_n(y, \kappa) = e^{-i\kappa n_y L} \int_O^L \tilde{H}(\tilde{y}, \tilde{x}_d) \Phi_n(\tilde{x}_d, \kappa) d\tilde{x}_d \qquad (35)$$

Taking this covariance for $y = y'$ represents the spectral energy emitted by the moving source at point y. Many simplified expressions may be derived from (35). For example stating that v is small compared to the wave speed in the soil, and y is far from the source, the correlation length is small (resp. large) compared to the period L.

3.4. Random incident fields

We now account for a random incident field modeled as a second order homogeneous and stationary random process $U_i(x, t)$ defined on $S_o \times \mathbb{R}$ where S_o is the free surface. It is characterized by its power spectral density $C_i(x, \omega)$ (see [18] and [13] for either theoretical or experimental expressions). Unfortunately formula (28) cannot be used directly here because although using the inverse transform (4), the modal synthesis (12), and the equation (13) where the force is given by (14), one can come up with a rather complex expression of the response $u(y)$ at any point y in the structure. This expression depends linearly on $\tilde{u}_i$ and $t_s(\tilde{u}_i)$ on $\tilde{\Sigma}$ but not depending on U_i on the free surface as it is given. Thus one has to first find $\tilde{u}_i$ and its associated covariance $\tilde{C}_i$ on $\tilde{\Sigma}$ as a function of U_i and C_i on the free surface and then a simple expression of the response as a function of $\tilde{u}_i$ on $\tilde{\Sigma}$ that can be practically computed.

The stochastic deconvolution The first step is the classical deconvolution process used in earthquake engineering. Given a deterministic ground motion at the free surface one makes some assumptions of a hypothesis on the incident field (usually a vertical incident plane wave) to find its amplitude, this motion being given. The same kind of process can be applied for random incident field [16]. Although the incident field is supposed to be random it is still assumed to satisfy the Navier equation in the half-space D. Then for an horizontally layered medium one can perform a stochastic deconvolution as follows :

1. compute $\hat{\Phi}_l(k, z_o)$ the three eigenvectors of $\tilde{C}_i(k)$ the Fourier transform of C_i on the free surface $z = z_o$ with respect to the horizontal space variables,

2. compute $\hat{\Phi}_l(k, z)$ performing the deterministic deconvolution of $\hat{\Phi}_l(k, z_o)$ for the given horizontal wavenumber k,

3. compute $\tilde{C}_i(k,z) = \sum_l \hat{\Phi}_l(k,z) \otimes \hat{\Phi}_l^*(k,z')$ and computing $\tilde{C}_i(\tilde{x}, \tilde{x}', \kappa)$ with $\tilde{x} = (\tilde{x}_h, z)$, $\tilde{x}' = (\tilde{x}_h', z')$ on $\tilde{\Sigma}$ and $k = (k_\xi, \kappa + 2n\pi/L)$, using the Fourier to Floquet transform (3) in the periodic horizontal direction and an inverse Fourier transform in the other one.

4. find the Karhunen-Loeve modes $\Phi_n(\tilde{x}, \kappa)$ of $\tilde{C}_i(\tilde{x}, \tilde{x}', \kappa)$ on $\tilde{\Sigma}$ for any κ.

Covariance on the response We propose here a simple method to compute the transfer function between the incident field at a given point $\tilde{x}$ of the soil-structure interface and any point y in the structure, so that equation (33) can be directly applied. As in the case of a moving load the basic idea is to define the Fundamental solution of the global problem (in the present case for a given κ), denoted $\tilde{U}_g^G(x, y)$, defined and continuous over every domain and which can be computed using section 2 methodology. In addition, let $\tilde{u}_d$ be the response in the structure for a given incident field $\tilde{u}_i$ and the diffracted field $\tilde{u}_s - \tilde{u}_i$ in the soil. This field satisfies all prescribed equations over the different domains except on the interface $\tilde{\Sigma}$ where it satisfies the following jump relationships:

$$[\tilde{u}_d] = \tilde{u}_i \quad , [\tilde{t}(\tilde{u}_d)] = t_s(\tilde{u}_i) \quad \text{on } \tilde{\Sigma} \tag{36}$$

As a consequence, using the representation theorem one directly gets:

$$\tilde{u}_s(\tilde{y}) = \tilde{u}_d(\tilde{y}) = \int_{\tilde{\Sigma}} \tilde{U}_g^G . t_s(\tilde{u}_i) - t(\tilde{U}_g^G).\tilde{u}_i dS \tag{37}$$

Now, noticing that either $\tilde{u}_i$ or any single layer $\tilde{\mathcal{U}}_s^G(\tilde{q}_s)$ defined by (20) satisfies the Navier equation in $\tilde{\Omega}_s^* = D \backslash \tilde{\Omega}_s$ and thus satisfies the reciprocity theorem in $\tilde{\Omega}_s^*$, leads to:

$$\int_{\tilde{\Sigma}} \tilde{\mathcal{U}}_s^G(\tilde{q}_s).t_s(\tilde{u}_i) - t_s(\tilde{\mathcal{U}}_s^G(\tilde{q}_s)).\tilde{u}_i dS = 0 \tag{38}$$

as integrals over the periodic interfaces or the free surface of the half-space vanish and where the traction vectors are defined using the outer normal of $\tilde{\Omega}_s^*$. Then using (37), the continuity of $\tilde{u}_i$, $t_s(\tilde{u}_i)$ and $\tilde{\mathcal{U}}_s^G(\tilde{q}_s)$ across $\tilde{\Sigma}$ and the classical jump relationship $[t_s(\tilde{\mathcal{U}}_s^G(\tilde{q}_s))] = -\tilde{q}_s$, one gets:

$$\tilde{u}_s(\tilde{y}) = \tilde{u}_d(\tilde{y}) = \int_{\tilde{\Sigma}} \tilde{Q}_s^g(\tilde{y}, \tilde{x}_d).\tilde{u}_i(\tilde{x}_d) dS(\tilde{x}_d) \tag{39}$$

where $\tilde{Q}_s^g$ is the source density on $\tilde{\Sigma}$ such that $\tilde{\mathcal{U}}_s^G(\tilde{Q}_s^g) = \tilde{U}_g^G$ in $\tilde{\Omega}_s$ which is computed when solving equation (21). Coming back to the stochastic analysis one then has using the same notation as in (35):

$$C_u(y, y') = \int_{-\pi/L}^{\pi/L} \sum_{n=-\infty}^{\infty} \lambda_n^2(\kappa) H_n(y, \kappa) \otimes H_n^*(y', \kappa) d\kappa$$

$$H_n(y, \kappa) = e^{-i\kappa n_y L} \int_{\tilde{\Sigma}} \tilde{Q}_s^g(\tilde{y}, \tilde{x}_d).\Phi_n(\tilde{x}_d, \kappa) dS(\tilde{x}_d) \tag{40}$$

4. Conclusion

We have shown in this paper that deterministic dynamic analyses on long periodic structures embedded in an infinite half-space may be carried out combining Domain Decomposition, BEM and standard FEM even for non periodic loadings (see [14] for numerical results). Moreover it has been shown that these deterministic tools combined with the Karhunen-Loeve expansion technique easily allow a stochastic analysis for various kinds of random loads. One has to remark that this methodology also applies for non periodic cases as presented in [8]. The extension of this analysis to random soil characteristics can be found in [9]. Further developments to account for the coupling between periodic and bounded structures throughout a propagation media can be found in [10] using Boundary Integral techniques together with asymptotic analysis.

References

[1] T. Abboud, V. Mathis, J.-C. Nedelec: Diffraction of an electromagnetic travelling wave by a periodic structure. In: G. Cohen et al.(editors): *Third International Conference on Mathematical and Numerical Aspects of Wave Propagation.* SIAM, 1995.

[2] D. Aubry: Sur une approche intégrée de l'interaction sismique sol-structure. *Revue Française de Géotechnique*, **38** (1986), 5–24.

[3] D. Aubry, D. Clouteau: A regularized boundary element method for stratified media. In: G. Cohen et al.(editors): *First International Conference on Mathematical and Numerical Aspects of Wave Propagation.* SIAM, 1991.

[4] A. Bensoussan, J. L. Lions, G. Papanicolaou: *Asymptotic analysis for periodic structures.* aubry.ps line 1/16057 0

[5] P. Bisch, editor: *11th European Conference on Earthquake Engineering*, Rotterdam, September 1998. Balkema.

[6] M. Bonnet: *Méthode des Equations Intégrales.* CNRS Editions/Eyrolles, 1995.

[7] D. Clouteau: *Propagation d'ondes dans des milieux hétérogènes, Application à la tenue d'ouvrages sous séismes.* PhD thesis, Ecole Centrale de Paris, 1990.

[8] D. Clouteau, D. Aubry, E. Savin: Influence of free field variability on soil-structure interaction. In: Bisch [5].

[9] D. Clouteau, D. Aubry, E. Savin: Influence of soil variability on soil-structure interaction. In: Bisch [5].

[10] D. Clouteau, A. Baroni, D. Aubry: Boundary integrals and ray method coupling for seismic borehole modeling. In: J. A. DeSanto, editor: *Fourth International Conference on Mathematical and Numerical Aspects of Wave Propagation*, page 768. SIAM, 1998.

[11] D. Colton, R. Kress: Integral equation methods in scattering theory, *Pure and applied Mathematics.* Wiley and Sons, 1983.

[12] R. Craig, M. Bampton: Coupling of substructures for dynamic analysis. *A.I.A.A. J.*, **6** (7) (1968), 1313–1319.

[13] A. Der Kiureghian: A coherency model for spatially varying ground motion. *Earthquake Engineering and Structural Dynamics*, **25** (1996), 99–111.

[14] M.-L. Elhabre, D. Clouteau, D. Aubry: Seismic behavior of diaphragm walls. In: Bisch [5].

[15] M. G. Floquet: Sur les equations differentielles linéaires a coefficients periodiques. *Annales de l'Ecole Normale* **12**, 1883.

[16] E. Kausel, A. Pais: Stochastic deconvolution of earthquake motions. *Journal of Engineering Mechanics ASCE*, **113** (2) (1987), 266–277.

[17] P. Kree, Ch. Soize: *Mathematics of random phenomena*. MIA, Reidel Publishing, Boston, 1986.

[18] J. E. Luco, H. L. Wong: Response of a rigid foundation to spatially random ground motion. *Earthquake Engineering and Structural Dynamics* **14** (1986), 583–596.

[19] K. Mahadevan: Edge–based finite element analysis of singly– and doubly– periodic scatterers using absorbing and periodic boundary conditions. *Electromagnetics* **16** (1996), 1–16.

[20] C. Pozrikidis: Computation of periodic Green's functions of Stokes flow. *Journal of engineering Mathematics* **30** (1996), 79–96.

[21] J. Sanchez-Hubert, N. Turbe: Ondes élastiques dans une bande périodique. *Mathematical Modelling and Numerical Analysis* **30** (3) (1986), 539–561.

[22] P. D. Spanos, R. Ghanem: *Stochastic Finite Elements: a Spectral Approach.* Springer–Verlag, 1991.

LMSSM, Ecole Centrale de Paris-CNRS/URA 850, 92295 Chatenay-Malabry, France

T. A. CRUSE and J. D. RICHARDSON

Self-regularized hypersingular BEM for Laplace's equation

1. Introduction

Recent work has established the computational effectiveness of the self-regularized, hypersingular boundary integral equation formulation for elastic stress analysis [1]. The results are fully competitive with standard, strongly singular BIE models but require the use of quartic rather than quadratic boundary support functions. The self-regularized formulation allows one to use standard non-conforming boundary elements and low order Gaussian integrations for *all* boundary elements. Convergence of the solution for non-trivial problems using the hypersingular formulation matches that of the strongly singular case based on the nodal degrees of freedom. The conclusion is that the self-regularized BIE formulation and computational algorithms are established as robust alternatives to strongly singular BIE formulations.

The self-regularized formulation of the BIE also provides the most direct basis for weakly-singular formulations of crack problems. The self-regularized potentials on the crack surface retain their weekly-singular nature. The formulation explicitly provides the Stokes' terms which account for the singular behavior of the potentials at the crack tip.

The purpose of the current note is to demonstrate the self-regularized formulation strategy using the Green's identity and its gradient form for Laplace's equation. The self-regularized formulation leads naturally to a highly effective BEM algorithm which utilizes standard, conforming boundary elements and low order Gaussian integrations. The boundary element algorithm utilizes a "relaxed continuity" interpretation, which has been accepted for the elasticity problem [2]. The note then extends the self-regularized approach to crack problems.

2. Self–Regularized Potential for Laplace's Equation

The material in this section is intended to illustrate the ideas behind self-regularization of potentials using Green's identity for Laplace's equation. The kernels are decomposed into their essential parts: discontinuous kernels or normal derivatives of the Newtonian potential; continuous but singular or hypersingular terms which are seen to be tangential derivatives of surface potentials; and weakly singular terms associated with the curvature of the body. The result will be seen to be the same as the use of simple state subtraction regularization as used by Rudolphi [3]. However, the self-regularized approach provides the basis for extending the weakly singular formulation to fracture problems. The approach also emphasizes the fact that the use of singular potentials for Laplace's equation or elasticity is naturally formulated using weakly singular potentials. There is no intrinsic need to treat Cauchy principal value or finite part concepts. Stated otherwise, what starts as a regular problem remains regular using self-regularized potential formulations.

Green's second identity for solutions to Laplace's equation is written using the three dimensional fundamental solutions, integrated over the closed surface $S = \Sigma_{i=1,\cdots,N} S_i$ where each S_i is taken to be smooth in the sense of Liapunov [4, p. 79]. The surface bounds the finite region R with outward normal $\vec{n}$. We will take the points (y, s) to be the interior free point and the surface integration point, respectively. The fixed surface point x may be an intersection point for two or more surfaces. The interior $y \in R$ form of the Green's identity is given by the following combination of a double and single layer potential.

$$4\pi T(y) = -\int_S T(s)\vec{\nabla}\left(\frac{1}{r(s,y)}\right)\cdot \vec{n}(s)dS + \int_S \vec{\nabla}T(s)\cdot \vec{n}(s)\frac{1}{r(s,y)}dS \quad \forall y \in R \quad (1)$$

The gradient in the integrals is denoted as $\vec{\nabla}T(s)$ and is the gradient as evaluated at the surface point s unless otherwise noted by a subscript. The field $T(y)$ satisfies Laplace's equation and is to satisfy the boundary conditions for the finite region. Of course, $\vec{\nabla}\left(\frac{1}{r(s,y)}\right)\cdot \vec{n}(s) = d[1/r(s,y)]/dn$. The more complicated form of the boundary condition derivative is retained for later clarity in obtaining the self–regularized form of the potentials.

The integral identity in Eq. 1 for the field variable $T(y)$ can be extended to points $y \in \bar{R}$, the exterior domain. This identity is discontinuous at the boundary and is given by the following identity.

$$0 = -\int_S T(s)\vec{\nabla}\left(\frac{1}{r(s,y)}\right)\cdot \vec{n}(s)dS + \int_S \vec{\nabla}T(s)\cdot \vec{n}(s)\frac{1}{r(s,y)}dS \quad \forall y \in \bar{R} \quad (2)$$

We can regularize the Green's identity for problems in which the field $T(y)$ is continuous in the sense of Hölder [5]. The regularization removes the discontinuity in the potential representation. We denote that continuity condition for the field variable as $T(y) \in C^{0,\alpha}$ where $\alpha > 0$.

$$4\pi T(y) = -T(x)\int_S \vec{\nabla}\left(\frac{1}{r(s,y)}\right)\cdot \vec{n}(s)dS - \int_S [T(s) - T(x)]\vec{\nabla}\left(\frac{1}{r(s,y)}\right)\cdot \vec{n}(s)dS$$
$$+ \int_S \vec{\nabla}T(s)\cdot \vec{n}(s)\frac{1}{r(s,y)}dS \quad (3)$$

The first integral is the normal derivative of the Newtonian potential $1/r(s,y)$ and is discontinuous as $y \to x$. The normal derivative of the Newtonian potential is most usefully written in terms of the solid angle Θ variable. The solid angle integral of the three dimensional closed surface is 4π for any point $y \in R$.

$$-\int_S \vec{\nabla}\left(\frac{1}{r(s,y)}\right)\cdot \vec{n}(s)dS = -\int_S \frac{d}{dn}\left(\frac{1}{r(s,y)}\right)dS = \int_S \frac{d\Theta}{dS}dS = \int_S d\Theta \equiv 4\pi \quad (4)$$

The self–regularized Green's identity is then given by the following combination of the double and single layer potentials.

$$4\pi T(y) = 4\pi T(x) - \int_S [T(s) - T(x)]\vec{\nabla}\left(\frac{1}{r(s,y)}\right)\cdot \vec{n}(s)dS + \int_S \vec{\nabla}T(s)\cdot \vec{n}(s)\frac{1}{r(s,y)}dS \quad (5)$$

The result is called self–regularized in that the mathematical properties of the double–layer potential and the Hölder continuity of the harmonic function $T(y)$ as $y \to s$ are all that is required to modify the original potential formulation. The result is the natural form of the potential using only weakly–singular potentials.

Eq. 5 is continuous as $y \to x(s), \forall x \in S$ including corners and edges in three dimensions. The boundary integral equation is given for all boundary points as follows.

$$0 = - \int_S [T(s) - T(x)] \frac{d}{dn}\left(\frac{1}{r(s,x)}\right) dS + \int_S \frac{dT}{dn}|_s \frac{1}{r(s,x)} dS \qquad (6)$$

Following the same steps, the self–regularized Green's identity for exterior points $y \in \bar{R}$ is given from Eq. 2 by the following double and single–layer potentials.

$$0 = - \int_S [T(s) - T(x)] \vec{\nabla}\left(\frac{1}{r(s,y)}\right) \cdot \vec{n}(s) dS + \int_S \vec{\nabla} T(s) \cdot \vec{n}(s) \frac{1}{r(s,y)} dS \qquad (7)$$

Again, the identity in Eq. 7 is continuous as $y \to x(s), \forall x \in S$ and gives the same boundary integral equation as the interior form, given in Eq. 6. The potential representation is therefore continuous for all boundary points at which the field variable $T(y) \in C^{0,\alpha}$ as $y \to s$.

2.1. Formulating the Gradient Potential

We will now compute the gradient of the potential $T(y)$ at the interior point y. The gradient operator for the field will be denoted by $\vec{\nabla}_y$. The gradient operator on the potential $r(s,y)$ has the property that $\vec{\nabla}_y r(s,y) = -\vec{\nabla}_s r(s,y) = -\vec{\nabla} r(s,y)$.

$$\begin{aligned}4\pi \vec{\nabla}_y T(y) &= - \int_S [T(s) - T(x)] \left[\frac{d}{dn} \vec{\nabla}_y \left(\frac{1}{r(s,y)}\right)\right] dS \\ &+ \int_S \vec{\nabla} T(s) \cdot \vec{n}(s) \vec{\nabla}_y \left(\frac{1}{r(s,y)}\right) dS \end{aligned} \qquad (8)$$

The indicial form of Eq. 8 is given as follows where the derivatives of the potentials are all in terms of the point s and the sign change has been taken.

$$4\pi T(y)_{,i} = \int_S [T(s) - T(x)] \left(\frac{1}{r(s,y)}\right)_{,ij} n_j(s) dS - \int_S \vec{\nabla} T(s) \cdot \vec{n}(s) \left(\frac{1}{r(s,y)}\right)_{,i} dS \qquad (9)$$

Each of the two integrals contains a strong singularity resulting in a discontinuous potential representation of the gradient of the field variable $T(y)$. In the elasticity problem [6], the first integrand is hypersingular as the elasticity kernel contains both tangential and normal derivatives of the usual Newtonian potential.

The continuous potential representation for the gradient field is obtained by another self–regularization of the two kernel functions in Eq. 9. In the first step, we regularize at a point on the surface x where there is a unique tangent surface. The tangential derivatives are denoted in the two independent directions ζ_1, ζ_2 as $\partial(\cdot)/\partial\zeta_\alpha$ for $\alpha = 1, 2$. The origin of the local expansion is $\zeta_\alpha = 0$.

$$\begin{aligned} 4\pi T(y)_{,i} &= \int_S [T(s) - T(x) - \frac{\partial T}{\partial \zeta_\alpha}|_x \zeta_\alpha] \left(\frac{1}{r(s,y)}\right)_{,ij} n_j(s) dS \\ &+ \frac{\partial T}{\partial \zeta_\alpha}|_x \int_S \zeta_\alpha \left(\frac{1}{r(s,y)}\right)_{,ij} n_j(s) dS \\ &- \int_S [\vec{\nabla} T(s) \cdot \vec{n}(s) - \vec{\nabla} T(x) \cdot \vec{n}(x)] \left(\frac{1}{r(s,y)}\right)_{,i} dS \\ &- \frac{dT}{dn}|_x \int_S \left(\frac{1}{r(s,y)}\right)_{,i} dS \end{aligned} \quad (10)$$

The first regularization requires a unique tangent vector for the point x while the second regularization is based on a unique value for the normal derivative of the field variable. For each of the integrands to be weakly–singular at $s = x$ requires that the field be one degree smoother than before; that is, $T(y) \in C^{1,\alpha}$ for $y \to x$. Such unique tangential and normal derivatives of the field variable at the boundary are conditions required to keep the potential representation of $T(y)$ continuous as $y \to x$. Corners and edges can be approached under these conditions but the potential in this form does not exist at the corner or edge.

We can now generalize the result to non–smooth boundary points. To do this, we use the analytical results for the "free" integrals (the second and fourth) in Eq. 10 taken from [6]. While a direct interpretation of the "free" integrals in Eq. 10 is possible, analytical integration confirms the non–singular but discontinuous nature of both.

The first free integral from Eq. 10 is converted by using the results in Eq. (52) and Eq. (53) from [6], as follows.

$$\nabla^2 r_{,ij} \zeta_\alpha n_j = 2\left(\frac{1}{r(s,y)}\right)_{,ij} \zeta_\alpha n_j = M_{(j\alpha)ij} + \delta_{i3} M_{()\alpha 3} + R_{i\alpha} + 2\delta_{i\alpha}\frac{d\Theta}{dS} \quad (11)$$

$$R_{i\alpha} = \nabla^2 r_{,\alpha} n_\beta \delta_{i\beta} + \delta_{i3} \nabla^2_{,3} n_\alpha \quad (12)$$

$$M_{(j\alpha)ij} = (\nabla^2 r_{,j} \zeta_\alpha)_{,i} n_j - (\nabla^2 r_{,j} \zeta_\alpha)_{,j} n_i \quad (13)$$

The terms denoted by $M_{()ij}$ are singular path integrals which are identically zero for closed surfaces, by making use of Stokes' theorem. Such Stokes' terms are, of course, tangential derivatives of potentials. The $R_{i\alpha}$ terms are weakly singular at all surface points on a Liapunov surface. The last term is the normal derivative of the Newtonian potential which is the discontinuous solid angle relative to the evaluation point y. The terms are written in the local coordinate system as previously defined. The ζ_3 direction corresponds to the unique normal at x. Details of the integral derivations are given in [7].

Next, we write the result for the second free integral from Eq. 10 by using the results in Eq. (66) and Eq. (67) from [6], as follows.

$$\nabla^2 r_{,i} = 2\left(\frac{1}{r(s,y)}\right)_{,i} = R_{3i} + M_{()i3} - 2\delta_{i3}\frac{d\Theta}{dS} \tag{14}$$

$$R_{3i} = \nabla^2 r_{,i}(1-n_3) + \delta_{i\alpha}\nabla^2 r_{,3} n_\alpha + \delta_{i3}[\nabla^2 r_{,3} n_3 - \nabla^2 r_{,m} n_m] \tag{15}$$

$$M_{()i3} = \nabla^2 r_{,i} n_3 - \nabla^2 r_{,3} n_i \tag{16}$$

Again, the strongly–singular Stokes' terms denoted by $M_{()i3}$ integrate to zero for closed surfaces. The R_{3i} terms are weakly singular for any point within a Liapunov surface. The source point can not yet be taken to a corner in the surface using this local decomposition.

2.2. Solid Angle Terms

The Stokes' terms are now taken to be zero. Crack problems are special cases with open surfaces and require that the Stokes' terms be included for the crack front path. In the following, only the weakly–singular terms and the discontinuous integrals are included.

$$\begin{aligned}
4\pi T(y)_{,i} &= \int_S [T(s) - T(x) - \frac{\partial T}{\partial \zeta_\alpha}|_x \zeta_\alpha]\left(\frac{1}{r(s,y)}\right)_{,ij} n_j(s) dS \\
&+ \frac{\partial T}{\partial \zeta_\alpha}|_x \left[\int_S \frac{1}{2} R_{i\alpha} dS + 4\pi \delta_{i\alpha}\right] \\
&- \int_S [\vec{\nabla}T(s)\cdot \vec{n}(s) - \vec{\nabla}T(x)\cdot \vec{n}(x)]\left(\frac{1}{r(s,y)}\right)_{,i} dS \\
&- \frac{dT}{dn}|_x \left[\int_S \frac{1}{2} R_{3i} dS - 4\pi \delta_{i3}\right]
\end{aligned} \tag{17}$$

The solid angle terms in Eq. 17 are written in the local coordinates. The terms can be combined and written in the global coordinates without loss of generality. The combined terms produce the gradient of the field at the surface point x, a result which is valid for all points $y \in R$. We therefore obtain the following self–regularized form of the gradient potential.

$$\begin{aligned}
4\pi T(y)_{,i} &= 4\pi T(x)_{,i} + \int_S [T(s) - T(x) - \frac{\partial T}{\partial \zeta_\alpha}|_x \zeta_\alpha]\left(\frac{1}{r(s,y)}\right)_{,ij} n_j(s) dS \\
&+ \frac{\partial T}{\partial \zeta_\alpha}|_x \int_S \frac{1}{2} R_{i\alpha} dS \\
&- \int_S [\vec{\nabla}T(s)\cdot \vec{n}(s) - \vec{\nabla}T(x)\cdot \vec{n}(x)]\left(\frac{1}{r(s,y)}\right)_{,i} dS \\
&- \frac{dT}{dn}|_x \int_S \frac{1}{2} R_{3i} dS
\end{aligned} \tag{18}$$

The result contains only weakly–singular integrals but they cannot be evaluated at corners or edges, owing to the original surface expansion that has been used. However, in the following section, we will generalize the form of the result which is valid at all surface points.

2.3. Weakly Singular Terms

Define a linear field given by $T^L(s) = T(x) + T_{,i}|_x[x_i(s) - x_i(x)]$. The gradient of the linear field is given as follows.

$$\vec{\nabla}T^L(s) = \vec{\nabla}T(x) \tag{19}$$

Next we subtract and add integrals of this linear term in Eq. 10. Combining the terms conveniently, we obtain the following relation.

$$\begin{aligned} 4\pi T(y)_{,i} &= 4\pi T(x)_{,i} + \int_S [T(s) - T^L(s)] \left(\frac{1}{r(s,y)}\right)_{,ij} n_j(s)dS \\ &+ \int_S [T^L(s) - T(x) - \frac{\partial T}{\partial \zeta_\alpha}|_x \zeta_\alpha] \left(\frac{1}{r(s,y)}\right)_{,ij} n_j(s)dS \tag{20} \\ &+ \frac{\partial T}{\partial \zeta_\alpha}|_x \int_S \frac{1}{2} R_{i\alpha}dS - \int_S \left[\vec{\nabla}T(s) - \vec{\nabla}T(x)\right] \cdot \vec{n}(s) \left(\frac{1}{r(s,y)}\right)_{,i} dS \\ &- \int_S \left[\vec{\nabla}T(x) \cdot \vec{n}(s) - \vec{\nabla}T(x) \cdot \vec{n}(x)\right] \left(\frac{1}{r(s,y)}\right)_{,i} dS \\ &- \frac{dT}{dn}|_x \int_S \frac{1}{2} R_{3i}dS \end{aligned}$$

There are two sets of cancelling, weakly singular integrals in Eq. 20. Each of these is valid for all surface points, except at the corner points given that each is derived at a point with a unique tangent. The following identities can be verified by expanding each of the first integrals.

$$\begin{aligned} 0 &= \int_S [T_{,i}|_x[x_i(s) - x_i(x)] - \frac{\partial T}{\partial \zeta_\alpha}|_x \zeta_\alpha] \left(\frac{1}{r(s,y)}\right)_{,ij} n_j(s)dS + \frac{\partial T}{\partial \zeta_\alpha}|_x \int_S \frac{1}{2} R_{i\alpha}dS \\ 0 &= -\int_S \vec{\nabla}T(x) \cdot [\vec{n}(s) - \vec{n}(x)] \left(\frac{1}{r(s,y)}\right)_{,i} dS - \frac{dT}{dn}|_x \int_S \frac{1}{2} R_{3i}dS \tag{21} \end{aligned}$$

The identities in Eq. 21 contain the weakly singular surface curvature effects on the original integrations. Each identity has been derived for a point on the interior of a smooth surface segment defined by a unique normal at x. However, the identity holds for all limits to the boundary for points arbitrarily close to corners.

We then apply the identities to Eq. 20. The following simple expression is obtained and is the self–regularized interior Green's integral identity for Laplace's equation.

$$\begin{aligned} 4\pi T(y)_{,i} &= 4\pi T(x)_{,i} + \int_S [T(s) - T^L(s)] \left(\frac{1}{r(s,y)}\right)_{,ij} n_j(s)dS \tag{22} \\ &- \int_S [\vec{\nabla}T(s) - \vec{\nabla}T(x)] \cdot \vec{n}(s) \left(\frac{1}{r(s,y)}\right)_{,i} dS \end{aligned}$$

The final result for the self-regularized BIE is seen to be identical to the method of Rudolphi in subtracting simple solutions [3]. However, by decomposing the kernels in the potential formulations, we have direct access to the self-regularized form for crack problems. The crack problem formulation is discussed in Section 4.

The result in Eq. 22 is weakly singular for all interior point limits to the boundary, including limits to the boundary at corners, at surface points where the continuity condition $T(y) \in C^{1,\alpha}$ is satisfied. The following weakly singular boundary integral equation is obtained at all boundary points satisfying the field continuity condition $T(y) \in C^{1,\alpha}$ for $y \to s$ at any given boundary point $s = x$.

$$
\begin{aligned}
0 = &\int_S [T(s) - T^L(s)] \left(\frac{1}{r(s,x)}\right)_{,ij} n_j(s) dS \\
&- \int_S [\vec{\nabla}T(s) - \vec{\nabla}T(x)] \cdot \vec{n}(s) \left(\frac{1}{r(s,x)}\right)_{,i} dS
\end{aligned}
\quad (23)
$$

It is of interest to note that in 3D Eq. 23 is three equations, one for each coordinate direction. The resulting BIE is therefore over-specified and one is left to ponder what subset of equations to use. The usual formulation, and the one that gives the best numerical results for the elasticity problem [1], is obtained by operating on the system of equations with the local normal $n_i(x)$. The resulting BIE is called a traction–BIE in elasticity. We will call this the gradient–BIE Green's identity for potential theory.

$$
\begin{aligned}
0 = &n_i(x) \int_S [T(s) - T^L(s)] \left(\frac{1}{r(s,x)}\right)_{,ij} n_j(s) dS - n_i(x) \int_S [\vec{\nabla}T(s) \\
&- \vec{\nabla}T(x)] \cdot \vec{n}(s) \left(\frac{1}{r(s,x)}\right)_{,i} dS
\end{aligned}
\quad (24)
$$

The collocation of Eq. 24 along edges or at corners requires that the surfaces be defined from each side so that discontinuities in the normal at the source point x can be allowed. Such collocations in elasticity may generally be done with single nodes or element edges; Laplace's equation is a scalar formulation and requires double nodes or edges to allow jumps in the surface normal and/or boundary flux.

3. Computational BEM Algorithm

The following development of a BEM algorithm for the self–regularized integral identity for Laplace's equation is based on the elastic BEM algorithm demonstrated in [1]. The algorithm uses standard conforming boundary elements of the C^0 class of support functions. These boundary elements do not preserve the $C^{1,\alpha}$ continuity of the tangential derivatives that is required for the analytical validity of the self-regularized BIE in Eq. 24. However, as discussed in [11], the BEM algorithm matches the weakly singular nature of the BIE in a two-sided sense without invalidating the underlying continuity requirement of the potential formulation. Such a BEM approach has been called a "relaxed continuity" BEM algorithm [2]. That algorithm is briefly summarized below.

The usual C^0 continuous support for the boundary data is applied in two or three dimensions using standard, isoparametric representations, as follows

$$T(s) \approx T(\xi) = \sum_{i=1}^{m} N_i(\xi) T^i \tag{25}$$

$$F(s) \approx F(\xi) = \sum_{i=1}^{m} N_i(\xi) F^i \tag{26}$$

where T^i and F^i are the nodal values of the boundary potential and flux, respectively.

The self-regularized potential formulation requires an explicit representation of the gradient of the potential, as evaluated at the boundary. The gradient is obtained for each unique boundary element by a vector combination of the local tangential derivative of the surface potential and the local flux. The tangential derivatives of the potential are computed for each boundary element in terms of the intrinsics for each element. The derivation of the tangential derivative is given in the following form for the two dimensional case

$$\frac{dT}{dS} \approx \frac{1}{J} \sum_{i=1}^{m} N'_i(\xi) T^i \tag{27}$$

where the Jacobian (J) is obtained in the usual manner from the isoparametric model of the element geometry. The gradient of the potential at any boundary point is given by the mapping of the local tangential and normal derviatives of the potential into the global coordinates.

$$\vec{\nabla} T(s) \approx \vec{\nabla} T(\xi_s) = \vec{A}(\xi_s) \sum_{i=1}^{m} N'_i(\xi) T^i + \vec{B}(\xi_s) \sum_{i=1}^{m} N_i F^i \tag{28}$$

The self-regularization terms for the BEM algorithm are given for element-M shared at a common collocation point x^I, as follows.

$$\begin{aligned} T^L(s) &= T(x^I) + \vec{\nabla} T(x^I) \cdot (\vec{s}^M - \vec{x}^I) \\ \vec{\nabla} T^L(s) &= \vec{\nabla} T(x^I) \end{aligned} \tag{29}$$

That is, the gradient for the Mth boundary element is computed using the form given in Eq. 28 and the coordinate expansion on that element. The potential is continuous at the collocation point. It is easily verified that the density functions

$$\begin{aligned} T(s) &- T^L(s) \\ \vec{\nabla} T(s) &- \vec{\nabla} T^L(s) \end{aligned} \tag{30}$$

are of order $r(s,x)^2$ and $r(s,x)$, respectively. As discussed in [11], the BEM algorithm matches the analytical regularity condition required by the weakly-singular BIE even though the $C^{1,\alpha}$ continuity of the potential is not met. This is the essence of the "relaxed continuity" BEM model as discussed in greater detail in [2].

Using the element-based regularizations in Eq. 29, we obtain the resulting BEM for the gradient-BIE as follows.

$$0 = \sum_{M=1}^{N} n(x^I) \int_{\Delta S_M} [T(\xi) - T^L(\xi)] \left(\frac{1}{r(\xi, x^I)}\right)_{,ij} n_j(\xi) dS(\xi)$$

$$- \sum_{M=1}^{N} n(x^I) \int_{\Delta S_M} [\vec{\nabla} T(\xi) - \vec{\nabla} T^L(\xi)] \cdot \vec{n}^M(\xi) \left(\frac{1}{r(\xi, x^I)}\right)_{,i} dS(\xi) \quad (31)$$

The regularizing gradients in Eq. 31 are single-valued in the analytical model (unique) at the collocation point x^I. The values are not unique in the "relaxed continuity" BEM algorithm. The element-based values are used locally for the $s = x$ integrals. As discussed in [1], the average nodal value at x^I is used for regularization of the integrals for boundary elements which do not share the collocation node x^I. This BEM algorithm recovers the correct analytical state for the local gradient in the limit as $N \to \infty$.

It is easily established on analytical grounds that the self-regularized BEM has the same matrix as the usual non-singular formulation for Green's identity. Theoretically, then, the self-regularized BEM should have the same numerical convergence characteristics as the standard, non-regularized BEM. The numerical results for the elasticity case have been published [1].

The elasticity results show that equivalent convergence based on the numbers of degrees of freedom is only achieved if the above integrals for the $s = x$ elements use two-degrees higher interpolants. That is, to get convergence rate parity in terms of degrees of freedom with quadratic elements in the non-singular BEM requires quartic boundary elements in the self-regularized formulation. The requirement for higher order interpolants appears to be associated with the local approximations and retained polynomial degrees of freedom on the $s = x$ integrals. The numerical basis for this interesting result is still under study.

4. Self-regularized Traction-BIE for Cracks

The use of boundary integral equations for the fracture mechanics analysis problem is still of great interest, particularly for growing cracks and multiple cracks. The advantages of the traction-BIE formulation for this problem is the ability to model only the surface of displacement discontinuity for each crack. One can than use an alternating algorithm to interact each crack with the uncracked geometry.

This note has developed the weakly-singular, self-regularized BEM for the gradient-BIE for Laplace's equation. In what follows, we extend the earlier formulation of the elasticity problem in [6] to crack problems. The development below is easily applied to the potential theory formulation. The collocation point is taken to be $P(x)$ and the integration point is $Q(y)$. The elasticity kernels are for the tractions due to a point load $S_{kij}(P,Q)$ and the displacements due to the point load $D_{kij}(P,Q)$. These kernels are hypersingular and strongly singular, respectively. The self-regularized BIE for the

Somigliana stress identity (SSI) is given from [6] as follows:

$$\begin{aligned} 0 = & -\int_S [u_k(Q) - u_k^L(Q)] S_{kij}(P,Q) dS(Q) \\ & + \int_S [t_k(Q) - t_k^L(Q)] D_{kij}(P,Q) dS(Q) \end{aligned} \quad (32)$$

where $t_i(Q)$ are the surface tractions and $u_i(Q)$ are the surface displacements. The self regularized equation is analogous to Eq. 23.

The crack is taken in the usual way to be an arbitrary, smooth surface Γ with upper and lower sides and self-equilibrating tractions.

$$t_i(Q^+) + t_i(Q^-) = 0 \text{ for } Q \in \Gamma \quad (33)$$

The displacement discontinuity across the crack surface is denoted by $v_j(Q) = u_j(Q^+) - u_j(Q^-)$. The crack surface is taken to be the open surface $\Gamma^+ = \Gamma$. The SSI can be written in the usual way [8] with an open surface Γ as follows.

$$\begin{aligned} 0 = & -\int_S [u_k(Q) - u_k^L(Q)] S_{kij}(P,Q) dS(Q) \\ & + \int_S [t_k(Q) - t_k^L(Q)] D_{kij}(P,Q) dS(Q) \\ & - \int_\Gamma v_k(Q) S_{kij}(P,Q) dS(Q) \end{aligned} \quad (34)$$

The integrals for the crack surface integrals of the linear state regularizations are identically zero.

The Somigliana stress-BIE can then be written as a traction–BIE for collocation points on the crack surface. In this case, we regularize only on the crack surface. Following the process in [6] and applying the results in Appendix A in that paper, the regularization terms multiply integrals that are discontinuous (involving the solid angle terms), weakly singular (involving the curvature of the crack surface), and continuous (both hyper and strongly singular). The continuous terms are all converted by Stokes' theorem into contour integrals at the crack front, as that is an open surface. Otherwise these terms would be zero. The discontinuous terms multiplying the tangential displacement derivatives can easily be shown to add to zero when operating on them with the normal at the collocation point, P. Therefore, after some manipulation, we obtain the following result for the hypersingular traction-BIE on the crack.

$$\begin{aligned}
t_i(P) = &-n_j(P)\int_S u_k(Q)S_{kij}(P,Q)dS(Q) \\
&+n_j(P)\int_S t_k(Q)D_{kij}(P,Q)dS(Q) \\
&-n_j(P)\int_\Gamma [v_k(Q)-v_k(P)-v_{k,\alpha}(P)]S_{kij}(P,Q)dS(Q) \quad (35)\\
&-n_j(P)v_k(P)\oint_{\partial\Gamma} S^*_{kijm}(P,Q)dx_m(Q) \\
&-n_j(P)v_{k,\alpha}(P)\oint_{\partial\Gamma} S^*_{kij\alpha m}(P,Q)dx_m(Q) \\
&+\text{weakly singular crack curvature terms}
\end{aligned}$$

The last terms are zero for a flat crack.

It is interesting to compare this regularized hypersingular traction-BIE to the regularized strongly-singular traction-BIE of Polch, et al. [12]. In that work, the hypersingularity is first removed by integration by parts. However, satisfactory numerical results were only achieved using "whole crack" regularization. We write the earlier formulation consistent with the current formulation, as follows.

$$\begin{aligned}
t_i(P) = &-n_j(P)\int_S u_k(Q)S_{kij}(P,Q)dS(Q) \\
&+n_j(P)\int_S t_k(Q)D_{kij}(P,Q)dS(Q) \\
&-\int_\Gamma [v_{j,\alpha}(Q)-v_{j,\alpha}(P)]D_{ij}(P,Q)dS(Q) \quad (36)\\
&-v_{j,\alpha}(P)\oint_{\partial\Gamma} D^*_{ij\alpha m}(P,Q)dx_m(Q) \\
&+\text{weakly singular crack curvature terms}
\end{aligned}$$

In fact, if one integrates Eq. 35 by parts prior to regularization and then regularizes, one obtains Eq. 36.

The new computational results for the elasticity problem suggest that both of the above formulations deserve some attention in terms of the goal of achieving highly accurate and efficient BEM fracture mechanics algorithms. Based on this experience, one can assume that the following will apply to the traction-BIE formulations:

- Quartic interpolations are likely to be needed for both of the above traction-BIE formulations. The earlier work by Polch et al. [12] was based on quadratic interpolations.

- Relaxed-continuity modeling using standard $C^{0,\alpha}$ displacement interpolants should apply to both formulations. Again, in Polch et al. and in all other traction-BIE applications to the fracture problem, special algorithms or non-conforming BEM algorithms were used.

- Special attention needs to be focused on the crack front integrals in order to collocate at the crack front. All earlier computational algorithms for the fracture problem have only collocated up to the row of nodes behind the crack tip. Local errors near the crack tip are therefore certain to be present.

The third point above can be more clearly seen by taking one of the traction-BIE equations for the normal loading on a flat crack.

$$t_3(P) = \frac{\mu}{4\pi(1-\mu)} \left\{ \int_\Gamma \frac{r_{,\alpha}}{r^2} [v_{3,\alpha}(Q) - v_{3,\alpha}(P)] dS(Q) \right.$$
$$\left. - \epsilon_{3\alpha\beta} v_{3,\alpha}(P) \oint_{\partial\Gamma} \frac{dx_\beta}{r} \right\} \qquad (37)$$

As shown by Polch et al. [12], the total integral in Eq. 37 is finite for points taken near the crack tip. In fact, on theoretical grounds, the result must be finite for all collocation points up to and including points on the crack front. The theoretical and computational results hold for the case of square-root singular values of the tangential derivative (normal to the crack front) of the crack opening displacement field. Given that the area integral is regularized and the path integral has a Cauchy principal value for crack front collocation, and given that the coefficient of the path integral itself is unbounded at the crack front, more work is required to fully understand the computational issues of collocating at the crack front. The focus of that work must be to assure a finite result for the total set of integrals in Eq. 37.

5. Conclusions

The intent of the current note is to review issues of formulation and computational implementation of self-regularization of boundary integral equation formulations. The formulation approach was reviewed by applying it to Laplace's equation. Numerical results for the two dimensional problem in elastostatics are cited which show that numerical results with desired and expected convergence characteristics requires higher order interpolants than locally-regularized algorithms. More work needs to be done to analytically and numerically explore the inter-relationships of the various BEM algorithms in this regard.

The note verifies an earlier fracture mechanics formulation based on a self-regularized traction-BIE. The traction-BIE formulation applications in [12] make use of linear state regularizations with quadratic interpolants. An pseudo-crack displacement function was used to "smooth" displacement derivative discontinuities. The recent experience in elastostatics [1] suggests the the computational approach embodied in "relaxed continuity" collocation with standard $C^{0,\alpha}$ boundary elements should be tried for the traction-BIE. However, it is also clear that higher order interpolations should also be applied to this problem. Finally, the numerical integrations of both area and path terms in the traction-BIE need to be carefully studied in order to allow collocation at the crack front nodes.

References

[1] J. D. Richardson, T. A. Cruse: Weakly Singular Stress–BEM for 2D Elastostatics. *Int. J. Numer. Meth. Engrg.*, in press.

[2] P. A. Martin, F. J. Rizzo, and T. A. Cruse: Smoothness-relaxation strategies for singular and hypersingular integral equations. *Int. J. Numer. Meth. Engrg.*, **42** 1998, 885–906.

[3] T. J. Rudolphi: The use of simple solutions in the regularization of hypersingular boundary integral equations. *Mathl. Comput. Modelling*, **15** 1991, 269–278.

[4] V. D. Kupradze: *Three–Dimensional Problems of the Mathematical Theory of Elasticity and Thermoelasticity*, North–Holland Publishing Company, 1979.

[5] O. D. Kellogg: *Foundations of Potential Theory*, Dover, 1953.

[6] T. A. Cruse and J. D. Richardson: Non–singular Somigliana stress identities, *Int. J. Numer. Meth. Engrg.*, **39** 1996, 3273–3304.

[7] T. A. Cruse and J. D. Richardson: *Traction BEM Integration Formulae*, Report 96-02, Vanderbilt University, 1996.

[8] T. A. Cruse: *Boundary Element Analysis in Computational Fracture Mechanics*, Martinus Nijhoff Publishers, Dordrecht, The Netherlands, 1988.

[9] A. E. H. Love: *A Treatise on the Mathematical Theory of Elasticity*, Dover, New York, 1944.

[10] Q. P. Huang and T. A. Cruse: On the nonsingular traction–BIE in elasticity, *Int. J. Numer. Meth. Engrg.* **37** 1994, 2041–2072.

[11] J. D. Richardson, T. A. Cruse, and Q. Huang: On the validity of conforming BEM algorithms for hypersingular boundary integral equations, *Int. J. Numer. Meth. Engrg.*, **42** 1997, 213–220.

[12] E. Z. Polch, T. A. Cruse, and C. J. Huang: Traction BIE solutions for flat cracks, *Comp. Mechs.*, **2** 1987, 253–267.

Vanderbilt University, Department of Mechanical Engineering,
Nashville, TN 37235, USA.
Email: Thomas.A.Cruse@Vanderbilt.Edu

C. ECK and W.L. WENDLAND
An Adaptive Boundary Element Method for Contact Problems

1. Introduction

Contact problems have a wide range of applications in many different areas of solid mechanics. Particular examples arise in fracture mechanics, where the contact of two opposite crack faces must be modelled, and in machine dynamics, where different parts of a machine may hit each other. At the contacting boundaries, friction forces may occur which often are too large to be neglected. The stresses generated by contact and friction can be very high and therefore often lead to fatigue and crack of the material. For a corresponding simulation it is decisive to have reliable numerical algorithms available. Such algorithms should be adaptive in order to enhance effectivity and reliability. This is especially important for contact problems, because the area of contact where large forces are transmitted is usually rather small and may also be time dependent. For the solution of contact problems the boundary element method is particularly well suited, because the nonlinearities of the contact and friction models are located on the bodies' boundaries. Then the boundary element method leads to a discrete system of nonlinear equations with much fewer degrees of freedom than methods where the whole body is also discretized.

Here we present an adaptive boundary element method for the solution of static frictional contact problems. The problem — which can be interpreted as one time step of a time discretization of the dynamic contact problem — can be formulated as an elliptic variational inequality. Application of the penalty method and a regularization of the generalized Coulomb law of friction leads to an approximation by a variational equation. This equation is discretized by Galerkin's method and then solved with a modified Newton method. For estimating the discretization error, a residual–based local error indicator developed in [13] for linear pseudodifferential equations is adapted to the nonlinear contact problem.

We begin with the description of the problem. For simplicity let us consider the contact of an elastic body with a rigid foundation. Let $\Omega \subset \mathbb{R}^d$, $d = 2, 3$, be the domain occupied by the elastic body in some reference configuration. Its boundary Γ is supposed to consist of three mutually disjoint, measurable parts Γ_U, Γ_F and Γ_C. On Γ_U and Γ_F, respectively, we prescribe boundary displacements g and boundary tractions b, respectively; on Γ_C may occur contact with the rigid foundation. The distance of a point $x \in \Gamma_C$ to the rigid foundation, measured in the direction of a given vector $N(x)$, is denoted by $g_N(x)$. Let $u = (u_i)_{i=1}^d$ denote the displacement field, $(\sigma_{ij})_{i,j=1}^d$ the stress tensor field, $n = (n_i)_{i=1}^d$ the outer normal vector on the boundary and $\sigma^{(n)} := (\sigma_{ij} n_j)_{i=1}^d$ the boundary stress vector. Throughout this paper we use the Einstein summation convention. For a vector field v defined on Γ_C we denote by $v_N = v_i N_i$ and $v_T = v - v_N N$ its components parallel and perpendicular to the vector N, respectively, and for simplicity we write $\sigma_N := \sigma_N^{(n)}$ and $\sigma_T = \sigma_T^{(n)}$. Moreover, with $v_{,i} := \frac{\partial v}{\partial x_i}$ the derivative of a vector field v with respect to the space variable

---

This work was supported by the German Research Foundation (DFG) under grant We 659/30-1 and Sonderforschungsbereich 404.

x_i is denoted. The stress tensor is given in terms of the (linearized) strain tensor with the components $e_{ij}(u) = \frac{1}{2}(u_{i,j} + u_{j,i})$ by Hooke's law $\sigma_{ij}(u) = a_{ijk\ell}e_{k\ell}(u)$ with a symmetric, elliptic and bounded tensor $\{a_{ijk\ell}\}$. Then the classical formulation of the problem is given by the following relations:

$$-\sigma_{ij,j}(u) = f_i \text{ in } \Omega, \tag{1}$$

$$u = g \text{ on } \Gamma_U, \tag{2}$$

$$\sigma^{(n)}(u) = b \text{ on } \Gamma_F, \tag{3}$$

$$u_N \leq g_N,\ \sigma_N \leq 0,\ \sigma_N(u_N - g_N) = 0 \text{ on } \Gamma_C, \tag{4}$$

$$\begin{aligned}(\delta u)_T = 0 &\Rightarrow |\sigma_T| \leq \mathcal{F}|\sigma_N|, \\ (\delta u)_T \neq 0 &\Rightarrow \sigma_T = -\mathcal{F}|\sigma_N|\tfrac{(\delta u)_T}{|(\delta u)_T|}\end{aligned} \text{ on } \Gamma_C. \tag{5}$$

Here $\delta u := u - w$ denotes some time increment, the function w depends on the time discretization scheme and on the solution of the previous time steps. For instance, for a backward Euler discretization we have $w(t^\ell) = u(t^{\ell-1})$ at the time step t^ℓ. Observe that the time increment does not appear explicitly in the formulation of the problem. In the quasistatic frictional contact problem the time has only the role of a parameter and not of a true variable. Problem (1) to (5) is interpreted as a given problem with the function w as part of the given data. In particular we do not care for the type and the convergence properties of the particular time discretization. Our aim is the construction of an efficient numerical method for the problem of one time step.

The weak formulation of the frictional contact problem is given by the following variational inequality:

Find a function $u \in \mathcal{K} := \{v \in H^1(\Omega; \mathbb{R}^d) \,|\, v = g \text{ on } \Gamma_U \text{ and } v_N \leq g_N \text{ on } \Gamma_C\}$, such that

$$\begin{aligned}\int_\Omega \sigma_{ij}(u)e_{ij}(v-u)\,dx &+ \int_{\Gamma_C} \mathcal{F}|\sigma_N(u)|\left(|(\delta v)_T| - |(\delta u)_T|\right) ds_x \\ \geq \mathcal{L}(v-u) &:= \int_\Omega f_i(v_i - u_i)\,dx + \int_{\Gamma_F} b_i(v_i - u_i)\,ds_x \quad \text{for all } v \in \mathcal{K}.\end{aligned} \tag{6}$$

This formulation is derived from the differential equations (1) by multiplication with $(v_i - u_i)$ and summation over i, by using the Green's formula and by employing the weak formulations

$$u_N \leq g_N,\ \sigma_N(v_N - u_N) \geq 0 \quad \text{for all } v_N \leq g_N,$$
$$\sigma_T(v_T - u_T) + \mathcal{F}|\sigma_N|\left(|(\delta v)_T| - |(\delta u)_T|\right) \geq 0 \quad \text{for all } v_T,\ v \in \mathcal{K},$$

corresponding to the Signorini contact condition (4) and the Coulomb law of friction (5) (cf. [9]).

2. Approximation with a variational equation

In order to approximate problem (6) with a variational equation we first use the penalty method and replace the Signorini condition (4) by the nonlinear compliance boundary condition

$$\sigma_N(u) = -\frac{1}{\lambda}[u_N - g_N]_+ \quad \text{with } [\cdot]_+ = \max\{0, \cdot\} \tag{7}$$

depending on a small positive approximation parameter λ. This yields an approximate problem which is obtained from variational inequality (6) by changing the set of admissible functions $\mathcal{K}$ to the affine space $g \oplus \mathcal{V}$ with

$$\mathcal{V} := \left\{ v \in H^1(\Omega; \mathbb{R}^d) \mid v = 0 \text{ on } \Gamma_U \right\},$$

by adding the penalty functional $\int_{\Gamma_C} \frac{1}{\lambda} [u_N - g_N]_+ (v_N - u_N) \, ds_x$ to the left–hand side, and by replacing $|\sigma_N(u)|$ in the friction term by $\frac{1}{\lambda}[u_N - g_N]_+$. In the next step the non–differentiable norms $|(\delta u)_T|$, $|(\delta v)_T|$ in the friction functional are replaced by differentiable approximations $\Psi_\varepsilon : \mathbb{R}^d \to \mathbb{R}$. The function Ψ_ε should be convex, continuously differentiable, satisfy the approximation property $|\Psi_\varepsilon(x) - |x|| \leq \varepsilon$ for all x and should have its minimum at the origin. After replacing the norms in the friction functional we choose $\tilde{v} = u + \eta v$ as the test function, divide the resulting inequality by η and perform the limit $\eta \to 0$. Then we obtain the following variational equation:

Find a displacement field $u \in g \oplus \mathcal{V}$ such that for all $v \in \mathcal{V}$ there holds

$$\int_\Omega \sigma_{ij}(u) \, e_{ij}(v) \, dx + \int_{\Gamma_C} \frac{1}{\lambda} [u_N - g_N]_+ \, v_N \, ds_x$$
$$+ \int_{\Gamma_C} \mathcal{F} \frac{1}{\lambda} [u_N - g_N]_+ \nabla \Psi_\varepsilon((\delta u)_T) \cdot v_T \, ds_x = \mathcal{L}(v) \quad \text{for all } v \in \mathcal{V}. \quad (8)$$

This problem can also be seen as a frictional contact problem where the contact is described by the nonlinear relation (7) and the friction is formulated by the law

$$\sigma_T(u) = -\mathcal{F} \frac{1}{\lambda} [u_N - g_N]_+ \nabla \Psi_\varepsilon((\delta u)_T). \quad (9)$$

The penalization of the contact–condition and the smoothing of the friction therefore correspond to an approximation of the multi–valued, discontinuous Signorini condition and Coulomb law of friction by single–valued continuous relations (7) and (9). In the approximate problem (8), the contact– and friction functional defines a completely continuous perturbation of the bilinear form of linear elasticity. Therefore the existence of solutions to this problem can be obtained with standard functional analytical arguments under appropriate assumptions (see e.g. in [4] and [5]). There it is also proved that a subsequence $u_{\lambda_k, \varepsilon_k}$ of solutions of the approximate problems with approximation parameters $\lambda_k, \varepsilon_k \to 0$ converges in $H^1(\Omega, \mathbb{R}^d)$ to a solution of the original contact problem (6). For the validity of this result it is essential that the support of the coefficient of friction $\mathcal{F}$ - which is supposed to depend also on $x \in \Gamma_C$ - is contained in the interior of the contact part of the boundary Γ_C and that its L_∞-norm is strictly smaller than a particular constant given in [4]. However, up to now it is not possible to derive precise estimates for the rate of convergence in the form $\|u_{\lambda_k, \varepsilon_k} - u\|_{H^1(\Omega; \mathbb{R}^d)} \leq c(\lambda_k, \varepsilon_k)$. Currently, such results are available only for contact without friction, where the estimate

$$\|u_{\lambda_k} - u\|_{H^1(\Omega; \mathbb{R}^d)} \leq c\lambda_k \|u\|_{H^2(\Omega; \mathbb{R}^d)}$$

can be proved, cf. [9]. The uniqueness of solutions for both, the approximate problem (8) and the original problem (6) is established also for the frictionless case only.

3. Boundary element method

The solution u of the elasticity equations with volume force f and corresponding boundary traction τ satisfies the two well-known boundary integral equations

$$\frac{1}{2}u = V\tau - Ku + N_0 f, \tag{10}$$

$$\frac{1}{2}\tau = K'\tau + Du + N_1 f \tag{11}$$

on the boundary Γ where the boundary integral operators are defined as

$$(V\tau)_i(x) = \int_\Gamma E_{ij}(x,y)\tau_j(y)\,ds_y \text{ (single layer operator)},$$

$$(Ku)_i(x) = \int_\Gamma (T_{(y)jk}E_{ik})(x,y)u_j(y)\,ds_y \text{ (double layer operator)},$$

$$(K'\tau)_i(x) = \int_\Gamma (T_{(x)ij}E_{jk})(x,y)\tau_k(y)\,ds_y \text{ (adjoint double layer operator)},$$

$$(Du)_i(x) = -T_{(x)ij}\int_\Gamma (T_{(y)k\ell}E_{j\ell})(x,y)u_k(y)\,ds_y \text{ (hypersingular operator)}$$

and the Newton potentials are given by

$$(N_0 f)_i(x) = \int_\Omega E_{ij}(x,y)f_j(y)\,dy, \quad (N_1 f)_i(x) = \int_\Omega (T_{(x)ij}E_{jk})(x,y)f_k(y)\,dy.$$

Here $E_{ij}(x,y)$ denotes the ij-th component of the fundamental solution of the elasticity equations — the Kelvin tensor — and $T_{(y)ij} = \frac{E}{2+2\nu}\left[\frac{2\nu}{1-2\nu}n_i\frac{\partial}{\partial y_j} + n_j\frac{\partial}{\partial y_i} + \delta_{ij}n_k\frac{\partial}{\partial y_k}\right]$ is the boundary stress operator with respect to the variable y. The single layer operator is given by a weakly singular integral, the double layer operator and its adjoint are defined by Cauchy singular integrals, and the operator D is calculated by a hypersingular integral which is defined in the sense of Hadamard's finite part. The boundary integral operators have the following properties (see [1, 7]).

Theorem 1. *Let Ω be a bounded domain with Lipschitz boundary Γ. Then for every $s \in (-\frac{1}{2}, \frac{1}{2})$ the operators $V: H^{-\frac{1}{2}+s}(\Gamma; \mathbb{R}^d) \to H^{\frac{1}{2}+s}(\Gamma; \mathbb{R}^d)$, $K: H^{\frac{1}{2}+s}(\Gamma; \mathbb{R}^d) \to H^{\frac{1}{2}+s}(\Gamma; \mathbb{R}^d)$, $K': H^{-\frac{1}{2}+s}(\Gamma; \mathbb{R}^d) \to H^{-\frac{1}{2}+s}(\Gamma; \mathbb{R}^d)$ and $D: H^{\frac{1}{2}+s}(\Gamma; \mathbb{R}^d) \to H^{-\frac{1}{2}+s}(\Gamma; \mathbb{R}^d)$ are continuous. The operator V is self-adjoint and positive definite,*

$$\langle V\tau, \tau \rangle_{L_2(\Gamma)} \geq c_1^V \|\tau\|^2_{H^{-\frac{1}{2}}(\Gamma; \mathbb{R}^d)}$$

for all $\tau \in H^{-\frac{1}{2}}(\Gamma; \mathbb{R}^d)$ with $c_1^V > 0$; in the two-dimensional case one needs to exclude specific cases (similar to the conformal radius 1 case in potential theory) which is guaranteed by scaling if $\operatorname{diam}(\Omega)$ is sufficiently small (see [2]). The hypersingular operator is self-adjoint and positive semi-definite satisfying

$$\langle Du, u \rangle_{L_2(\Gamma)} \geq c_1^D \|u\|^2_{H^{\frac{1}{2}}(\Gamma; \mathbb{R}^d)} \quad \textit{for all } u \in H^{\frac{1}{2}}(\Gamma; \mathbb{R}^d)/\ker(D)$$

with $c_1^D > 0$. The kernel $\ker(D)$ is given by the set of rigid motions, $\ker(D) = \mathcal{R}$.

Since V and D are positive (semi–) definite, the first boundary integral equation (10) is especially well suited for the solution of Dirichlet–problems where the boundary tractions τ are the unknowns, whereas the second equation (11) is better suited for Neumann–problems with unknown displacements u. For mixed boundary value problems and also for contact problems both integral equations must be used.

In order to transform the domain variational equation (8) to a boundary variational equation it is sufficient to replace the domain integral $\int_\Omega (\sigma_{ij}(u)e_{ij}(v) - f_i v_i)\,dx$ by a boundary integral by using the Green's formula

$$\int_\Omega \left(\sigma_{ij}(u)e_{ij}(v) + \sigma_{ij,j}(u)v_i\right) dx = \int_\Gamma \sigma^{(n)}(u) \cdot v\,ds_x \tag{12}$$

and the differential equation $f_i = -\sigma_{ij,j}(u)$. Here, we need a representation of the boundary stress $\tau = \sigma^{(n)}(u)$ in terms of the boundary displacements. The operator, which maps a given boundary displacement field to the corresponding boundary traction field for a *homogeneous* Dirichlet problem, is called the *Steklov–Poincaré operator* and denoted by the symbol S. This operator has different representations in terms of the integral operators V, K, K' and D. For the solution of a boundary value problem with inhomogeneous volume data f, the relation between boundary displacement u and boundary stress τ is given by the formula

$$\tau = Su + Nf = Su - V^{-1}N_0 f.$$

Here we realize the Dirichlet–Neumann map with the help of both integral equations. Adding $\frac{1}{2}\tau$ to the second equation (11) yields

$$\tau = \left(\frac{1}{2}I + K'\right)\tau + Du + N_1 f.$$

This equation is used in the Green's formula (12) which is then inserted into the approximate contact problem (8). For the determination of the contact stress τ the first integral equation is added. Thus, the following boundary variational formulation of the contact problem is obtained:

Find $(u, \tau) \in \left(g \oplus H^{\frac{1}{2}}_{\Gamma_U}(\Gamma; \mathbb{R}^d)\right) \times H^{-\frac{1}{2}}(\Gamma; \mathbb{R}^d)$ with

$$H^{\frac{1}{2}}_{\Gamma_U}(\Gamma; \mathbb{R}^d) := \{v \in H^{\frac{1}{2}}(\Gamma; \mathbb{R}^d) \,|\, v = 0 \text{ on } \Gamma_U\}$$

such that there holds

$$\int_\Gamma \begin{pmatrix} V & -(\frac{1}{2}I + K) \\ \frac{1}{2}I + K' & D \end{pmatrix} \begin{pmatrix} \tau \\ u \end{pmatrix} \cdot \begin{pmatrix} \varphi \\ v \end{pmatrix} ds_x + \int_\Gamma B_{\lambda,\varepsilon}(u) \cdot v\,ds_x$$
$$= \int_\Gamma (-N_0 f \cdot \varphi + (b - N_1 f) \cdot v)\,ds_x \tag{13}$$

for all $(v, \varphi) \in H^{\frac{1}{2}}_{\Gamma_U}(\Gamma; \mathbb{R}^d) \times H^{-\frac{1}{2}}(\Gamma; \mathbb{R}^d)$ with the nonlinear contact- and friction function

$$B_{\lambda,\varepsilon}(u) = \frac{1}{\lambda}[u_N - g_N]_+ N + \mathcal{F}\frac{1}{\lambda}[u_N - g_N]_+ \nabla \Psi_\varepsilon((\delta u)_T).$$

Note that the linear operator on the left–hand side of (13) is positive semi-definite and has a block skew–symmetric structure. Let us also remark that the first boundary integral equation (10) is used on the whole boundary whereas the second equation (11) is employed only on the parts Γ_F and Γ_C.

Equation (13) is discretized with a conforming Galerkin method. Therefore we use a partition $\Gamma = \bigcup_{i=1}^{M} \Gamma_i$ of the boundary into segments in the 2D–case or approximate Γ by interpolating plane triangles in the 3D–case. Let h_j be the diameter of Γ_j and $h = \max_{j=1,\ldots M} h_j$. Let $S_h^{r,t}$, $r \in \mathbb{N}$, $t \in \mathbb{N}_0 := \mathbb{N} \cup \{0\}$, be the space of piecewise polynomials of degree $r - 1$ associated with the elements $\{\Gamma_j\}$ which are in $C^{t-1}(\Gamma)$ globally ($S_h^{r,0}$ is the space of discontinuous splines). The boundary displacements are discretized with smoothest splines from the space $S_h^D := \{\varphi \in S_h^{r_D, r_D - 1} \,|\, \varphi = 0 \text{ on } \Gamma_U\}$ with respect to the parametric variables. The approximate solution is sought in the form $g + u_h$ with $u_h \in S_h^D$. The discretization of the boundary traction is supposed to be a discontinuous spline from the space $S_h^N = S_h^{r_N, 0}$ of piecewise constants. The test functions are elements of the same spaces, $v \in S_h^D$ and $\varphi \in S_h^N$, correspondingly. The discrete system of equations is then solved with Newton's method. However, in order to obtain a discrete linearized equation with symmetric structure in every Newton step, a symmetric approximation of the tangent matrix for the nonlinear functional $B_{\lambda,\varepsilon}$ is employed: In step k of the Newton iteration, the derivative of

$$\widetilde{B}_{\lambda,\varepsilon}(u) = \frac{1}{\lambda}[u_N - g_N]_+ N + \mathcal{F}\frac{1}{\lambda}\left[u_N^{(k-1)} - g_N\right]_+ \nabla\Psi_\varepsilon\bigl((\delta u)_T\bigr)$$

with the solution $u^{(k-1)}$ of the previous Newton step is used instead of that of the function $B_{\lambda,\varepsilon}$. This can be seen as an approximation of the friction functional by a friction functional with given contact stress. The solution of contact problems with an iteration over approximate problems with given contact stress in the friction functional was proposed first by Panagiotopoulos in [11]. In [10] and [8] the convergence of such an iterative process was proved for the original contact problem. In our case of the penalized and smoothed contact problem, the convergence of such an iteration is investigated in [6].

4. An error estimator for boundary element methods

For the a–posteriori estimate of the discretization- and approximation error we extend a residual–based error indicator derived in [13] for linear pseudodifferential equations to the nonlinear frictional contact problem. Let us first outline briefly the error indicator. Consider the equation

$$Au = f, \qquad (14)$$

with a positive definite and bounded pseudodifferential operator $A : H^\alpha(\Gamma) \to H^{-\alpha}(\Gamma)$ of order 2α, a linear functional $f \in H^{-\alpha}(\Gamma)$ and a sufficiently smooth manifold $\Gamma \subset \mathbb{R}^d$ of dimension $d - 1$. Let us also assume that u_h is any approximate solution and χ_i is a cut-off function associated with an element Γ_i from a partition of Γ. Let $e_h := u - u_h$ denote the error and $r_h := Au_h - f$ the residual. Then in [12] the error estimate

$$c_1\|\chi_i r_h\|_{H^{s-2\alpha}(\Gamma)} - \varepsilon_h^{(i)}(u) \leq \|\chi_i e_h\|_{H^s(\Gamma)} \leq C_1\|\chi_i r_h\|_{H^{s-2\alpha}(\Gamma)} + \varepsilon_h^{(i)}(u) \qquad (15)$$

is proved where

$$\varepsilon_h^{(i)}(u) \leq c\|u - u_h\|_{H^{s-1}(\Gamma)}. \tag{16}$$

If the approximate solution u_h is computed with a conforming Galerkin method, where the trial- and test space $S_h^{r,t}$ is defined as in the previous section, then the well-known a-priori error estimate

$$\|u_h - u\|_{H^p(\Gamma)} \leq ch^{q-p}\|u\|_{H^q(\Gamma)} \tag{17}$$

is valid for all Sobolev indices p, q satisfying the conditions $2\alpha - r \leq p \leq \alpha \leq q \leq r$ and $\alpha < t + \frac{1}{2}$ for the 2D–case or $\alpha \leq t$ in the 3D–case. Using (17) in (15) and in (16) it can be seen that the perturbation $\varepsilon_h^{(i)}(u)$ in the error estimate is of higher order than the expected error with respect to the mesh size h. Therefore these terms can be neglected in an asymptotically correct error estimator.

However, this estimate is valid for fixed cut–off functions χ_i only. In particular it is not possible to vary the support of the cut–off function according to the size of the associated element Γ_i which limits the applicability of this error indicator for the construction of adaptive methods considerably. Therefore in [13] an improved estimate of the terms $\varepsilon_h^{(i)}(u)$ was derived which allows a variation of the supports of the cut–off functions. Let us first collect the corresponding assumptions.

Assumption 1. Let the partition of the boundary Γ be β–regular, this means $h_{\min} \geq ch_{\max}^\beta$ with $\beta > 1$. The cut–off function χ_i and a parameter ε_i which are associated with an element Γ_i are supposed to satisfy the conditions $0 \leq \chi_i \leq 1$, $\chi_i = 1$ on Γ_i, $\mathrm{supp}(\chi_i) \subset \tilde{\Gamma}_i(\varepsilon_i) := \{x \in \Gamma \mid \mathrm{dist}(x, \Gamma_i) < \varepsilon_i\}$ and $|D^a \chi_i| \leq c\varepsilon_i^{-|a|}$ for all $a \in \mathbb{N}_0^N$. The approximate solution u_h shall satisfy the a-priori estimate (17).

Then H. Schulz showed in [13] the following result.

Theorem 2. *Under Assumption 1, the perturbation $\varepsilon_h^{(i)}(u)$ of the a–posteriori error estimate (15) is bounded by*

$$\varepsilon_h^{(i)}(u) \leq ch^{p-s+\delta}\|u\|_{H^p(\Gamma)}. \tag{18}$$

The parameter δ is defined by $\delta = 1 - \beta(\ell + k + 1)\gamma$, where the constant γ is calculated from the condition $\varepsilon_i \geq c\tilde{h}_i^\gamma$ with $\tilde{h}_i := \max\{h_j \mid \Gamma_j \cap \tilde{\Gamma}_i(\varepsilon_i) \neq \emptyset\}$ and where for the parameters k, ℓ there holds either $\ell > d+|s-1|$ and $k > d+|s-2\alpha|$ or $\ell > d+|s-1-2\alpha|$ and $k > d+|s|$.

The parameter ε_i in Assumption 1 indicates how small the support of the cut–off function can be made in dependence on the mesh size $\tilde{h}_i$. In an adaptive procedure it is desirable that the cut–off function associated with an element has a size which is proportional to the size of that element. This corresponds to a relation $\varepsilon \sim h$, or to an exponent $\gamma = 1$. Unfortunately this case is not covered by the above theorem. One can easily see that it is not possible to have γ close to 1 if the gain δ in the order of convergence for the remainder $\varepsilon_h^{(i)}(u)$ is positive. Nevertheless Theorem 2 is a remarkable progress compared to the older results where no variation of the size of the support of the cut-off function was allowed at all.

The described error estimator for linear pseudodifferential equations is now adjusted to the special situation of contact problems. We first consider the case of frictionless contact. Let (u,τ) denote the displacement field and the boundary stress of the exact solution of the contact problem. Let (u_h, τ_h) be an approximate solution. It is supposed that these solutions satisfy optimal convergence properties, i.e.

$$\|u - u_h\|_{H^p(\Gamma;\mathbb{R}^d)} \leq Ch^{q-p} \|u\|_{H^q(\Gamma;\mathbb{R}^d)}, \tag{19}$$

$$\|\tau - \tau_h\|_{H^{p'}(\Gamma;\mathbb{R}^d)} \leq Ch^{q'-p'} \|\tau\|_{H^{q'}(\Gamma;\mathbb{R}^d)} \tag{20}$$

with $-\frac{1}{2} \leq p \leq \frac{1}{2} \leq q \leq q_{\max}$ and $-\frac{3}{2} \leq p' \leq -\frac{1}{2} \leq q' \leq q_{\max} - 1$. The maximal value $q_{\max}$ is determined by the test spaces $S_h^D = S_h^{r_D, r_D - 1}$ and $S_h^N = S_h^{r_N, 0}$ as described after relation (17) and by the regularity of the exact solution (u, τ). For simplicity it is assumed that the approximate solutions u_h and τ_h satisfy the Dirichlet condition on Γ_U and the Neumann condition on Γ_F exactly. This is no major restriction, since the error of the approximation of the boundary conditions can be computed exactly from the approximate solution and the boundary condition.

Let S denote the Steklov–Poincaré operator, see Section 3. Then the (semi-) norms $\|\cdot\|_+$ and $\|\cdot\|_-$ can be defined by

$$\|u\|_+ := (\langle Su, u\rangle)^{\frac{1}{2}} \text{ and } \|\tau\|_- := (\langle S^{-1}\tau, \tau\rangle)^{\frac{1}{2}}. \tag{21}$$

The norm $\|\cdot\|_+$ is equivalent to the $H^{\frac{1}{2}}(\Gamma;\mathbb{R}^d)$–norm in the space $H^{\frac{1}{2}}(\Gamma;\mathbb{R}^d)/\mathcal{R}$ and $\|\cdot\|_-$ is equivalent to the norm of $H^{-\frac{1}{2}}(\Gamma;\mathbb{R}^d) \cap \mathcal{R}^\perp$. The local discretization error of the contact problem associated with the element Γ_i is measured by the quantity

$$E_i(u, \tau; u_h, \tau_h) := \|\chi_i(u - u_h)\|_+ + \|\chi_i(\tau - \tau_h)\|_-. \tag{22}$$

This error measure is composed of the error in the displacements and the error in the stresses, both errors measured in their corresponding energy norm. Then we have the following estimate:

Theorem 3. *Let the Assumption 1 and the a-priori estimates (19, 20) be valid. Then we have*

$$E_i(u, \tau; u_h, \tau_h) \leq c_1 \left\|\chi_i r_h^D\right\|_+ + c_2 \left\|\chi_i r_h^N\right\|_- + c_3 \left\|\chi_i [u_{hN} - g_N]_+\right\|_+$$

$$+ c_4 \left\|\chi_i [\tau_{hN}]_+\right\|_- + c_5 \left(\int_{\Gamma_C} [\chi_i \tau_{hN}]_- \cdot \chi_i [u_{hN} - g_N]_- \, ds_x\right)^{\frac{1}{2}} + \varepsilon_h^{(i)}(u, \tau) \tag{23}$$

where $r_h^D := V\tau_h - \left(\frac{1}{2}I + K\right)u_h + N_0 f$ and $r_h^N := Du_h - \left(\frac{1}{2}I - K'\right)\tau_h + N_1 f$. The remainder $\varepsilon_h^{(i)}(u, \tau)$ is bounded by

$$\varepsilon_h^{(i)}(u, \tau) \leq ch^{p+\delta}\left(\|u\|_{H^{\frac{1}{2}+p}(\Gamma;\mathbb{R}^d)} + \|\tau\|_{H^{-\frac{1}{2}+p}(\Gamma;\mathbb{R}^d)}\right) \tag{24}$$

for every $0 \leq p \leq q_{\max} - \frac{1}{2}$ *with* $\delta > 1 - \beta(2d+3)\gamma$.

Remark. The expression r_h^D can be interpreted as the residual of a Dirichlet problem with the given boundary datum u_h and the approximate solution τ_h. The residual corresponds to the first boundary integral equation (10). Analogously, r_h^N corresponds to the residual of a Neumann problem with boundary datum τ_h and the approximate solution u_h. Here the residual is calculated with the help of the second boundary integral equation (11).

At first sight, the error estimator given in Theorem 3 seems unsuitable for practical applications, because the evaluation of the norms $\|\cdot\|_+$ and $\|\cdot\|_-$ requires the solution of a boundary value problem. However, the norm $\|\cdot\|_+$ is spectrally equivalent to $\|\cdot\|_D := \langle D\cdot,\cdot\rangle^{\frac{1}{2}}$ and $\|\cdot\|_-$ is spectrally equivalent to $\|\cdot\|_V := \langle V\cdot,\cdot\rangle^{\frac{1}{2}}$. Hence in the error estimate $\|\cdot\|_+$ can be replaced by $\|\cdot\|_D$ and $\|\cdot\|_-$ by $\|\cdot\|_V$ and the estimate remains valid with possibly modified constants $c_1, \ldots, c_5$. The numerical evaluation of $\|\cdot\|_D$ and $\|\cdot\|_V$ is readily obtained if the Galerkin discretizations of the operators V and D are available. The constants c_j in Theorem 3 depend on various quantities as e.g. the norms of A and A^{-1}, and they are not explicitly computable, unfortunately. On the other hand, the error inclusions estimate the true local error in the contrary to normal global error inclusions.

For frictional contact problems it is much harder to derive mathematically rigorous error indicators. This is due to the non–monotonicity of the friction functional and due to the lack of uniqueness of the solution. Therefore all error indicators for frictional problems are based on some heuristic approach. Here we use the contact problem with given contact stress in the friction functional in order to derive an error indicator as we have mentioned already in Section 3. This problem is obtained from the variational inequality (6) if the contact stress $\mathcal{F}|\sigma_n(u)|$ in the friction functional is replaced by some given function $\mathcal{G}$. For the resulting problem, existence and uniqueness of the solution is well established, see e.g. [3]. Moreover, it is possible to define the following error indicator:

Theorem 4. *Let (u,τ) and (u_h,τ_h) be the exact and approximate solutions of the contact problem with given friction $\mathcal{G} \in L_2(\Gamma_C)$ and let the assumptions of Theorem 3, respectively, be valid. Then the approximation error can be estimated by*

$$E_i(u,\tau;u_h,\tau_h) \leq c_1 \left\|\chi_i r_h^D\right\|_+ + c_2 \left\|\chi_i r_h^N\right\|_- + c_3 \left\|\chi_i [u_{hN} - g_N]_+\right\|_+$$

$$+ c_4 \left\|\chi_i [\tau_{hN}]_+\right\|_- + c_5 \left(\int_{\Gamma_C} [\chi_i \tau_{hN}]_- \cdot \chi_i [u_{hN} - g_N]_- \, ds_x\right)^{\frac{1}{2}}$$

$$+ c_6 \left(\int_{\Gamma_C} \chi_i^2 \left(\tau_{hT}(\delta u_h)_T + \mathcal{G}|(\delta u_h)_T| + [|\tau_{hT}| - \mathcal{G}]_+ |(\delta u_h)_T|\right) ds_x\right)^{\frac{1}{2}}$$

$$+ c_7 \left\|\chi_i [|\tau_{hT}| - \mathcal{G}]_+\right\|_- + \varepsilon_h^{(i)}(u,\tau). \qquad (25)$$

The remainder $\varepsilon_h^{(i)}(u,\tau)$ satisfies the same estimate as in Theorem 3.

Observe that the local error estimator E_i in (25) is given by the error estimator of Theorem 3 plus additional terms describing the residual in the friction law.

The error indicator of Theorem 4 is applied also to the original contact problem. There the given contact stress $\mathcal{G}$ is replaced by the computed approximation $\mathcal{F}|\tau_{hN}|$. This approach is heuristic, and due to the difficulties mentioned above we are not able to prove a rigorous estimate in the general case. Nevertheless, in the numerical example presented below the results with this error indicator in the frictional case are as good as in the frictionless case.

5. Example

In this section we present the performance of the adaptive method driven by our estimators for a simple 2D example. We consider the contact of an elastic half disk pressed upon a rigid foundation as illustrated in Figure 1.

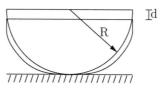

Figure 1: Example

The data of the problem are $R = 0.4$, $d = 0.02$; the elasticity modulus is $E = 1$ and the Poisson ratio $\nu = 0.3$. The problem is solved by the boundary element method and the error indicator is evaluated. Then all the elements are refined into two elements of equal size where the value of the local relative error indicator was bigger than a given tolerance. This procedure is iterated until the desired accuracy is obtained in every element. In order to estimate the accuracy of the error indicator, the difference of the approximate solution to a reference solution computed on an extremely fine mesh was calculated in the energy norm. We present first the results for contact without friction. In Table 1 the l_2-norm of the relative error indicators and the reference error is given for different refinement levels. Figures 2 and 4 show the grid of the finest level and the computed boundary stress on the contact part of the boundary. For the frictional contact problem with friction coefficient $\mathcal{F} = 1$, the error indicator and the reference error are compared in Table 2. Figure 3 shows the grid on its finest level and in Figure 5, both the normal and tangential component of the boundary stress on the contact part of the boundary are presented.

nodes	estim. error	refer. error
9	0.4157	0.5067
18	0.2099	0.2506
30	0.1191	0.1160
46	0.0744	0.0545
72	0.0471	0.0230
108	0.0302	0.0096

Table 1: Error estimator for contact without friction

nodes	estim. error	refer. error
9	0.4327	0.5101
18	0.2173	0.2533
28	0.1309	0.1225
44	0.0814	0.0590
70	0.0501	0.0247
104	0.0333	0.0105

Table 2: Error estimator for contact with friction

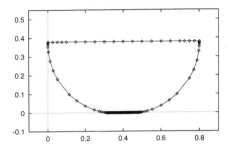

Figure 2: Contact without friction
Grid with 108 nodes

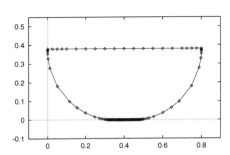

Figure 3: Contact with friction
Grid with 104 nodes

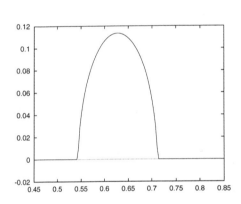

Figure 4: Contact without friction
Stress on contact boundary

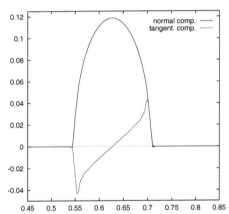

Figure 5: Contact with friction
Stress on contact boundary

References

[1] M. Costabel: Boundary integral operators on Lipschitz domains: elementary results. *SIAM J. Math. Anal.* **19** (1988), 613–626.

[2] M. Costabel, M. Dauge: Invertibility of the biharmonic single layer potential operator. *Integr. Equat. Th.* **24** (1996), 46–67.

[3] G. Duvaut, J.L. Lions: *Inequalities in Mechanics and Physics.* Springer–Verlag, Berlin, Heidelberg, New York, 1976.

[4] C. Eck, J. Jarušek: Existence results for the static contact problem with Coulomb friction. *Math. Mod. Meth. Appl. Sci.* **8** (1998), 445–468.

[5] C. Eck, J. Jarušek: Existence results for the semicoercive static contact problem with Coulomb friction. *Preprint 97/45, SFB 404,* University of Stuttgart, 1997.

[6] C. Eck, O. Steinbach, W.L. Wendland: A symmetric boundary element method for contact problems with friction. In preparation.

[7] G.C. Hsiao, W.L. Wendland: A finite element method for some integral equation of the first kind. *J. Math. Anal. Appl.* **58** (1977), 449–481.

[8] J. Jarušek: Contact problems with bounded friction, coercive case. *Czechosl. Math. J.* **33** (1983), 237–261.

[9] N. Kikuchi, J.T. Oden: *Contact Problems in Elasticity: A Study of Variational Inequalities and Finite Element Methods.* SIAM, Philadelphia, 1988.

[10] J. Nečas, J. Jarušek, J. Haslinger: On the solution of the variational inequality to the Signorini problem with small friction. *Bollettino U.M.I.* **17** (1980), 796–81.

[11] P.D. Panagiotopoulos: A nonlinear programming approach to the unilateral contact problem with friction. *Ingenieur–Archiv* **44** (1975), 421–432 .

[12] J. Saranen, W.L. Wendland: Local residual-type error estimates for adaptive boundary element methods on closed curves. *Applicable Analysis* **48** (1993), 37-50.

[13] H. Schulz: Über lokale und globale Fehlerabschätzungen für adaptive Randelementmethoden. Doctoral–thesis, University of Stuttgart, 1997.

[14] O. Steinbach: Gebietszerlegungsmethoden mit Randintegralgleichungen und effiziente numerische Lösungsverfahren für gemischte Randwertprobleme. Doctoral–thesis, University of Stuttgart, 1996.

[15] W.L. Wendland, D. Yu: Adaptive boundary element methods for strongly elliptic integral equations. *Numer. Math.* **53** (1988), 539-558.

[16] W.L. Wendland, D. Yu: Local error estimates of boundary element methods for pseudo-differential-operators of non-negative order on closed curves. *J. Comp. Math.* **10** (1990), 273-289.

C. Eck: Inst. Appl. Math. 1, Univ. Erlangen, 91058 Erlangen, Germany.
Email: `eck@am.uni-erlangen.de`
W.L. Wendland: Math. Inst. A, Univ. Stuttgart, 70695 Stuttgart, Germany.
Email: `wendland@mathematik.uni-stuttgart.de`

Y. FU, J. R. OVERFELT and G. J. RODIN
Fast Summation Methods and Integral Equations

1. Introduction

Fast summation methods (FSMs) allow one to solve large electrostatics (or gravitational) problems in $\mathcal{O}(N \log^p N)$ operations, where N is the number of charges and p is a small integer independent of N. FSMs have firmly established themselves in the computational physics community and have become standard tools in molecular dynamics, astrophysics, and atomistic simulations (see references in Greengard and Rokhlin, 1997). Also it has been well recognized that FSMs lend themselves to fast iterative strategies for integral equations in which FSMs perform dense-matrix-vector multiplication using only $\mathcal{O}(N \log^p N)$ memory and operations.

Currently, the majority of FSM applications involve particle simulations rather than integral equations. This situation exists because in particle simulations FSMs are used to solve the same, conceptually simple, summation problem: determine the attraction forces among given charges. Thus, as far as particle simulations are concerned, FSM codes can be treated as black boxes with simple interfaces, and currently there are many FSM codes available in the public domain. In contrast, for integral equations, the summation problems are usually more complicated and diverse. For those problems, there are two main approaches: either to represent the summation problem associated with the integral equation as a finite number of the electrostatics summation problems or to develop a FSM for the integral equation of interest.

In this paper, we consider FSMs applicable to a very broad class of integral equations associated with a general vector-valued elliptic partial differential equation with constant coefficients

$$C_{ijkl} u_{k,lj}(\boldsymbol{x}) + b_i(\boldsymbol{x}) = 0. \tag{1}$$

In this equation, C_{ijkl} is a constant positive definite fourth rank tensor such that

$$C_{ijkl} = C_{jikl}, \quad C_{ijkl} = C_{jilk}, \quad \text{and} \quad C_{ijkl} = C_{klij},$$

$u_k(\boldsymbol{x})$ is the unknown vector field, and $b_i(\boldsymbol{x})$ is the prescribed vector field. Equation (1) describes the elastic deformation of anisotropic solids and it is well-known that in general the fundamental solution of (1) cannot be expressed in closed-form (Bacon et al., 1978; Mura, 1982). Particular cases of (1) include Laplace's equation, Navier's equations of isotropic linear elasticity, and Stokes' equations of low-Reynolds number hydrodynamics.

Existing applications of FSMs to three-dimensional problems include those of Nabors et al. (1994) for Laplace's equation, Sangani and Mo (1996) and Fu and Rodin (1998) for Stokes' equations of low-Reynolds-number hydrodynamics, and Hayami and Sauter (1997), Nishimura et al. (1998), Yoshida et al. (1998) and Fu et al. (1998)

This work was supported by the National Science Foundation through the National Grand Challenges program grant ECS-9422707 and by the National Partnership for Advanced Computational Infrastructure.

for equations of isotropic linear elasticity. There are many applications of FSMs to two-dimensional problems (for references see Greengard and Moura, 1994) because two-dimensional integral equation solvers and FSMs are much easier to implement than their three-dimensional counterparts.

The solution strategies for integral equations can also be improved using matrix compression methods based on wavelets (Lage and Schwab, 1997; von Petersdorff and Schwab, 1997). In those methods, the dense system matrix is approximated with a sparse one, which then can be inverted using methods that take the sparsity into account. In general, matrix compression methods may involve several matrices whose structure lends itself to fast inversion.

The rest of our presentation is organized as follows. In Section 2, we describe FSMs for the electrostatics summation problem. Our description emphasizes various aspects of those FSMs that allow one to adopt them for solving integral equations. In Section 3, we develop an expression for the fundamental solution of (1) that lends itself to fast summation. In Section 4, to demonstrate the power of FSMs, we present a case study that involves a problem with more than one million unknowns.

2. Fast Summation Methods

In this section, we consider three FSMs for solving the electrostatics problem. Our objective is to understand to what extent these methods have to be modified in order to be applicable to integral equations corresponding to (1).

The electrostatics problem is stated for N charges q_i located at positions z_i inside a unit cube. The task is to compute the potential

$$\Phi_i = \Phi(z_i) = \sum_{j=1, j\neq i}^{N} \frac{q_j}{|z_i - z_j|}$$

for each charge.

To distinguish between the electrostatics summation problem and those associated with integral equations corresponding to (1), we suppose that for the latter the potential has the form

$$\Phi(z_i) = \sum_{j=1, j\neq i}^{N} \frac{q_j}{|z_i - z_j|} \Psi\left(\frac{z_i - z_j}{|z_i - z_j|}\right),$$

where Ψ is an implicitly prescribed function. The validity of this functional form for Φ follows from dimensional analysis, and at this stage we are not concerned with the structure of Ψ. To the contrary, we treat Ψ as an arbitrary smooth function, and evaluate the FSMs for the electrostatics problem in terms of their ability to handle such a function.

The naive computation of the electrostatics problem involves a nested loop structure in which the outer loop is over the target charges, located at z_i, and the inner loop is over the source charges, located at z_j. Consequently the required number of operations is equal to

$$\mathcal{O}(\text{number of target charges}) \times \mathcal{O}(\text{number of source charges}) = \mathcal{O}(N^2).$$

One can say that the $\mathcal{O}(N^2)$ operation count occurs because the target charges and the source charges are processed *simultaneously*. In this regard, Greengard (1994) observed that FSMs can be viewed as approximation methods in which the target and source charges are processed *separately*, and as a result the operation count becomes

$$\mathcal{O}(\text{number of target charges}) + \mathcal{O}(\text{number of source charges}) = \mathcal{O}(N).$$

It has been demonstrated by Fu and Rodin and their co-workers (Fu, 1998; Fu et al., 1998; Fu and Rodin, 1998; Rodin, 1998) that the notion that FSMs are separation methods is particularly useful for applying FSMs to integral equations.

To fix the idea of separation, we consider a *non-physical* summation problem. In this problem, the potential for the target charge at x_i is an exponential sum,

$$\Phi(x_i) = \sum_{j=1}^{N} q_j \exp\left[\boldsymbol{\lambda} \cdot (x_i - y_j)\right],$$

in which $\boldsymbol{\lambda}$ is a constant vector. To draw a clear distinction between the target and source charges, here and in the rest of the paper, we denote the positions of the former by x and the positions of the latter by y. To compute this problem in $\mathcal{O}(N)$ operations, first, we compute the moment (process the source charges),

$$m = \sum_{j=1}^{N} q_j \exp\left(-\boldsymbol{\lambda} \cdot y_j\right),$$

and then compute the potential for each target charge (process the target charges)

$$\Phi_i = m \exp\left(\boldsymbol{\lambda} \cdot x_i\right), \quad i = 1, ..., N.$$

This example clearly demonstrates that exponential approximations naturally lend themselves to fast summation. For the electrostatics problem, such approximations have been developed by Yarvin and Rokhlin (1996) and subsequently used as the basis for an efficient FSM by Greengard and Rokhlin (1997). An important feature of the Yarvin-Rokhlin approximations is that they exhibit an exponential decay and therefore can capture reasonably well the decay of $|x - y|^{-1}$. In contrast, exponential approximations based on the trigonometric Fourier series of $|x - y|^{-1}$ are unsuitable for FSMs. In the next section, we briefly describe the Yarvin-Rokhlin approximations and establish an integral representation that allows one to develop exponential approximations for the integral equation summation problem.

Besides the idea of separation, FSMs for the electrostatics problem rely on hierarchical partitioning of the unit cube into 8^1, 8^2, 8^3, etc. cubic cells. Hierarchical partitioning is needed for two reasons. First, it allows one to reduce the number of terms involved in the approximation of $|x - y|^{-1}$ by using local (with respect to the cells) rather than global (with respect to the unit cube) expansions. Second, hierarchical partitioning allows one to recycle the source moments so that the total number of moments is $\mathcal{O}(N \log N)$ and the number of moments used to compute the potential for each target charge is $\mathcal{O}(\log N)$. For summation problems associated with boundary integral equations, it may be advantageous to use hierarchical partitioning schemes that take advantage of the fact that the charges are located on the surface, as in the panel clustering method of Hackbusch and Nowak (1989).

With the exception of the FSM proposed by Greengard and Rokhlin (1997), existing FSMs for the electrostatics problem rely on non-exponential approximations derived either through Taylor's or spherical harmonics expansions of $|\boldsymbol{x} - \boldsymbol{y}|^{-1}$. Such approximations are easy to develop and they do not exhibit oscillatory behavior of exponential approximations. In particular, Taylor's expansion can be adopted for developing approximations for the integral equation summation problem.

Now let us demonstrate the algorithm of one of the simplest and very frequently-used FSM due to Barnes and Hut (1986).

Source processing:
Step 1: Partition the unit cube so that there is approximately one charge per cell at the finest partitioning level; accordingly the number of partitioning levels is an integer close to $\log_8 N$.
Step 2: Compute the charge (monopole) Q and dipole $\boldsymbol{D}$ for each cell at the finest partitioning level,

$$Q = \sum_{\text{cell charges}} q_j, \quad \boldsymbol{D} = \sum_{\text{cell charges}} q_j (\boldsymbol{y}_j - \boldsymbol{c}),$$

where $\boldsymbol{c}$ is the cell center.
Step 3: Compute the charges and dipoles for the remaining cells using the relationships

$$Q^p = \sum_{k=1}^{8} Q_k^c \tag{2}$$

and

$$\boldsymbol{D}^p = \sum_{k=1}^{8} \boldsymbol{D}_k^c + \sum_{k=1}^{8} Q^c (\boldsymbol{c}_k^c - \boldsymbol{c}^p), \tag{3}$$

In these equations, the superscripts p and c denote the parent and children cells, respectively; each parent has eight children.

Target processing:
For each target charge:
Step 1: Start with the source cells at the first partitioning level. For each of these cells, compute the error associated with replacing the actual charges in the cell with their monopole and dipole.
Step 2: If the error is bellow the tolerance, compute the potential induced by the monopole and dipole, and disregard the children and grandchildren of this cell. If the error is above the tolerance, disregard the cell and repeat Step 1 for its children.

Let us make the following remarks in connection with the Barnes-Hut FSM:

- Equations (2) and (3) follow directly from the definitions. Following Greengard and Rokhlin (1987) and Greengard (1988), we refer to such relationships as translation theorems. Their extension to higher-order moments is straightforward for both Taylor's and spherical harmonics expansions for the kernel $|\boldsymbol{x} - \boldsymbol{y}|^{-1}$.

- The operation count for the method should be at least $\mathcal{O}(N \log^2 N)$, provided that the charge distribution is quasi-uniform and the error control is properly implemented. The log-square rather than log dependence exists because the error tolerance for the individual cells must be $\mathcal{O}(\log^{-1} N)$ in order to keep the cumulative error constant. To this end, we observe that the issue of error control is open and difficult to resolve because the error for each target charge is accumulated from *a priori* unknown source cells, and therefore it is unclear how to distribute the error among those cells during the computation. In practice, the issue of error control is resolved through numerical experiments.

- The method is very simple and can be extended without major difficulties to the integral equation summation problem, especially if one chooses to use lower-order approximations based on Taylor's expansion. Further, the partitioning scheme can be defined on a two-dimensional manifold as it is done in the panel clustering method of Hackbusch and Nowak (1989).

- Among the many implementations of the method available in the public domain, we single out the one due to Warren and Salmon (1992, 1993) that is designed for large-scale distributed computing.

The principal difference between the Barnes-Hut method and the Fast Multipole Method (FMM) of Greengard and Rokhlin (1987) is in the target processing stage. In the FMM, this stage is carried out by introducing target cells so that the interaction scheme has the form

$$\text{source charges} \to \text{source cells} \to \text{target cells} \to \text{target charges}$$

The interactions among target cells and charges are computed based on solid spherical harmonics (or local) expansions. Implementation of these interactions requires an additional translation theorem. Another translation theorem is needed to compute the interactions among the source and target cells.

The additional translation theorems endow the FMM with a well-defined geometric structure, which allows one to implement the method as a sequence of matrix multiplications (Schmidt and Lee, 1991) and improve the error control. However, the translation theorems require lengthy calculations even for $|\boldsymbol{x} - \boldsymbol{y}|^{-1}$. In fact the only set of translation theorems, other than that developed by Greengard and Rokhlin (1987), is due to Sangani and Mo (1996) for the single layer potential of low-Reynolds-number hydrodynamics. For the kernels associated with (1), the translation theorem calculations would be so complicated that it is difficult to expect that they would ever be carried out. For this reason, the FMM should be used only for particular cases of (1) and one such case is considered in Section 4.

3. Exponential Representations

In this section, we develop an integral representation for the fundamental solution of (1) that can be used as the basis for constructing exponential approximations. This integral representation generalizes the well-known representation (Morse and Feshbach, 1953) used by Yarvin and Rokhlin (1996) for constructing exponential approximations for $|\boldsymbol{x}-\boldsymbol{y}|^{-1}$.

Yarvin and Rokhlin (1996) developed exponential approximations for $|\boldsymbol{x}-\boldsymbol{y}|^{-1}$ using as the starting point the well-known integral representation (Morse and Feshbach, 1953)

$$\frac{1}{|\boldsymbol{x}-\boldsymbol{y}|} = \frac{1}{2\pi} \int_0^\infty d\rho \int_0^{2\pi} d\lambda \, \exp\left(-\rho|x_3-y_3|\right) \cos\left[\rho(x_1-y_1)\cos\lambda + \rho(x_2-y_2)\sin\lambda\right]. \tag{4}$$

Those approximations are obtained as quadrature rules in which particular attention is given to the location of the integration points with respect to ρ:

$$\frac{1}{|\boldsymbol{x}-\boldsymbol{y}|} \approx \sum_m \sum_n W_{mn} \exp\left(-\rho_m|x_3-y_3|\right) \cos\left[\rho_m(x_1-y_1)\cos\lambda_n + \rho_m(x_2-y_2)\sin\lambda_n\right].$$

Here W_{mn} are the integration weights corresponding to the quadrature points ρ_m and λ_n. It is important to mention that in the FSM of Greengard and Rokhlin (1997), the Yarvin-Rokhlin approximations are constructed such that their domain of validity reflects the geometry of interacting cells. Also Greengard and Rokhlin (1997) had to consider separately pairs of cells with $x_3 - y_3 > 0$ and $x_3 - y_3 < 0$.

In the remainder of this section, we develop a representation similar to (4) but for the fundamental solution of (1) rather than Laplace's equation. Our derivation follows closely that of Wang and Achenbach (1995) who exploited the Radon transform to obtain several useful representations for the fundamental solution of (1). Alternatively, one could follow the derivation of (4) given by Morse and Feshbach (1953). At this moment, the work on the integration scheme for the new representation is in progress and its results will be reported elsewhere.

We define the Fourier image $\bar{f}(\boldsymbol{s})$ of a function $f(\boldsymbol{x})$ as

$$\bar{f}(\boldsymbol{s}) = \int_{R^3} f(\boldsymbol{x}) \exp\left(-i\boldsymbol{s}\cdot\boldsymbol{x}\right) d\boldsymbol{x},$$

so that the inverse Fourier transform is defined as

$$f(\boldsymbol{x}) = \frac{1}{8\pi^3} \int_{R^3} \bar{f}(\boldsymbol{s}) \exp\left(i\boldsymbol{s}\cdot\boldsymbol{x}\right) d\boldsymbol{s}.$$

The fundamental solution of (1) is a second rank tensor $G_{ij}(\boldsymbol{x}-\boldsymbol{y})$ that satisfies the equation

$$C_{ijkl} G_{kr,lj}(\boldsymbol{x}-\boldsymbol{y}) + \delta(\boldsymbol{x}-\boldsymbol{y})\delta_{ir} = 0, \tag{5}$$

where $\delta(\boldsymbol{x}-\boldsymbol{y})$ is the delta function and δ_{ir} is the second rank identity tensor. Upon application of the Fourier transform to (5), we obtain

$$-s_l s_j C_{ijkl} \bar{G}_{kr}(\boldsymbol{s}) + \delta_{ir} = 0.$$

Thus the Fourier image $\bar{G}_{ij}(s)$ can be calculated as

$$\bar{G}_{ij}(s) = M_{ij}^{-1}(s) = \frac{A_{ij}(s)}{D(s)},$$

where

$$M_{ij}(s) = C_{ikjl} s_k s_l, \quad D(s) = \det[M_{ij}(s)], \quad \text{and} \quad A_{ij}(s) = \text{adj}[M_{ji}(s)]$$

As a result we obtain

$$G_{ij}(x-y) = \frac{1}{8\pi^3} \int_{R^3} \frac{A_{ij}(s)}{D(s)} \exp[is \cdot (x-y)] \, ds. \tag{6}$$

To obtain the desired representation for $G_{ij}(x-y)$ we evaluate (6) in cylindrical coordinates (ρ, λ, ξ) of the Fourier space:

$$s_1 = \rho\cos\lambda, \quad s_2 = \rho\sin\lambda, \quad s_3 = \xi.$$

In these coordinates, we can rewrite (6) as

$$\begin{aligned} G_{ij}(x-y) &= \frac{1}{8\pi^3} \int_0^\infty \rho \, d\rho \int_0^{2\pi} d\lambda \int_{-\infty}^\infty d\xi \, \frac{A_{ij}(\rho\cos\lambda, \rho\sin\lambda, \xi)}{D(\rho\cos\lambda, \rho\sin\lambda, \xi)} \\ &\quad \times \exp\{i[\rho(x_1-y_1)\cos\lambda + \rho(x_2-y_2)\sin\lambda + (x_3-y_3)\xi]\}. \end{aligned} \tag{7}$$

Next we integrate (7) with respect to ξ by observing that $D(\rho\cos\lambda, \rho\sin\lambda, \xi)$ is a polynomial of degree six in ξ and therefore the integration can be carried out using the residue theory. To simplify the calculations, we observe that $A_{ij}(s)$ and $D(s)$ are homogeneous polynomials in the Cartesian components of s of degree four and six, respectively. Thus we write

$$A_{ij}(s) = A_{ij}(\rho\cos\lambda, \rho\sin\lambda, \xi) = \rho^4 \hat{A}_{ij}(\lambda, \eta),$$

$$D(s) = D(\rho\cos\lambda, \rho\sin\lambda, \xi) = \rho^6 \hat{D}(\lambda, \eta),$$

with

$$\eta = \frac{\xi}{\rho}.$$

Here $\hat{A}_{ij}(\lambda, \eta)$ and $\hat{D}(\lambda, \eta)$ are polynomials of degree four and six in η, respectively; the coefficients of those polynomials are trigonometric functions of λ. Hence the roots of $D(\rho\cos\lambda, \rho\sin\lambda, \xi)$ with respect to ξ can be easily expressed in terms of the roots of $\hat{D}(\lambda, \eta)$ with respect to η. Wang and Achenbach (1993) have shown that for positive definite C_{ijkl} the roots of $\hat{D}(\lambda, \eta)$ cannot be real. Here we assume that the roots of $\hat{D}(\lambda, \eta)$ are distinct and denote those of them with positive imaginary parts by α^1,

α^2, and α^3. Now by replacing integration with respect to ξ with contour integration about these roots, we obtain the main result

$$G_{ij}(\boldsymbol{x} - \boldsymbol{y}) = \frac{\mathrm{i}}{4\pi^2} \int_0^\infty \mathrm{d}\rho \int_0^{2\pi} \mathrm{d}\lambda \, \exp\{\mathrm{i}\rho[(x_1 - y_1)\cos\lambda + (x_2 - y_2)\sin\lambda]\}$$
$$\times \sum_{m=1}^3 \frac{\hat{A}_{ij}(\lambda, \eta)}{\partial_\eta \hat{D}(\lambda, \eta)}\bigg|_{\eta=\alpha^m} \exp\left(\alpha^m \mathrm{i}\rho |x_3 - y_3|\right). \tag{8}$$

In (8) the absolute value appears because the contour integration involves the half-space $x_3 - y_3 > 0$. It is clear that the representation in (8) exhibits an exponential decay thanks to the imaginary parts of the roots α^m.

4. Applications of FSMs to Integral Equations of Isotropic Elasticity

Equations of isotropic elasticity can be obtained from (1) by imposing the restriction

$$C_{ijkl} = \frac{2\mu\nu}{1-2\nu}\delta_{ij}\delta_{kl} + \mu\delta_{ik}\delta_{jl} + \mu\delta_{il}\delta_{jk},$$

where μ and ν are material constants. In this case, the fundamental solution can be calculated in closed-form. Further, as it has been shown by Fu et al. (1998), the fundamental solution of isotropic elasticity can be related to $|\boldsymbol{x}-\boldsymbol{y}|^{-1}$ in such a way that the integral equation summation problem can be expressed as sixteen electrostatics summation problems. Therefore, in this case, one can choose any FSM for the electrostatics problem.

In this section we present computational results obtained with the code FLEMS (Fu et al., 1998) which relies on the FMM for matrix-vector multiplication. The problems we choose to solve involve identical spherical particles embedded in an infinite binder. In each problem, the spheres form a simple cubic array in which the number of spheres is between $3^3 = 27$ and $11^3 = 1331$ and the spacing between the two nearest spheres is equal to 2.5 radii (Fig. 1). The material properties are chosen such that $\mu^B = 5\mu^P$ and $\nu^B = \nu^P = 0.25$, where B represents the binder and P represents the particle. The binder is subjected to uniform tension at infinity. The principle objective in this section is to demonstrate that even these large problems can be quickly solved on a modest parallel computer.

In our computations we used the boundary element mesh which allows us to compute stresses on the surface of the particles to within 3% accuracy. The mesh is formed by 9-node isoparametric elements and the number of degrees of freedom per sphere is 1158 as shown in Figure 1. Within each element, integration is performed using 4×4 Gaussian and singular integration schemes. We used an FMM expansion depth of ten and the partitioning depth was chosen such that the number of charges in the smallest cells was close to one hundred. The iterations were performed using the GMRES algorithm, for which we set the stopping criterion equal to 5×10^{-5} in the relative l_2 norm. All problems were solved using 32 nodes of an IBM SP2, 160 MHz (640MFLOPs), 256 MB per node. In Figure 2, the upper and lower curve represent the CPU time needed to perform one matrix-vector multiplication using the $\mathcal{O}(N^2)$ method and the FMM. To this end, we observe that the number of matrix-vector multiplications for all problems was independent of the problem size. However, the $\mathcal{O}(N^2)$ method required

135

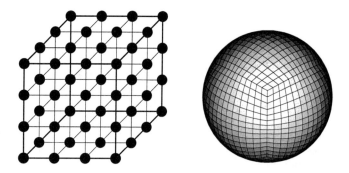

Figure 1: A simple 4x4x4 cubic lattice of spheres and the surface discretization for a single sphere.

six multiplications whereas the FMM required nine multiplications. The three additional multiplications were performed in order to compute the singularity of the double layer kernel (Fu, 1998). Nevertheless the additional multiplications are insignificant if one compares the performance of the two methods. Let us comment that the kinks in the lower curve represent the discrete nature of the partitioning strategy in which the number of charges per cell at the finest partitioning level is kept as close to one hundred as possible. Accordingly, each kink signifies a switch in the number of the partitioning levels.

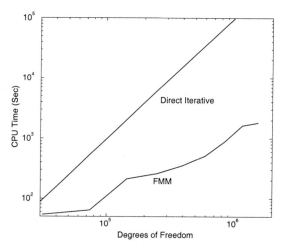

Figure 2: Matrix-vector multiplication for the $\mathcal{O}(N^2)$ method and the FMM.

It is expedient to mention that we used only 32 out of 128 processors available. If all the processors were used, the problem with 1000 spheres that involves more than one million unknowns could be solved in about an hour. Note that such a large problem can be very efficiently distributed across all processors. In comparison, the same problem requires about three weeks, if solved with the $\mathcal{O}(N^2)$ method on 128 processors. Perhaps we should not even theorize how long it would take to solve this problem using Gaussian elimination.

The fact that the chosen test problems required few iterations to achieve the desired degree of accuracy does not affect the superiority of FSMs over the $\mathcal{O}(N^2)$ method. To the contrary, the overhead associated with the three additional multiplications becomes even more insignificant as the number of iteration increases. Also it appears that there are preconditioning techniques for this class of problems that allow one to keep the number of iterations below one hundred even for rather challenging problems such as those involving long cylindrical particles (Fu, 1998).

5. Summary

In this paper, we considered fast summation methods applicable to integral equations corresponding to a general system of elliptic equations with constant coefficients. We identified two fast summation methods: one is a generalization of the Barnes and Hut (1986) method and the other is a generalization of the Greengard and Rokhlin (1997) method. We do not regard the proposed methods as competitors of each other – to the contrary we believe that their advantages and disadvantages are problem-dependent and therefore both methods deserve consideration. Also, using a conceptually simple example problem, we demonstrated the superiority of fast summation methods over conventional methods of linear algebra. In the general setting considered here, fast summation methods are particularly useful to problems arising in mechanics of anisotropic composite materials and mechanics of lattice defects in polycrystals.

6. Acknowledgments

We wish to thank our colleagues at TICAM, James C. Browne, Emery Berger, and Robert van de Geijn, for their contributions to the development of the code FLEMS. We are particularly grateful to Dr. C. Y. Wang of Schlemberger Research Laboratories for his advice in connection with the main result in Section 3.

References

[1] Barnes, J., Hut, P.: A hierarchical $O(NlogN)$ force-calculation algorithm, *Nature*, **324**, 446-449, (1986).

[2] Bacon, D. J., Barnett, D. M, Scattergood, R. O.: Anisotropic continuum theory of lattice defects, *Prog. Mater. Sci.*, 51-262, (1978).

[3] Fu, Y., Klimkowski, K. J., Rodin, G. J., Berger, E., Browne, J. C., Singer, J., van de Geijn, R. A., Vemaganti, K. S.: A fast solution method for three-dimensional many-particle problems of linear elasticity, *Int. J. Numer. Meth. Engng.*, **42**, 1215-1229, (1998).

[4] Fu, Y.: *Rapid Solution of Large-Scale Three-Dimensional Micromechanics Problems,* Ph.D. Dissertation, The University of Texas at Austin (1998).

[5] Fu, Y., Rodin, G. J.: A rapid solution method for Stokesian many-particle problem, submitted for publication (1998).

[6] Greengard, L., Moura, M.: On the numerical evaluation of elastostatic fields in composite materials, *Acta Numerica*, 379-410, (1994).

[7] Greengard, L., Rokhlin, V.: A fast algorithm for particle simulations, *J. Comput. Phys.*, **73**, 325-348, (1987).

[8] Greengard, L.: *The Rapid Evaluation of Potential Fields in Particle Systems*, The MIT Press (1988).

[9] Greengard, L., Rokhlin, V.: A new version of the fast multipole method for the Laplace equation in three dimensions, *Acta Numerica*, **6**, 229-270, (1997).

[10] Hackbusch, W., Nowak, Z.: On the fast matrix multiplication in the boundary element method by panel clustering, *Numer. Math.*, **54**, 463-491, (1989).

[11] Hayami, K., Sauter, S. A.: Application of the panel clustering method to the three-dimensional elastostatic problem, *Proceedings of the International Conference on Boundary Element Method*, Univ. of Tokyo, 625-634, (1997).

[12] Lage, C., Schwab, C.: On the implementation of a fully discrete multiscale Galerkin BEM, presented at *BEM 19 Conference*, Rome (1997).

[13] Morse, P. F., Feshbach, H.: *Methods of Theoretical Physics*, McGraw-Hill, New York, 1255-1256, (1953).

[14] Mura, T.: *Micromechanics of Defects in Solids*, Martinus Nijhoff, Dordrecht (1982).

[15] Nabors, K., Korsmeyer, F. T., Leighton, F. T., White, J.: Preconditioned, adaptive, multipole-accelerated iterative methods for three-dimensional first-kind integral equations of potential theory, *SIAM, J. Sci. Stat. Comput.*, **15**, 714-731, (1994).

[16] Nishimura, N., Yoshida, K., Kobayashi, S.: A fast multipole boundary integral equation method for crack problems in 3D, to appear in *Engng. Analy. Boundary Elem.*, (1998).

[17] Rodin, G. J.: Toward rapid evaluation of the elastic interactions among three-dimensional dislocations, *Phil. Mag. Letters A*, **77**, No. **4**, 187-190, (1998).

[18] Schmidt, K. E., Lee, M. A.: Implementing the fast multipole method in three dimensions, *J. Statist. Phys.*, **63**, 1223-1235, (1991).

[19] Sangani, A. S., Mo, G.: An $\mathcal{O}(N)$ algorithm for Stokes and Laplace interactions of particles, *Phys. Fluids*, **8**, 1990-2010, (1996).

[20] von Petersdorff, T., Schwab, C.: Fully discrete multiscale Galerkin BEM, *Multiscale Wavelet Methods for Partial Differential Equation*, edited by Dahmen, W., Kurdila, A. and Oswald, P., Academic Press, 287-346, (1997).

[21] Wang, C. Y., Achenbach, J. D.: Three-dimensional time-harmonic elastodynamic Green's functions for anisotropic solid, *Proc. R. Soc. Lond. A*, 441-458, (1995).

[22] Wang, C. Y., Achenbach, J. D.: New method to obtain 3D Green's functions for anisotropic solids, *Wave Motion*, **18**, 273-289, (1993).

[23] Warren, M. S., Salmon, J. K.: Astrophysical N-body simulations using hierarchical tree data structures, In: *Supercomputing '92*, Minneapolis, MN, IEEE Comput. Soc. Press, 570-576, (1992).

[24] Warren, M. S., Salmon, J. K.: A parallel hashed oct-tree N-body algorithm, In: *Supercomputing '93*, Washington, DC, IEEE Comput. Soc. Press, 12-21, (1993).

[25] Yarvin, N., Rokhlin V.: Generalized Gaussian quadratures and singular value decompositions of integral operators, research report YALEU/DCS/RR-1109, (1996).

[26] Yoshida, K., Nishimura, N., Kobayashi, S.: Analysis of three-dimensional elastostatic crack problems with fast multipole boundary integral equation method, to appear in *J. Appl. Mech.*, JSCE, **1** (1998).

Y. Fu, J. R. Overfelt:
Department of Aerospace Engineering and Engineering Mechanics
The University of Texas at Austin
Austin, Texas 78712

G. J. Rodin:
Texas Institute for Computational and Applied Mathematics
The University of Texas at Austin
Austin, Texas 78712, USA
Email: gjr@ticam.utexas.edu

I.G. GRAHAM, W. HACKBUSCH and S.A. SAUTER

Hybrid Galerkin Boundary Elements on Degenerate Meshes

1. Introduction

In recent work ([4],[5]) the authors have presented a new discretisation scheme for boundary integral equations which has the same energy norm stability and convergence properties as the Galerkin method but has a complexity comparable with discrete collocation or Nyström methods. The results in [4],[5] were for non-quasiuniform but nevertheless shape-regular meshes. Here we extend the theory to much more general meshes, including the degenerate meshes commonly used to handle singularities arising from corners and edges in 3D applications. As an application we give numerical results for the classical problem of computing the capacitance of a two-dimensional plate in $\mathbb{R}^3$. These show that the method is capable of attaining the same type of complexity reduction for singular problems as was already attained for smooth applications in [5].

Our new method is an approximate Galerkin method, where the integrals arising in the matrix are computed by quadrature. The important novel feature is that the matrix is (adaptively) partitioned into (i) a small "conventional Galerkin" part (near the diagonal), in which standard triangle-based quadrature rules are used to compute the singular or nearly singular integrals, and (ii) the remainder, which is done by novel node-based rules. In effect, in (ii), suitable approximations to the Galerkin matrix entries are computed from local weighted averages of elements of the "Nyström matrix" (formed from kernel evaluations only at nodal points of the mesh). This has a low complexity compared with the method in (i), since triangle-based rules typically use (often many) additional kernel evaluations at non-nodal points. The ability of the new method to compute most of the matrix using the cheaper method is the key to its competitiveness. The term "*hybrid*" is used because, in a certain way, the new method is a blend of two classical methods: Galerkin and Nyström. We note that the word "hybrid" is also used to describe various other methods, but, as far as we are aware, these have no connection with the "hybrid Galerkin method" proposed here.

Nevertheless we want to mention here that our work has connections to two earlier works [12, 1]. In particular, [12] considers 2D boundary integral equations with operator composed of the sum of a convolution and a (smooth) compact perturbation. The convolution part is discretised by the Galerkin method, yielding a Toeplitz matrix which can be assembled very efficiently in a semi-discrete manner ([12, p.257]). The smooth perturbation and the the right-hand side function are discretised by node-based quadrature analogous to that proposed below. In a different but related direction, [1] proves stability for nodal collocation onto splines of odd degree (again for integral equations on 2D contours) by utilising a discrete integration by parts argument to show that the collocation equations are equivalent to a non-standard Galerkin method. Thus [1] utilises the link between the discrete Galerkin and collocation methods. We also exploit this link, but our method considers in fact fully discrete quadrature errors and more precisely exploits an analogous link between the discrete Galerkin and Nyström methods. In addition our work concerns integral equations on 3D surfaces.

This work was supported by British Council / DAAD ARC Grant 869.

Thus, in general we shall consider (first or second kind) integral equations of the form

$$(\lambda I + \mathcal{K})u(\mathbf{x}) := \lambda u(\mathbf{x}) + \int_\Gamma k(\mathbf{x}, \mathbf{y})u(\mathbf{y})dy = f(\mathbf{x}), \quad \mathbf{x} \in \Gamma, \tag{1}$$

where $\lambda \in \mathbb{R}$, $\Gamma = \cup\{\overline{\Gamma}_\ell : \ell = 1, \ldots, L\}$ is the surface of a polyhedron in $\mathbb{R}^3$ with (open) polygonal faces Γ_ℓ. (The method and analysis given here extends to general piecewise smooth (curved) Lipschitz surfaces - as described in the case of non-degenerate meshes in [4, 5] - but for brevity's sake (only), we do not give that generality here.)

For (1), we assume that for some $\mu \in [-1, 1]$, $\lambda I + \mathcal{K} : H^\mu \to H^{-\mu}$. We also assume that the bilinear form $a(u, v) := ((\lambda I + \mathcal{K})u, v)$ is elliptic in H^μ and that $f \in H^{-\mu}$. (Here $(\cdot, \cdot)$ denotes the usual L_2 inner product on Γ.) Finally we assume also that the kernel k has the characteristic singularity which arises in fundamental solutions of elliptic PDEs (or derivatives of fundamental solutions), i.e., for some $\alpha \in \mathbb{R}$ and all $m \in \mathbb{N}_0$, $|D^m k(\mathbf{x}, \mathbf{y})| \leq B_m |\mathbf{x} - \mathbf{y}|^{-\alpha - m}$ holds, where B_m is a constant and D^m denotes any mth order partial differential operator with respect to $(\mathbf{x}, \mathbf{y}) \in \Gamma_\ell \times \Gamma_{\ell'}$. Note that this ensures that $k \in C^\infty(\Gamma_\ell \times \Gamma_{\ell'})$ when $\ell \neq \ell'$, but it allows k to be discontinuous across the edges of Γ (as happens, for example, with the double layer potential kernel).

2. Discrete Galerkin Methods

The main object of this paper is to extend the theoretical results of [4, 5] to more general (irregular) mesh sequences, including the anisotropic graded meshes typically used to handle edge and corner singularities.

We want to emphasise that the hybrid Galerkin method proposed here can in practice be applied directly to general boundary integral equations on curved surfaces, discretised using discontinuous or continuous elements on triangular or quadrilateral meshes. For illustrative purposes we restrict here to the model case of piecewise flat surfaces and continuous linear triangular elements. More general theory is given in [4, 5] and further extensions will be addressed in a forthcoming paper.

For the approximation of (1), we introduce a family of triangular meshes $\mathcal{T}_h$ ($h \to 0$) on Γ, each of which consists of open planar triangles $\tau \in \mathcal{T}_h$. (We assume that the edges of Γ consist of triangle edges.) The nodes of $\mathcal{T}_h$ are denoted $\{\mathbf{x}_p : p \in \mathcal{N}\}$, with $\mathcal{N}$ denoting a suitable index set. We set $N = \#\mathcal{N}$. For any $\tau \in \mathcal{T}_h$, we let $|\tau|$ denote its area, h_τ its diameter and r_τ its smallest perpendicular height:

$$r_\tau = 2|\tau|/h_\tau \tag{2}$$

For $p \in \mathcal{N}$, $\mathcal{T}_h(p) := \{\tau \in \mathcal{T}_h : \mathbf{x}_p \in \overline{\tau}\}$. The parameter h is identified with the maximum mesh diameter: $h = \sup\{h_\tau : \tau \in \mathcal{T}_h\}$. We also introduce the "local mesh diameters": $h_p = \max\{h_\tau : \tau \in \mathcal{T}_h(p)\}$. As in [4], if A and B are any two quantities which depend (implicitly or explicitly) on h, we use the notation $A \lesssim B$ to mean that $A \leq \text{const} B$, with const denoting a constant independent of h.

The chief new theoretical contribution of this paper is to develop the theory of hybrid Galerkin methods for meshes which satisfy only very weak regularity assumptions as $h \to 0$. Specifically we require (with l.q.u. an abbreviation for *locally quasiuniform*):

(A1) The number of triangles adjoining any node is bounded: $\max_{p \in \mathcal{N}}\{\#\mathcal{T}_h(p)\} \lesssim 1$.
(A2) The triangle areas are *l.q.u.*: $\max_{p \in \mathcal{N}} \max_{\tau, \tau' \in \mathcal{T}_h(p)}\{|\tau|/|\tau'|\} \lesssim 1$.

(**A3**) The minimal heights are *l.q.u.*: $\max_{p\in\mathcal{N}} \max_{\tau,\tau'\in\mathcal{T}_h(p)}\{r_\tau/r_{\tau'}\} \lesssim 1$.

It is easy to show (see, e.g. [4, §2.5]) that shape-regular meshes satisfy (**A1**)-(**A3**). However more general meshes are also allowed, in particular the degenerate meshes which result from anisotropic algebraic or geometric refinement used (respectively) in the $h-$ or $hp-$approximations of solutions with edge singularities (e.g. [7, 3, 10]). Using these meshes, we define $\mathcal{S}_h$ to be the space of continuous piecewise linear functions on Γ with respect to $\mathcal{T}_h$, spanned by the usual basis $\{\phi_p : p \in \mathcal{N}\}$ with $\phi_p(\mathbf{x}_q) = \delta_{p,q}$, for all $p, q \in \mathcal{N}$.

The *Galerkin method* for (1) seeks $U \in \mathcal{S}_h$ such that

$$a(U, V) = (f, V), \quad \text{for all } V \in \mathcal{S}_h, \tag{3}$$

With the expansions $U = \sum_{q\in\mathcal{N}} U_q \phi_q$ and $V = \sum_{p\in\mathcal{N}} V_p \phi_p$, we have

$$a(U, V) = \sum_{p\in\mathcal{N}}\sum_{q\in\mathcal{N}} U_q(\lambda M_{p,q} + K_{p,q}) V_p, \tag{4}$$

where $M_{p,q} = (\phi_q, \phi_p)$ and

$$K_{p,q} = (\mathcal{K}\phi_q, \phi_p) = \int_\Gamma \int_\Gamma k(\mathbf{x}, \mathbf{y}) \phi_q(\mathbf{y}) \phi_p(\mathbf{x}) dy dx, \tag{5}$$

the exact Galerkin method (3) reduces to the solution of a linear system with coefficient matrix $(\lambda M_{p,q} + K_{p,q})_{p,q\in\mathcal{N}}$.

Discrete Galerkin methods arise when the matrix elements are approximated by quadrature. The (sparse) M is relatively trivial to compute, but in almost all 3D engineering applications it is necessary to approximate K by quadrature. In general we let $\tilde{a}$ denote the result of replacing $K_{p,q}$ in (4) by some approximation $\tilde{K}_{p,q}$:

$$\tilde{a}(U, V) = \sum_{p\in\mathcal{N}}\sum_{q\in\mathcal{N}} U_q(\lambda M_{p,q} + \tilde{K}_{p,q}) V_p. \tag{6}$$

Then the corresponding *discrete Galerkin* solution $\tilde{U}$ is defined to be the solution of (3) with a replaced by $\tilde{a}$. (Related semidiscrete methods are discussed in [11].)

To define suitable $\tilde{K}_{p,q}$, let us first restrict attention to the (most common) case when $\text{supp}\phi_p \cap \text{supp}\phi_q = \emptyset$. Then in (5) the kernel k is smooth on each of the faces of Γ but may be discontinuous across edges. We shall build approximations to $K_{p,q}$ from fundamental quadrature rules of the form:

$$\int_\Gamma F\phi_p \approx \sum_{j\in J_p} w_j F(\mathbf{z}_j), \quad \text{for each } p \in \mathcal{N}. \tag{7}$$

Here J_p is a finite index set, w_j are weights and $\mathbf{z}_j$ are quadrature points on Γ which are "near" supp ϕ_p. The integral (5) can then be computed by applying such rules for the integrals over $\mathbf{x}$ and $\mathbf{y}$, yielding

$$\tilde{K}_{p,q} = \sum_{j\in J_p}\sum_{j'\in J_q} w_j k(\mathbf{z}_j, \mathbf{z}_{j'}) w_{j'}. \tag{8}$$

Because of the regularity of k, (7) has to make sense for functions F which are discontinuous across edges of Γ. When any $\mathbf{z}_j$ lies on such an edge, the notation $F(\mathbf{z}_j)$ in (7) should be understood as an abbreviation for the *limit* of $F(\mathbf{x})$ as $\mathbf{x} \to \mathbf{z}_j$ from within *one* particular face of Γ. With this convention the quantity $F(\mathbf{z}_j)$ may appear more than once in (7), but with different weights according to which face the limiting value is taken from. The implicit notation (7) is sufficient for theoretical purposes. More precise practical details are given in [4].

Let $\mathcal{P}_d(\Gamma)$ denote the discontinuous functions on Γ which reduce to bivariate polynomials of degree d on each of the faces Γ_ℓ. Then we will be interested in rules (7) which satisfy the assumptions (for some $d \in \mathbb{N}_0$ and $\sigma, \delta \geq 1$ independent of h):

(**A4**) *"Degree of Precision"*: Equality holds in (7) for all $F \in \mathcal{P}_d(\Gamma)$;

(**A5**) *"Stability"*: $\sum_{j \in J_p} |w_j| \leq \sigma \sum_{j \in J_p} w_j$, for all p ;

(**A6**) *"Compactness"*: $\sup_{j \in J_p} |\mathbf{z}_j - \mathbf{x}_p| \leq \delta h_p$, for all p .

Examples of rules satisfying (**A4**) - (**A6**) for any $d \in \mathbb{N}$, with $\sigma = \delta = 1$ are furnished by separating the left-hand side of (7) into a sum of integrals over the triangles $\tau \in \mathcal{T}_h(p)$ and then applying suitable rules to integrate $F\phi_p$ over triangles. An alternative (with reduced complexity, as mentioned above) is to construct (7) using only quadrature points chosen exclusively from the set of nodal points $\{\mathbf{x}_p : p \in \mathcal{N}\}$ of the mesh and which have the exactness property (**A4**) for some required d.

To see how to do this, suppose, for example, that supp ϕ_p lies entirely within one $\overline{\Gamma}_\ell$. Then to achieve (**A4**) for $d = 1$ we would choose 3 nodes $\{\mathbf{x}_j\}$ near $\mathbf{x}_p$ and require (7) to hold for all functions F which are polynomials of degree 1 on Γ_ℓ. This leads to a 3 × 3 system to be solved for the weights w_j. Increasing d leads to larger systems for the weights but, for any fixed d, the cost of doing this for all $p \in \mathcal{N}$ is still only $O(N)$. For these "node-based" rules, assumptions (**A4**), (**A6**) can in principle be satisfied directly by construction, but nonsingularity of the systems to be solved for the weights (and the consequent size of the stability constant σ) is, in general, unknown *a priori*. In [5] (see also §5 below), we give a very detailed theory on this question, which shows that for $d \leq 2$, all "reasonable" choices of quadrature points lead to a stability constant σ which is bounded independently of h.

Remark 1. If supp ϕ_p does not lie entirely within one face of Γ then it will be necessary to split the support up into components belonging to each of the smooth faces and construct appropriate quadrature rules on each face (see also [4]).

It should now be clear why the use of node-based rules can decrease the complexity of computing the matrix K: If kernel evaluations within triangles are used then this will automatically be more expensive than methods which confine the evaluations to the nodes and in many cases a reduction in kernel evaluations by a factor of between 10 and 70 is possible using node-based rules (see [5]). (For quadrilateral elements the same principle applies, although the complexity achievable depends on the individual choices of quadrature rules.)

So far we have not discussed the fundamental question of how the degree of precision of the rules used for computing each $K_{p,q}$ should be chosen in order to preserve the optimal order of convergence of the underlying Galerkin method. We show below that in order to achieve this, d needs only to be quite low (e.g. 1 or 2) in the "far field" (where $\mathbf{x}_p$ and $\mathbf{x}_q$ are well-separated), and d needs to increase modestly as the distance between $\mathbf{x}_p$ and $\mathbf{x}_q$ decreases. In the *singular case* supp $\phi_p \cap$ supp $\phi_q \neq \emptyset$, regularising transforms and triangle-based rules must be used [8, 10]. We now give an algorithm in which efficient quadrature rules which maintain optimal convergence rates are determined automatically. It has a first phase in which node-based rules satisfying (**A1**)-(**A3**) for some (small) choices of d are computed. Then in the second phase these are used to compute (5), provided this integral is sufficiently far from being singular. In determining what degree of precision should be employed the following quantities are useful: $h_{p,q} := \max\{h_p, h_q\}$, and

$$\rho_{p,q} = \min\{|\mathbf{x} - \mathbf{y}| : \mathbf{x} \in \text{supp } \phi_p \cup \{\mathbf{z}_j : j \in J_p\}, \ \mathbf{y} \in \text{supp } \phi_q \cup \{\mathbf{z}_{j'} : j' \in J_q\}\}.$$

The algorithm requires the user to choose the following parameters: real numbers $\delta, \sigma, C^* \geq 1$ and positive integers $d_{\min} \leq d_{\max}$. Also set $\chi = 1 - \mu - \min\{\mu, 0\}$, where μ is as defined in §1.

Hybrid Galerkin Algorithm
procedure generate_node_based_quadrature_rules;
begin
    for all $p \in \mathcal{N}$ **do**
        for all integers $d \in [d_{\min}, d_{\max}]$ **do**
        (i) Select $(d+1)(d+2)/2$ nodes $\{\mathbf{x}_j : j \in J_p = J_p(d)\}$ on Γ near $\mathbf{x}_p$.
        (ii) Find $\{w_j : j \in J_p(d)\}$ such that (7) is exact for polynomials F of degree d.
        (iii) If the w_j satisfy (**A5**),(**A6**):
            set $B_p(d) =$ "admissible" and store the $J_p(d)$ and w_j.
        (iv) If the w_j do not satisfy (**A5**),(**A6**), set $B_p(d) =$ "inadmissible".
        end of loop over d
    end of loop over p
end;

Remark 2. If $\mathbf{x}_p$ is an edge node of Γ, the procedure above must be slightly modified (see Remark 1).

Remark 3. Step (ii) requires solution of a square system. It is also permissible to select more than $(d+1)(d+2)/2$ quadrature points, provided the resulting underdetermined system is solved by a minimal norm algorithm [5, §3.3].

The second phase of the algorithm then follows:
procedure generate_hybrid_system_matrix;
begin
    for all $p, q \in \mathcal{N}$ **do**
    begin
        • Compute d_{pq}, the smallest positive integer satisfying

$$d_{p,q} \geq \frac{\chi + (1+\alpha)\log(C^*\rho_{p,q})/\log h_{p,q}}{1 - \log(C^*\rho_{p,q})/\log h_{p,q}} \quad (9)$$

if $(d_{p,q} \in [d_{\min}, d_{\max}])$ and $(B_p(d_{pq}) =$ "admissible") and $(B_q(d_{pq}) =$ "admissible")
then
        • Compute $\tilde{K}_{pq}$ using the node-based rules obtained above.

else
- Compute $\tilde{K}_{pq}$ using triangle-based rules, and ensure that

$$|K_{p,q} - \tilde{K}_{p,q}| \leq C h_{p,q}^{\chi+1} s_p s_q \ . \tag{10}$$

end
end;

Remark 4. (10) is always possible, with C independent of h, using triangle-based rules (see [2, 8, 10]).

In the next section we prove the stability and convergence of this algorithm.

3. Convergence Theory

Since Galerkin stability follows from the continuity and ellipticity of a, discrete Galerkin stability can be obtained using the "Strang Lemma", provided the error $a - \tilde{a}$ is suitably small on $\mathcal{S}_h \times \mathcal{S}_h$. The estimation of this error often makes use of *inverse inequalities* (especially when $\mu < 0$) and for these, the meshes normally need to be quasiuniform. However in [4, §3] we obtained more general norm inequalities which play the role of inverse estimates in the theory but which hold true for meshes assumed only to be shape-regular. It turns out that these norm inequalities can be further generalised to include degenerate meshes and the results are as follows.

Recalling (2), we define, for $p \in \mathcal{N}$, $s_p = \int_\Gamma \phi_p = \sum_{\tau \in \mathcal{T}_h(p)} |\tau|/3$ and $r_p = \max\{r_\tau : \tau \in \mathcal{T}_h(p)\}$. Then, for any $t \in \mathbb{R}$, define the weighted ℓ_2-norms $\|\cdot\|_{\ell_2,t}$ by:

$$\|\boldsymbol{V}\|_{\ell_2,t}^2 = \sum_{p \in \mathcal{N}} r_p^{2t} s_p |V_p|^2, \quad \boldsymbol{V} = (V_p)_{p \in \mathcal{N}} \in \mathbb{R}^N \ .$$

We also define $\mathcal{I} : \mathbb{R}^N \to \mathcal{S}_h$, by: $\mathcal{I}\boldsymbol{V} = \sum_{p \in \mathcal{N}} V_p \phi_p$. The following result is a generalisation of [4, Theorem 3.5]. Its proof will be given in a forthcoming paper.

Theorem 1 *For all $t \in [0, 1]$ and $\boldsymbol{V} \in \mathbb{R}^N$,*

$$\|\boldsymbol{V}\|_{\ell_2,t} \lesssim \|\mathcal{I}\boldsymbol{V}\|_{H^{-t}} \lesssim \|\boldsymbol{V}\|_{\ell_2,0} \lesssim \|\mathcal{I}\boldsymbol{V}\|_{H^t} \lesssim \|\boldsymbol{V}\|_{\ell_2,-t} \ .$$

From this we can prove the following result, which estimates the error $a - \tilde{a}$ in the form required by the Strang Lemma.

Theorem 2 *Let $\tilde{K}$ be obtained by the hybrid Galerkin algorithm and define $\tilde{a}$ by (6). Suppose $h < 1$, $t \in [-1, 1]$ and set $t^* = \min\{t, 0\}$. Then*

$$\frac{|a(V,W) - \tilde{a}(V,W)|}{\|V\|_{H^t} \|W\|_{H^t}} \lesssim C h^{\chi+1} \sum_{\tau \in \mathcal{T}_h} h_\tau^{-2t^*} |\tau|^{1+2t^*} \tag{11}$$

$$\frac{|a(V,W) - \tilde{a}(V,W)|}{\|V\|_{H^0} \|W\|_{H^t}} \lesssim C h^{\chi+1} \left\{ \sum_{\tau \in \mathcal{T}_h} h_\tau^{-2t^*} |\tau|^{1+2t^*} \right\}^{1/2}, \tag{12}$$

with C denoting a generic constant which depends on C^*, δ and σ.

Proof. We start by estimating the errors $|K_{pq} - \widetilde{K}_{pq}|$. First, if $\widetilde{K}_{pq}$ is computed by a node-based rule with degree of precision d_{pq}, then $B_p(d_{pq}) = B_q(d_{pq}) =$ "admissible", $d_{p,q}$ satisfies (9) and $d_{p,q} \in [d_{\min}, d_{\max}]$. From these it follows that $\log(C^*\rho_{p,q})/\log h_{p,q} < 1$. Then, since $h_{p,q} \leq h < 1$, a simple rearrangement of (9) shows that

$$\rho_{p,q} \geq (C^*)^{-1} h_{p,q}^{(d_{p,q}-\chi)/(d_{p,q}+1+\alpha)} . \tag{13}$$

Now, a careful examination of the arguments in [4] shows that [4, Lemma 4.3], remains true even for the more general meshes satisfying (**A1**) - (**A3**), i.e., there exists $C = C(\delta, \sigma)$ such that

$$|K_{p,q} - \widetilde{K}_{p,q}| \leq C h_{p,q}^{d_{p,q}+1} \rho_{p,q}^{-(d_{p,q}+1+\alpha)} s_p s_q . \tag{14}$$

Combining (14) with a simple rearrangement of (13) yields

$$\left| K_{pq} - \widetilde{K}_{pq} \right| \leq C h_{p,q}^{\chi+1} s_p s_q . \tag{15}$$

In fact (15) holds even when triangle-based rules are used because of the explicit condition (10).

Now we can proceed with the proofs of (11), (12). First observe that by (15),

$$|a(V,W) - \tilde{a}(V,W)| \leq \sum_{p,q \in \mathcal{N}} |V_q| |K_{p,q} - \widetilde{K}_{p,q}| |W_p|$$

$$\lesssim C h^{\chi+1} \sum_{q \in \mathcal{N}} s_q |V_q| \sum_{p \in \mathcal{N}} s_p |W_p| . \tag{16}$$

To bound (16), observe that, for any $t \in [-1, 1]$ and $t^* = \min\{t, 0\}$, we have, using Cauchy-Schwarz and Theorem 1,

$$\sum_{p \in \mathcal{N}} s_p |W_p| = \sum_{p \in \mathcal{N}} r_p^{-t^*} s_p^{1/2} |W_p| r_p^{t^*} s_p^{1/2}$$

$$\lesssim \left\{ \sum_{p \in \mathcal{N}} r_p^{-2t^*} s_p |W_p|^2 \right\}^{1/2} \left\{ \sum_{p \in \mathcal{N}} r_p^{2t^*} s_p \right\}^{1/2} \lesssim \left\{ \sum_{p \in \mathcal{N}} r_p^{2t^*} s_p \right\}^{1/2} \|W\|_{H^{t^*}} . \tag{17}$$

Now note that, by definition of r_p, we have $r_p^{2t^*} \leq r_{\tau'}^{2t^*}$ for all $\tau' \in \mathcal{T}_h(p)$. Hence, using (**A3**), we have

$$r_p^{2t^*} s_p \leq r_{\tau'}^{2t^*} \sum_{\tau \in \mathcal{T}_h(p)} |\tau|/3 \lesssim \sum_{\tau \in \mathcal{T}_h(p)} r_\tau^{2t^*} |\tau| .$$

Now, using the definition of r_τ (2), it follows that

$$\sum_{p \in \mathcal{N}} r_p^{2t^*} s_p \lesssim \sum_{p \in \mathcal{N}} \sum_{\tau \in \mathcal{T}_h(p)} h_\tau^{-2t^*} |\tau|^{1+2t^*} . \tag{18}$$

Now substitution of (18) into (17) and noting that $\|V\|_{H^{t^*}} \le \|V\|_{H^t}$, yields

$$\sum_{p \in \mathcal{N}} s_p |W_p| \lesssim \left\{ \sum_{\tau \in \mathcal{T}_h} h_\tau^{-2t^*} |\tau|^{1+2t^*} \right\}^{1/2} \|W\|_{H^t} . \tag{19}$$

Using (19) twice in (16) yields (11). Also, since (19) holds for any t and W we have, in particular,

$$\sum_{q \in \mathcal{N}} \int \phi_q |V_q| \lesssim \left\{ \sum_{\tau \in \mathcal{T}_h} |\tau| \right\}^{1/2} \|V\|_{H^0} \lesssim \|V\|_{H^0} . \tag{20}$$

Combining (19), (20) with (16) yields (10).

Combining the result of Theorem 2 with the Strang Lemma (e.g. [4]) we have
Corollary 3 *Let $\mu^* = \min\{\mu, 0\}$. If the mesh sequence $\mathcal{T}_h$ satisfies*

$$h^{\chi+1} \sum_{\tau \in \mathcal{T}_h} h_\tau^{-2\mu^*} |\tau|^{1+2\mu^*} \to 0 , \quad \text{as } h \to 0 , \tag{21}$$

then the hybrid Galerkin solution $\tilde{U}$ is stable and satisfies the error bound:

$$\left\| u - \tilde{U} \right\|_{H^\mu} \lesssim \inf_{V \in S_h} \left\{ \|v - V\|_{H^\mu} + h^{\chi+1} \left\{ \sum_{\tau \in \mathcal{T}_h} h_\tau^{-2\mu^*} |\tau|^{1+2\mu^*} \right\}^{1/2} \right\} . \tag{22}$$

Remark 5. Recalling that $\chi = 1 - \mu - \mu^*$, it is easy to see that the stability condition (21) is automatically satisfied when $\mu = 0$ or $\mu = 1/2$ (double layer or hypersingular operator arising from the Laplacian), and that the stability condition when $\mu = -1/2$ (single layer) is $h^3 \sum_{\tau \in \mathcal{T}_h} h_\tau \to 0$. This is a very mild condition. Indeed we shall see in the next section that it is satisfied for strongly anisotropic meshes, and indeed the error estimate (22) turns out to be optimal in this case.

4. Numerical Experiment: Capacitance of a plate

In this section we report on some numerical experiments for the hybrid Galerkin algorithm applied to the single layer potential equation:

$$\mathcal{K}u(\mathbf{x}) := -\frac{1}{4\pi} \int_\Gamma \frac{1}{|\mathbf{x} - \mathbf{y}|} u(\mathbf{y}) dy = f(\mathbf{x}) , \quad \mathbf{x} \in \Gamma , \tag{23}$$

where $\Gamma = \{(x_1, x_2, 0)^T : x_1, x_2 \in [0,1]\}$ is the two-dimensional unit square in $\mathbb{R}^3$. We will be concerned with the special case $f \equiv 1$, in which case the capacitance κ of the plate is given by the formula:

$$\kappa = -\frac{1}{4\pi} \int_\Gamma u(\mathbf{x}) dx . \tag{24}$$

(See also [11].) For this problem the parameters μ (the Sobolev index of the energy norm) and α (the rate of blow-up of the kernel) are $-1/2$ and 1, respectively.

It is known that the solution u of this problem has corner and edge singularities. In fact, as $\mathbf{x} \in \Gamma$ approaches an interior point $\mathbf{x}_0$ of an edge (in the normal direction), then $u(\mathbf{x}) = O(|\mathbf{x} - \mathbf{x}_0|^{-1/2})$. (This edge singularity is more severe than any arising from problem (23), when Γ is any (closed) Lipschitz polyhedron - see, e.g., [7]. The optimal approximation of these singularities can be obtained only by appropriate mesh refinement near the edges of Γ. This may be implemented either *adaptively* or *a priori*. In this experiment we shall implement a sequence of *a priori* graded meshes which yield optimal approximations for this problem, with a complexity (in terms of numbers of degrees of freedom) which is the same as that of uniform meshes. This low complexity arises because the meshes are strongly *anisotropic* i.e., near the interior portions of edges they are highly refined in the direction orthogonal to the edge, but are more or less uniformly subdivided in the direction tangential to the edges. To define these meshes, for each integer $n \geq 1$, and a *grading exponent* $g \geq 1$, we introduce the points in $[0,1]$: $t_i = (i/n)^g/2$, and $t_{2n-i} = 1 - (i/n)^g/2$, for $i = 0, \ldots, n$. Then, for $i,j = 1, \ldots 2n+1$, we introduce the nodes of the tensor-product mesh on Γ: $\mathbf{x}_{i,j} = (t_i, t_j)^T$. From these we construct a triangular mesh by dividing each of the rectangles into two using its diagonal (drawn from bottom left to top right). As g increases these 2D meshes become more anisotropic, with the most extreme triangles having an aspect ratio of about gn^{g-1}. Nevertheless these meshes satisfy assumptions (**A1**)-(**A3**) above. In [7] it was shown that if $\mathcal{S}_n$ denotes the discontinuous piecewise linear basis functions defined on these meshes, then there exists $\xi_n \in \mathcal{S}_n$ such that $\|U - \xi_n\|_{-1/2} \leq C(1/n)^{\min(g-\epsilon,5)/2}$ (for arbitrary $\epsilon > 0$). In addition, in [3] it is shown that if U_n denotes the Galerkin solution to (23) in any subspace of $H^{-1/2}$, and if κ_n denotes the corresponding approximation to the capacitance obtained by substituting U_n into (24), then $|\kappa - \kappa_n| \leq C\|u - U_n\|_{-1/2}^2$. Combining these two inequalities, we expect that with discontinuous piecewise linear basis functions,

$$|\kappa - \kappa_n| \leq C(1/n)^{\min(g-\epsilon,5)}, \text{ for } g \geq 1 \text{ and } \epsilon > 0. \qquad (25)$$

Since the space of continuous piecewise linear functions is contained in the space of discontinuous piecewise linears we cannot expect a better rate than (25) for our discretisation. Our experiments indicate that (25) remains true in the discontinuous case. Assuming this, since $\mu = \mu^* = -1/2$ and $\chi = 2$, the second term on the right hand side of (22) can be shown to be $O(1/n)^{5/2}$ and condition (21) holds. Then, by Corollary 3, $\|u - \tilde{U}\|_{-1/2} = O(1/n)^{5/2}$ and so the capacitance $\tilde{\kappa}_n$ obtained by inserting the hybrid Galerkin solution $\tilde{U}$ into (24) also satisfies (25). Our experiments also indicate this to be true.

In the following experiments we implement the hybrid Galerkin algorithm with parameter choice and $d_{\min} = d_{\max} = 2$ and various C^*. For this choice of parameters the right-hand side of (9) is bounded above by 2 (as $n \to \infty$) for p, q satisfying

$$\rho_{p,q} \geq 1/C^* . \qquad (26)$$

Thus in our experiments the condition (9) is replaced by (26), i.e. node-based rules are used only for p, q satisfying (26).

In the most general version of the hybrid algorithm the parameters δ and σ in (**A5**), (**A6**) depend upon the choice of points in the node-based rules. In these experiments we choose quadrature points around each node (depicted below) which satisfy (**A6**) for $\delta = 2$. Moreover the theory in [5] can be used to show that for our choices of quadrature points, (**A5**) holds with σ independent of n (but dependent on g). We choose here not to stipulate in advance an upper bound for σ but rather we run the code and record the computed smallest value of σ for which (**A5**) holds.

To construct node-based quadrature rules of degree of precision 2 we need to choose at least 6 nodes near the reference node $\mathbf{x}_p$. In fact, for all nodes except the four corner nodes, we chose 7 nodes and computed the weights using a minimal norm method for underdetermined systems (see [5]). The choices of nodes used is given diagrammatically below. In each picture the circled dot denotes $\mathbf{x}_p$ and the other dots denote the nodes which (together with $\mathbf{x}_p$) are used to compute a rule of precision 2.

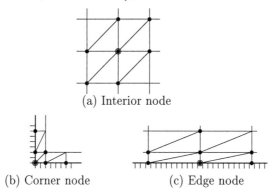

(a) Interior node

(b) Corner node   (c) Edge node

Finally, to fully describe the algorithm which we have implemented we must stipulate how we have computed the triangle-based Galerkin integrals in the "near field" when (26) does not hold. To do this we take the double integral (5) and expand it in the conventional way as a double sum of integrals over pairs of triangles in supp $\phi_p \times$ supp ϕ_q. For pairs of triangles which touch we use the regularising transformations in [2] to reduce these to smooth integrals over tensor products of a simplex in $\mathbb{R}^s$ and $[0,1]^{4-s}$, $s = 1, 2, 3$. The resulting integrals are computed using conical Gauss rules. For pairs of triangles which do not touch (but for which (26) does not hold), we use again conical Gauss rules in each triangle. The numbers of points in these Gauss rules were adjusted experimentally until we were confident that the corresponding Galerkin integrals were accurate to 5 figures. For triangles which have high aspect ratio, it was found necessary to choose highly accurate Gaussian rules in order to preserve the overall convergence of the method. In particular for pairs of triangles which touch, we chose the degree of precision increasing approximately linearly with the maximum of the aspect ratios of the pair. (For example, in Table 1 for $N = 169$, integrals over the most distorted intersecting triangles required up to 14 Gauss points in some coordinate directions.) A similar increase in precision was necessary for disjoint pairs of triangles whose distance apart was smaller than the maximum of the diameters of each of the triangles in the pair. In [3] a semi-analytic method was used for computing the Galerkin matrix, and so these difficulties could be somewhat reduced. (See also [11].) However in more general cases, fully discrete quadrature can not be avoided and so the experiments recorded here provide a guide to what may happen in larger applications.

In the tables presented below we use the following notation: $N = (2n+1)^2$ denotes the number of nodes in the mesh, σ denotes the computed stability constant in (**A5**), $\tilde{\kappa}_n$ denotes the capacitance computed using the hybrid algorithm and the column headed % denotes the percentage of the matrix entries which were computed using the node-based rules. Table 1 gives results for $g = 3$ in the "true Galerkin" case where $C^* = 0.5$ ($<$ diam$(\Gamma))^{-1}$. By (26), no entries of the matrix are computed with node-based rules. Based on (25) we expect $\tilde{\kappa}_n$ to converge with about $O(1/n)^3$ to κ. From this, extrapolation yields a true value $\kappa \cong 0.73363$, which coincides to 4 decimal places with the suggested true value in [3]. We use this to compute the errors and the estimated order of convergence (EOC). Table 2 repeats this experiment for $C^* = 2, 4$. As C^* increases a higher percentage of the matrix is computed using node-based rules. For $C^* = 2$ the asymptotic convergence rate predicted in (25) is clearly visible. This is less clear for the case $C^* = 4$, where finer meshes will be required before asymptotia is reached. This observation resonates with the theory, since the constant C in (14) is known to grow as C^* increases (see also [5, eq. (2.16)]). For $N = 169$ and $C^* = 2$, more than half the matrix is computed with a node-based rule of precision 2. This uses about 64 times fewer kernel evaluations than the cheapest triangle-based rules used in Table 1, but the increase in error is only 9×10^{-5}. To further illustrate the effect of the grading exponent g, we repeat this experiment with a uniform mesh $g = 1$. In Table 3 a convergence rate of only $O(1/n)$ is observed, but with a remarkably robust dependence on C^*.

N	$\tilde{\kappa}_n$	$|\kappa - \tilde{\kappa}_n|$	EOC	%
25	0.729115	4.5(-3)	2.7	0
49	0.732154	1.5(-3)	2.3	0
81	0.732848	7.8(-4)	3.3	0
121	0.733257	3.7(-4)	1.7	0
169	0.733364	2.7(-4)	3.0	0
225	0.733461	1.7(-4)		0

Table 1: $q = 3$, $C^* = 0.5$

		$C^* = 2$				$C^* = 4$			
N	σ	$\tilde{\kappa}_n$	$|\kappa - \tilde{\kappa}_n|$	EOC	%	$\tilde{\kappa}_n$	$|\kappa - \tilde{\kappa}_n|$	EOC	%
25	5.7	0.729129	4.5(-3)	2.7	9	0.729047	4.6(-3)	2.2	24
49	5.7	0.732168	1.5(-3)	2.2	34	0.731726	1.9(-3)	1.9	56
81	4.0	0.732835	8.0(-4)	2.5	44	0.732516	1.1(-3)	3.2	62
121	4.0	0.733167	4.6(-4)	3.1	49	0.733093	5.4(-4)	1.4	65
169	4.0	0.733375	2.6(-4)		53	0.733215	4.2(-4)		71

Table 2: $q = 3$, $C^* = 2, 4$

		$C^* = 2$				$C^* = 4$			
N	σ	$\tilde{\kappa}_n$	$|\kappa - \tilde{\kappa}_n|$	EOC	%	$\tilde{\kappa}_n$	$|\kappa - \tilde{\kappa}_n|$	EOC	%
25	2.0	0.718333	1.5(-2)	0.8	4	0.718833	1.5(-2)	1.0	30
49	2.0	0.723048	1.1(-2)	1.1	7	0.723221	1.0(-2)	0.7	37
81	2.0	0.725532	8.1(-3)	0.9	25	0.725580	8.1(-3)	1.0	59
121	2.0	0.727056	6.6(-3)	0.9	29	0.727127	6.5(-3)	0.9	59
169	2.0	0.728077	5.6(-3)		33	0.728083	5.5(-3)		66

Table 3: $q = 1$, $C^* = 2, 4$

References

[1] D.N. Arnold, W.L. Wendland: On the asymptotic convergence of collocation methods. *Math. Comp.* **41**, 349–381, 1983.

[2] S. Erichsen, S. A. Sauter: Efficient automatic quadrature in 3D Galerkin BEM. To appear in *Comp. Meth. Appl. Mech. Eng.*, 1998.

[3] V.J. Ervin, E.P. Stephan, S. Abou El-Seoud: An improved boundary element method for the charge density of a thin electrified plate in $\boldsymbol{R}^3$. *Math. Meth. Appl. Scis.* **13** (1990), 291-303.

[4] I. G. Graham, W. Hackbusch, S. A. Sauter: *Discrete boundary element methods on general meshes in 3D*. Bath Mathematics Preprint 97/19, 1997.

[5] I.G. Graham, W. Hackbusch, S.A. Sauter: *Hybrid boundary elements: Theory and implementation*. Preprint Number 98-6, University of Kiel, Feb. 1998.

[6] W. Hackbusch, S. A. Sauter: On the efficient use of the Galerkin method to solve Fredholm integral equations. *Applications of Mathematics*, **38** (1993), 301–322.

[7] T. von Petersdorff: Elasticity Problems in Polyhedra-Singularities and Approximation with Boundary Elements, Ph.D. Dissertation, T.H. Darmstadt, 1989.

[8] S. A. Sauter: Cubature Techniques for 3D Galerkin BEM. In: W. Hackbusch and G. Wittum, Eds., *BEM: Implementation and Analysis of Advanced Algorithms*, Vieweg Verlag, 1996.

[9] S. A. Sauter, A. Krapp: On the effect of numerical integration in the Galerkin boundary element method. *Numer. Math.* **74** (1997), 337-359.

[10] S.A. Sauter, C. Schwab: Quadrature for $hp-$ Galerkin BEM in 3D, *Numer. Math.* **78** (1997), 211-258.

[11] I.H. Sloan: Semidiscrete Galerkin method for piecewise-smooth curves and surfaces. Submitted for publication.

[12] W.L. Wendland: On Galerkin collocation methods for integral equations of elliptic boundary value problems. In: *Numerical Treatment of Integral Equations*, J. Albrecht and L. Collatz, Eds., ISNM 53, Birkhauser-Verlag, Basel, 1980.

I.G. Graham, Mathematical Sciences, University of Bath, Bath BA2 7AY, U.K.
Email: igg@maths.bath.ac.uk
W. Hackbusch, Praktische Mathematik, Universität Kiel, 24098 Kiel, Germany
Email: wh@numerik.uni-kiel.de
S.A. Sauter, Praktische Mathematik, Universität Kiel, 24098 Kiel, Germany
Email sas@numerik.uni-kiel.de

N. HEUER and E. P. STEPHAN

The Poincaré-Steklov Operator within Countably Normed Spaces

1. Introduction

When treating elliptic boundary value problems by domain decomposition methods the interaction between the boundary data of the subdomains plays an important role, see, e.g., [2] and the references therein. The mapping of the Dirichlet datum on the boundary of a subdomain to its Neumann datum is called Poincaré-Steklov operator and corresponds in its discrete form to a so-called Schur complement. However, often a reduction of this mapping to an operator which acts on the trace spaces exclusively is advantageous, cf. [9] and [12]. This reduction corresponds to solving the appearing interface problem by the boundary element method. Besides domain decomposition, the coupling of the finite element and the boundary element methods is an efficient way to solve various transmission problems, see, e.g., [13, 5]. When applying a Schur complement step to eliminate the unknown jump of the normal derivative across the interface, this reduces the whole problem to inverting the Poincaré-Steklov operator.

This work follows the lines of [8] where basic integral operators have been investigated. These integral operators appear also in this work but are combined in a specific way. Whereas the basic integral operators are continuous in the whole scale of countably normed spaces, we here obtain a restricted continuity of the Poincaré-Steklov operator, depending on the angles of the polygon under consideration. The continuity of the inverse of the Poincaré-Steklov operator requires analogous restrictions as the continuity of the operator itself.

Now let us define the Poincaré-Steklov operator. For a polygonal domain $\Omega \subset \mathbb{R}^2$ with boundary Γ we consider the Dirichlet problem for the Laplacian: *For given trace $u|_\Gamma$ on Γ find u in Ω with $\Delta u = 0$ in Ω.*

Once, the solution u to the Dirichlet problem is known, the Neumann datum $\partial u/\partial n|_\Gamma$ on Γ is computable. By this mapping we formally define the Poincaré-Steklov operator

$$S: u|_\Gamma \mapsto \frac{\partial u}{\partial n}\bigg|_\Gamma.$$

In variational form this operator is given by the solution $Su|_\Gamma$ of the problem

$$\int_\Gamma (Su|_\Gamma)v|_\Gamma \, ds = \int_\Omega \nabla u \nabla v \, dx \quad \text{for any } v \in H^1(\Omega). \tag{1}$$

This paper deals with the mapping properties of S purely within trace spaces on Γ. These properties ensure exponentially fast convergence of the h-p version of the Galerkin method for the inversion of S.

We note that sometimes the inverse of S is called Poincaré-Steklov operator and S is referred to as the Steklov-Poincaré operator. Further we note that, in what follows, more general elliptic differential operators than the Laplacian can be considered as well. The main restriction is that the differential operator possesses a fundamental solution such that representation formulae for the solution of the boundary value problem are available. Then, in principle, our analysis leads to analogous results.

As mentioned above, in order to invert the Poincaré-Steklov operator one can solve (1) directly by the finite element method. This corresponds, in its discrete form, to inverting the Schur complement of the block of the stiffness matrix which belongs to the basis functions attaching the boundary Γ. However, the problem (1) can also be efficiently solved by the boundary element method since S can be given explicitly in terms of boundary integral operators, which will be introduced in the following.

For sufficiently smooth ϕ we define the integral operators V (single layer operator), K (double layer potential), K' (the adjoint of the double layer potential), and D (the normal derivative of the double layer potential):

$$V\phi(x) = -\frac{1}{\pi}\int_\Gamma \phi(y)\log|x-y|\,ds_y \quad (x \in \Gamma), \tag{2}$$

$$K\phi(x) = -\frac{1}{\pi}\int_\Gamma \phi(y)\frac{\partial}{\partial n_y}\log|x-y|\,ds_y \quad (x \in \Gamma), \tag{3}$$

$$K'\phi(x) = -\frac{1}{\pi}\int_\Gamma \phi(y)\frac{\partial}{\partial n_x}\log|x-y|\,ds_y \quad (x \in \Gamma), \tag{4}$$

$$D\phi(x) = -\frac{\partial}{\partial n_x}K\phi(x) \quad (x \in \Gamma). \tag{5}$$

Now, the Poincaré-Steklov operator can be equivalently defined by

$$Sv := \bigl(D + (I+K')V^{-1}(I+K)\bigr)v \quad \text{for any } v \in H^{1/2}(\Gamma) \tag{6}$$

where I is the identity and $H^{1/2}(\Gamma)$ is the trace space of $H^1(\Omega)$ on Γ, see [11]. For a precise definition of the trace space we refer to the next section. For this formulation we have to assume that the single layer operator V is injective which is the case for domains Ω with conformal radius less than 1.

Relating to the original problem, there holds

$$Su|_\Gamma = \bigl(D + (I+K')V^{-1}(I+K)\bigr)u|_\Gamma = \frac{\partial u}{\partial n}\bigg|_\Gamma.$$

In variational form our problem is as follows. *For given $g \in \bigl(H^{1/2}(\Gamma)\bigr)'$ with $\int_\Gamma g\,ds = 0$, find $v \in H_0^{1/2}(\Gamma) := \{w \in H^{1/2}(\Gamma); \int_\Gamma w\,ds = 0\}$ such that*

$$\int_\Gamma w\bigl(D + (I+K')V^{-1}(I+K)\bigr)v\,ds = \int_\Gamma wg\,ds \quad \text{for any } w \in H_0^{1/2}(\Gamma). \tag{7}$$

This is the suitable form for introducing boundary elements for approximately inverting the Poincaré-Steklov operator. Since S is positive definite on $H^{\frac{1}{2}}(\Gamma)$, (7) is uniquely solvable. Note that, if we are considering the boundary value problem for the Laplacian, $g = \partial u/\partial n|_\Gamma$ is the Neumann data and, therefore,

$$\int_\Gamma wg\,ds = \int_\Omega \nabla v \nabla u\,dx \quad \text{for any } v \in H^1(\Omega).$$

The solution v of (7) behaves singularly at the corners of the polygon Γ. Therefore, the h- and p-versions of the Galerkin method converge only algebraically. On the other hand, the h-p version with geometrically graded meshes and appropriate degree distributions converges exponentially fast, see [1, 7]. This can be shown by approximation theory in the framework of countably normed Sobolev spaces, cf. [6]. In this paper we provide the needed regularity theory for the Poincaré-Steklov operator. For the continuity and regularity of certain boundary integral operators we also refer to [3, 10].

An outline of the paper is as follows. In §2 we introduce the needed Sobolev spaces and the countably normed spaces. Further, we define the Mellin transformation which is our tool to deal with the integral operators in these spaces. Section 3 presents the main results about the continuity and regularity of the Poincaré-Steklov operator.

2. Countably normed spaces and Mellin transformation

In this section we describe the technical tools for investigating the mapping properties of the Poincaré-Steklov operator on polygons. We introduce the needed Sobolev and countably normed spaces. After that we deal with the Mellin transformation which can be used to handle the norms of the countably normed spaces. The corresponding results are recalled from [8].

Let $\Gamma = \cup_{j=1}^{J} \bar{\Gamma}_j$ be the boundary of a polygonal domain $\Omega \subset \mathbb{R}^2$ with edges Γ_j. By t_j ($j = 1, \ldots, J$) we denote the corner points where Γ_j and Γ_{j+1} meet ($\Gamma_{J+1} = \Gamma_1$). The interior angle at t_j is denoted by ω_j. Each side Γ_j is bisected into two pieces Γ_1^j (containing the vertex t_{j-1}) and Γ_2^j (containing the vertex t_j).

The definition of the Sobolev spaces is as usual:

$$H^s(\Omega) = \{\phi|_\Omega\,;\ \phi \in H^s(\mathbb{R}^2)\} \quad (s \in \mathbb{R}),$$

$$H^s(\Gamma) = \begin{cases} \{\phi|_\Gamma\,;\ \phi \in H^{s+1/2}(\mathbb{R}^2)\} & (s > 0) \\ L^2(\Gamma) & (s = 0) \\ (H^{-s}(\Gamma))' \ \text{(dual space)} & (s < 0) \end{cases},$$

$$H^s(\Gamma_j) = \{\phi|_{\Gamma_j}\,;\ \phi \in H^s(\Gamma)\} \quad (s \geq 0,\ j = 1, \ldots, J),$$

$$\tilde{H}^s(\Gamma_j) = \{\phi \in H^s(\Gamma_j)\,;\ \tilde{\phi} \in H^s(\Gamma)\} \quad (s \geq 0,\ j = 1, \ldots, J).$$

Here, $\tilde{\phi}$ is the extension of ϕ by 0 outside Γ_j. Finally we define the dual spaces on an edge Γ_j:

$$H^s(\Gamma_j) = (\tilde{H}^{-s}(\Gamma_j))' \quad \text{and} \quad \tilde{H}^s(\Gamma_j) = (H^{-s}(\Gamma_j))' \quad (s < 0).$$

Our regularity investigations of Section 3 will base on the weighted Sobolev spaces and the countably normed spaces as introduced in the following.

Let $I = (0, 1)$. By $H_\beta^{m,l}(I)$ ($m \geq l \geq 1$, integers and $0 < \beta < 1$) we denote the completion of the set of all infinitely differentiable functions on $[0, 1]$ under the norm

$$\|\phi\|_{H_\beta^{m,l}(I)}^2 = \|\phi\|_{H^{l-1}(I)}^2 + \sum_{j=l}^{m} |\phi|_{H_\beta^{j,l}(I)}^2 \tag{8}$$

where
$$|\phi|_{H^{j,l}_\beta(I)} := \|x^{\beta+j-l}\phi^{(j)}\|_{L^2(I)}. \tag{9}$$

The countably normed spaces on I are defined as
$$\begin{aligned}B^l_\beta(I) &= \{\phi \in H^{m,l}_\beta(I), m = l, l+1, \ldots\,;\ \exists C \geq 0, d \geq 1\ \forall j = l, l+1, \ldots \\ &\quad |\phi|_{H^{j,l}_\beta(I)} \leq Cd^{(j-l)}(j-l)!\}\quad (l \geq 1,\ \text{integer})\end{aligned} \tag{10}$$

On Γ these spaces are defined as the product spaces
$$H^{m,l}_\beta(\Gamma) = \Pi^J_{j=1}\Pi^2_{k=1}H^{m,l}_\beta(\Gamma^j_k),\ B^l_\beta(\Gamma) = \Pi^J_{j=1}\Pi^2_{k=1}B^l_\beta(\Gamma^j_k) \cap H^{l-1}(\Gamma), \tag{11}$$

where each boundary piece Γ^j_k has to be mapped onto I such that the vertex t_{j+k-2} falls onto 0 in order to apply the definition (10). If we want to emphasize the dependence on the constants C and d we will write $B^l_{\beta,C,d}$ instead of B^l_β.

For technical reasons we need the following representation of the countably normed spaces:
$$B^l_{\beta,C,d}(I) = \bigcap_{L=l}^{\infty} B^{l,L}_{\beta,C,d}(I) \tag{12}$$

where
$$B^{l,L}_{\beta,C,d}(I) := \{\phi \in H^{L,l}_\beta(I)\,;\ |\phi|_{H^{j,l}_\beta(I)} \leq Cd^{(L-l)}(j-l)!,\ j = l, l+1, \ldots, L\}. \tag{13}$$

The spaces $B^{l,L}_\beta(\Gamma)$ are defined accordingly to (11). For localization techniques one needs to introduce cut–off functions. These turn out not to be comprised by the general countably normed spaces B^l_β. But, evidently, for the spaces $B^{l,L}_\beta$ there exist partitions of unity. Furthermore, they can be chosen such that the constants C and d of $B^{l,L}_{\beta,C,d}(\Gamma)$ do not depend on the parameter L.

Lemma 1. [8, Lemma 2.1] Let $U \subset \Gamma$ be an open set and $U_\delta := \{x \in \Gamma\,;\ \text{dist}(x,U) \leq \delta\}$ for $\delta > 0$. Let $\phi \in B^{l,L}_{\beta,C,d}(U_\delta)$ for all $L \geq l$. Then there exists for each $L \geq l$ a cut–off function $\chi_L \in C^\infty(\Gamma)$ such that
$$\chi_L|_U \equiv 1 \quad \text{and} \quad \chi_L|_{\Gamma\setminus U_\delta} \equiv 0$$

and
$$\chi_L\phi \in B^{l,L}_{\beta,\tilde{C},\tilde{d}}(\Gamma)$$

with constants $\tilde{C}$ and $\tilde{d}$ independent of L.

For $\phi \in C_0^\infty(0,\infty)$ the Mellin transformation is defined by

$$\mathcal{M}(\phi)(\lambda) := \hat{\phi}(\lambda) := \int_0^\infty x^{i\lambda-1} \phi(x)\, dx. \tag{14}$$

The seminorm $|\phi|_{H_\beta^{j,l}(I)}$ can be characterized by using this transformation.

Lemma 2. [8, Lemma 2.4] *Let $\phi \in C_0^\infty(I)$ and $0 < \beta < 1$. Then*

$$|\phi|^2_{H_\beta^{j,l}(I)} \simeq \int_{\mathrm{Im}(\lambda) = l-1/2-\beta} |f_j(\lambda)|^2 |\hat{\phi}(\lambda)|^2\, d\lambda \quad (j \geq l),$$

where $f_j(\lambda) := i\lambda \cdot (i\lambda + 1) \cdots (i\lambda + j - 1)$. The constants in the mutual estimates do not depend on j.

3. Mapping properties and regularity

In this section we present our main results concerning the Poincaré-Steklov operator in countably normed spaces. Before doing so we need to recall the respective results for standard Sobolev spaces.

For Lipschitz domains continuity and regularity of the integral operators (2)–(5) as mappings between usual Sobolev spaces have been investigated by Costabel, see [3]. Using these estimates and noting that the Poincaré-Steklov operator maps the Dirichlet data onto the Neumann data, we obtain the following proposition.

Proposition 1. [3] *For all $\sigma \in [0, 1/2]$ $S : H^{1/2+\sigma}(\Gamma) \to H^{-1/2+\sigma}(\Gamma)$ is continuous. For $\sigma \in [0, 1/2]$ let $v \in H^{1/2}(\Gamma)$ satisfy $Sv \in H^{-1/2+\sigma}(\Gamma)$. Then $v \in H^{1/2+\sigma}(\Gamma)$, and there holds the a priori estimate*

$$\|v\|_{H^{1/2+\sigma}(\Gamma)} \leq C \left(\|Sv\|_{H^{-1/2+\sigma}(\Gamma)} + \|v\|_{H^{1/2}(\Gamma)} \right).$$

Following Costabel and Stephan in [4] we use the method of Mellin transformation to investigate the Poincaré-Steklov operator acting on countably normed spaces, see also [8]. First we look at the local properties on the boundary of an infinite side Γ^ω. In a second step we apply these results to the boundary Γ of a polygonal domain. Let $\Gamma^\omega = \Gamma^- \cup \{0\} \cup \Gamma^+$ with $\Gamma^- = e^{i\omega}\mathbb{R}_+$ and $\Gamma^+ = \mathbb{R}_+$ ($\omega \in (0, 2\pi)$). A function ϕ on Γ^ω can be identified with the pair (ϕ_-, ϕ_+) of functions on $\mathbb{R}_+$ defined by $\phi_-(x) = \phi(xe^{i\omega})$, $\phi_+(x) = \phi(x)$ ($x > 0$). We will choose the representation of ϕ by its "even" and "odd" parts (in a formal sense) which are defined by

$$\phi^e = \frac{1}{2}(\phi_- + \phi_+), \qquad \phi^o = \frac{1}{2}(\phi_- - \phi_+).$$

This induces for any operator A acting on functions on Γ^ω a representation by a 2×2–matrix of operators acting on functions on $\mathbb{R}_+$:

$$A \hat{=} \mathcal{A} := \begin{pmatrix} A_{ee} & A_{oe} \\ A_{eo} & A_{oo} \end{pmatrix} \qquad \text{where} \qquad \begin{aligned} (A\phi)^e &= A_{ee}\phi^e + A_{oe}\phi^o, \\ (A\phi)^o &= A_{eo}\phi^e + A_{oo}\phi^o. \end{aligned}$$

We need the following operators acting on functions on $\mathbb{R}_+$:

$$V_\omega \phi(x) := -\frac{1}{\pi} \int_0^\infty \log\left|1 - \frac{x}{y} e^{-i\omega}\right| \phi(y)\, dy, \quad V_0 = V_\omega \text{ for } \omega = 0,$$

$$K_\omega \phi(x) := \frac{1}{\pi} \int_0^\infty \operatorname{Im}\left(\frac{1}{xe^{i\omega} - y}\right) \phi(y)\, dy,$$

$$K'_\omega \phi(x) := \frac{1}{\pi} \int_0^\infty \operatorname{Im}\left(\frac{e^{i\omega}}{xe^{i\omega} - y}\right) \phi(y)\, dy,$$

$$D_\omega \phi(x) := -\frac{1}{x} \frac{\partial}{\partial \omega} K_\omega \phi(x), \quad D_0 = \lim_{\omega \to 0} D_\omega.$$

Then, with the exception of finite dimensional operators which are negligible in our theory, the integral operators (2)–(5) can be represented by the following matrices (see [4]):

$$V \triangleq \mathcal{V} = \begin{pmatrix} V_0 + V_\omega & 0 \\ 0 & V_0 - V_\omega \end{pmatrix},$$

$$D \triangleq \mathcal{D} = \begin{pmatrix} D_\omega - D_0 & 0 \\ 0 & -(D_0 + D_\omega) \end{pmatrix},$$

$$K \triangleq \mathcal{K} = \begin{pmatrix} K_\omega & 0 \\ 0 & -K_\omega \end{pmatrix}, \quad K' \triangleq \mathcal{K}' = \begin{pmatrix} K'_\omega & 0 \\ 0 & -K'_\omega \end{pmatrix}.$$

Using these representations we also obtain the representation of the Poincaré-Steklov operator acting on even and odd functions on the infinite angle:

$$S \triangleq \mathcal{S} = \begin{pmatrix} S_{ee} & S_{oe} \\ S_{eo} & S_{oo} \end{pmatrix}$$

with

$$\begin{aligned} S_{ee} &= D_\omega - D_0 + (I + K'_\omega)(V_0 + V_\omega)^{-1}(I + K_\omega), \\ S_{oo} &= D_0 - D_\omega + (I - K'_\omega)(V_0 - V_\omega)^{-1}(I - K_\omega), \end{aligned} \qquad (15)$$

and $S_{eo} = S_{oe} = 0$.

The Mellin symbols of all the components are explicitly known (see [4]):

$$\mathcal{M}(V_\omega \phi)(\lambda) = \hat{V}_\omega(\lambda) \hat{\phi}(\lambda - i) := \frac{\cosh[(\pi - \omega)\lambda]}{\lambda \sinh \pi \lambda} \hat{\phi}(\lambda - i), \quad \operatorname{Im}(\lambda) \in (0, 1), \qquad (16)$$

$$\begin{aligned} \mathcal{M}(D_\omega \phi)(\lambda) &= \hat{D}_\omega(\lambda + i) \hat{\phi}(\lambda + i) \\ &:= -(\lambda + i) \frac{\cosh[(\pi - \omega)(\lambda + i)]}{\sinh[\pi(\lambda + i)]} \hat{\phi}(\lambda + i), \quad \operatorname{Im}(\lambda) \in (-2, 0), \end{aligned} \qquad (17)$$

$$\mathcal{M}(K_\omega \phi)(\lambda) = \hat{K}_\omega(\lambda)\hat{\phi}(\lambda) := -\frac{\sinh[(\pi-\omega)\lambda]}{\sinh \pi \lambda}\hat{\phi}(\lambda), \quad \text{Im}(\lambda) \in (-1,1), \qquad (18)$$

$$\mathcal{M}(K'_\omega \phi)(\lambda) = \hat{K}_\omega(\lambda+i)\hat{\phi}(\lambda), \quad \text{Im}(\lambda) \in (-2,0). \qquad (19)$$

On an infinite angle the continuity with respect to seminorms is as follows.

Lemma 3. *Let $\rho < \beta < 1$ for $\rho := 3/2 - \min\{\frac{\pi}{2\pi-\omega}, \frac{\pi}{\omega}\}$. Then there holds*

$$|S\phi|_{H^{j,1}_\beta(\Gamma^\omega)} \leq C|\phi|_{H^{j+1,2}_\beta(\Gamma^\omega)} \quad (j \geq 1)$$

The constant $C > 0$ does not depend on j.

Proof. We use the representation (15) of the Poincaré-Steklov operator and Lemma 2 to handle the norms of the weighted Sobolev spaces. To calculate the seminorms $|\cdot|_{H^{j,l}_\beta(\mathbb{R}_+)}$ it suffices to concentrate on test functions with compact support in $(0,\infty)$: Let $\phi \in C_0^\infty[0,\infty)$ with $\text{supp}(1-\phi) \subset (0,\infty)$. Then we automatically have

$$\text{supp}(\phi^{(j)}) \subset C_0^\infty(0,\infty), \quad j \geq l \geq 1,$$

and the values $\phi(x)$ for x near 0 are not taken into account for calculating the seminorms $|\cdot|_{H^{j,l}_\beta(\mathbb{R}_+)}$.

Therefore we take $\phi \hat{=} (\phi^e, \phi^o) \in C_0^\infty(\mathbb{R}_+)^2$. Then we have

$$\mathcal{S}\phi = (S_{ee}\phi^e + S_{oe}\phi^o, S_{eo}\phi^e + S_{oo}\phi^o)^T = (S_{ee}\phi^e, S_{oo}\phi^o)^T$$

and it remains to estimate the two components of the right hand side by means of the seminorm $|\phi|_{H^{j,l}_\beta(\mathbb{R}_+)}$.

Using the representations (16)–(19) we obtain

$$\mathcal{M}(S_{ee}\phi^e)(\lambda - i)$$
$$= \mathcal{M}(D_\omega - D_0)(\lambda)\phi^e(\lambda) + \mathcal{M}(I + K'_\omega)(\lambda)\mathcal{M}((V_0 + V_\omega)^{-1})(\lambda)\mathcal{M}(I + K_\omega)(\lambda)\hat{\phi}^e(\lambda)$$
$$= -\lambda\frac{\cosh(\pi-\omega)\lambda - \cosh\pi\lambda}{\sinh\pi\lambda}\hat{\phi}^e(\lambda) + \Big(1 - \frac{\sinh(\pi-\omega)\lambda}{\sinh\pi\lambda}\Big)\frac{\lambda\sinh\pi\lambda}{\cosh\pi\lambda + \cosh(\pi-\omega)\lambda}$$
$$\times\Big(1 - \frac{\sinh(\pi-\omega)\lambda}{\sinh\pi\lambda}\Big)\hat{\phi}^e(\lambda)$$
$$= \frac{\lambda\big(\cosh\pi\lambda - \cosh(\pi-\omega)\lambda\big)\big(\cosh\pi\lambda + \cosh(\pi-\omega)\lambda\big) + \lambda\big(\sinh\pi\lambda - \sinh(\pi-\omega)\lambda\big)^2}{\sinh\pi\lambda(\cosh\pi\lambda + \cosh(\pi-\omega)\lambda)}\hat{\phi}^e(\lambda)$$
$$= \lambda\frac{\cosh^2\pi\lambda - \cosh^2(\pi-\omega)\lambda + \sinh^2\pi\lambda - 2\sinh\pi\lambda\sinh(\pi-\omega)\lambda + \sinh^2(\pi-\omega)\lambda}{\sinh\pi\lambda(\cosh\pi\lambda + \cosh(\pi-\omega)\lambda)}\hat{\phi}^e(\lambda)$$
$$= 2\lambda\frac{\sinh\pi\lambda - \sinh(\pi-\omega)\lambda}{\cosh\pi\lambda + \cosh(\pi-\omega)\lambda}\hat{\phi}^e(\lambda) =: \mathcal{M}(S_{ee})(\lambda)\hat{\phi}^e(\lambda). \qquad (20)$$

Therefore, by Lemma 2,

$$|S_{ee}\phi^e|^2_{H_\beta^{j,1}(\mathbb{R}_+)} \simeq \int_{\text{Im}(\lambda)=1/2-\beta} |f_j(\lambda)|^2 |\mathcal{M}(S_{ee}\phi^e)(\lambda)|^2 \, d\lambda$$

$$= 4\int_{\text{Im}(\lambda)=1/2-\beta} |f_j(\lambda)|^2 \left|(\lambda+i)\frac{\sinh\pi(\lambda+i) - \sinh(\pi-\omega)(\lambda+i)}{\cosh\pi(\lambda+i) + \cosh(\pi-\omega)(\lambda+i)}\right|^2 |\phi^e(\lambda+i)|^2 \, d\lambda$$

$$= 4\int_{\text{Im}(\lambda)=1/2-\beta} |f_{j+1}(\lambda+i)|^2 \left|\frac{\sinh\pi(\lambda+i) - \sinh(\pi-\omega)(\lambda+i)}{\cosh\pi(\lambda+i) + \cosh(\pi-\omega)(\lambda+i)}\right|^2 |\phi^e(\lambda+i)|^2 \, d\lambda$$

$$= 4\int_{\text{Im}(\lambda)=3/2-\beta} |f_{j+1}(\lambda)|^2 \left|\frac{\sinh\pi\lambda - \sinh(\pi-\omega)\lambda}{\cosh\pi\lambda + \cosh(\pi-\omega)\lambda}\right|^2 |\phi^e(\lambda)|^2 \, d\lambda.$$

Here we used the relation $|f_j(\lambda)(\lambda+i)| = |f_{j+1}(\lambda+i)|$.

With $\lambda = x + iy$ there holds

$$\cosh\pi\lambda + \cosh(\pi-\omega)\lambda = 2\cosh(\pi - \frac{\omega}{2})\lambda \cosh\frac{\omega}{2}\lambda$$

$$= 2\left(\cosh(\pi - \frac{\omega}{2})x \cos(\pi - \frac{\omega}{2})y + i\sinh(\pi - \frac{\omega}{2})x \sin(\pi - \frac{\omega}{2})y\right)$$

$$\times \left(\cosh\frac{\omega}{2}x \cos\frac{\omega}{2}y + i\sinh\frac{\omega}{2}x \sin\frac{\omega}{2}y\right)$$

$$= 0 \quad \text{for} \quad y = \text{Im}(\lambda) = 3/2 - \beta \in (1/2, 1)$$

if and only if

$$x = 0 \quad \text{and} \quad y = \frac{\pi}{2\pi - \omega} \quad \text{and} \quad \omega < \pi$$

or

$$x = 0 \quad \text{and} \quad y = \frac{\pi}{\omega} \quad \text{and} \quad \omega > \pi.$$

Therefore, $\cosh\pi\lambda + \cosh[(\pi-\omega)\lambda]$ does not vanish if $\text{Im}(\lambda) = 3/2 - \beta < \min\{\frac{\pi}{2\pi-\omega}, \frac{\pi}{\omega}\}$, i.e. if $\beta > 3/2 - \min\{\frac{\pi}{2\pi-\omega}, \frac{\pi}{\omega}\} = \rho$.

Thus, since

$$\left|\frac{\sinh\pi\lambda - \sinh(\pi-\omega)\lambda}{\cosh\pi\lambda + \cosh(\pi-\omega)\lambda}\right|$$

is bounded for $|\text{Re}(\lambda)| \to \infty$ when avoiding the roots of the denominator, there holds

$$|S_{ee}\phi^e|^2_{H_\beta^{j,1}(\mathbb{R}_+)} \simeq \int_{\text{Im}(\lambda)=3/2-\beta} |f_{j+1}(\lambda)|^2 \left|\frac{\sinh\pi\lambda - \sinh(\pi-\omega)\lambda}{\cosh\pi\lambda + \cosh(\pi-\omega)\lambda}\right|^2 |\phi^e(\lambda)|^2 \, d\lambda$$

$$\leq C\int_{\text{Im}(\lambda)=3/2-\beta} |f_{j+1}(\lambda)|^2 |\phi^e(\lambda)|^2 \, d\lambda \simeq |\phi^e|^2_{H_\beta^{j+1,2}(\mathbb{R}_+)}$$

for $\beta > \rho$ and $j \geq 1$.

Analogously, we obtain

$$|S_{oo}\phi^o|^2_{H^{j,1}_\beta(\mathbb{R}_+)} \simeq \int_{\text{Im}(\lambda)=3/2-\beta} |f_{j+1}(\lambda)|^2 \left|\frac{\sinh \pi\lambda + \sinh(\pi-\omega)\lambda}{\cosh \pi\lambda - \cosh(\pi-\omega)\lambda}\right|^2 |\phi^o(\lambda)|^2\, d\lambda$$

$$\leq C\int_{\text{Im}(\lambda)=3/2-\beta} |f_{j+1}(\lambda)|^2 |\phi^o(\lambda)|^2\, d\lambda \simeq |\phi^o|^2_{H^{j+1,2}_\beta(\mathbb{R}_+)}$$

for $\beta > \rho$ and $j \geq 1$. Therefore,

$$|S\phi|^2_{H^{j,1}_\beta(\Gamma^\omega)} \leq C(|\phi^e|^2_{H^{j+1,2}_\beta(\mathbb{R}_+)} + |\phi^o|^2_{H^{j+1,2}_\beta(\mathbb{R}_+)})$$

$$\leq c(|\phi_-|^2_{H^{j+1,2}_\beta(\mathbb{R}_+)} + |\phi_+|^2_{H^{j+1,2}_\beta(\mathbb{R}_+)}) = c|\phi|^2_{H^{j+1,2}_\beta(\Gamma^\omega)}$$

and the assertion of the lemma is proved. □

Using this lemma we obtain the continuity of S within countably normed spaces on the whole polygon Γ.

Theorem 1. *For $\rho < \beta < 1$ with $\rho := 3/2 - \min\{\frac{\pi}{2\pi-\omega}, \frac{\pi}{\omega}\}$ let $\phi \in B^2_\beta(\Gamma)$. Then there holds $S\phi \in B^1_\beta(\Gamma)$.*

Proof. Let $\phi \in B^2_{\beta,C,d}(\Gamma)$. Due to the definition of the countably normed spaces in (10), (11) and the respective norms (8) we have to consider the global norm $\|S\phi\|_{L^2(\Gamma)}$ and the seminorms $|S\phi|_{H^{j,1}_\beta(\Gamma^i_k)}$ ($j \geq 1$) for $i = 1,\ldots,J$ and $k = 1,2$. The continuity with respect to the L^2-norm is proved by Proposition 1. To show the continuity with respect to the seminorms (9) we need a partition of unity which exists due to Lemma 1 for each $B^{2,L}_{\beta,C,d}(\Gamma)$ ($L \geq 2$), cf. (12). Let $\chi_i \in C^\infty(\Gamma)$, $i = 1,\ldots,J$, such that supp $\chi_i \subset \Gamma_i \cup \{t_i\} \cup \Gamma_{i+1}$ and $\sum_{i=1}^J \chi_i = 1$ and

$$\chi_i\phi \in B^{2,L}_{\beta,\tilde{C},\tilde{d}}(\Gamma_i \cup \{t_i\} \cup \Gamma_{i+1})$$

($\chi_i\phi$ is supposed to be extended by 0 outside supp χ_i). Then for $\beta > \rho$, Lemma 3 yields

$$|S\chi_i\phi|_{H^{j,1}_\beta(\Gamma_i\cup\{t_i\}\cup\Gamma_{i+1})} \leq c|\chi_i\phi|_{H^{j+1,2}_\beta(\Gamma_i\cup\{t_i\}\cup\Gamma_{i+1})} \tag{21}$$

$$\leq c\tilde{C}\tilde{d}^{L-2}(j-1)!, \quad j=1,\ldots,L-1,$$

where the second inequality is caused by the regularity of $\chi_i\phi$ and definition (13). Now, the already known boundedness of $\|S\chi_i\phi\|_{L^2(\Gamma)}$ and the estimate (21) yield due to (13) the local regularity

$$S\chi_i\phi \in B^{1,L-1}_{\beta,c\tilde{C},\tilde{d}}(\Gamma_i \cup \{t_i\} \cup \Gamma_{i+1}). \tag{22}$$

Again, there exists a partition of unity $\{\zeta_i;\ i = 1,\ldots, J\}$ for $S\chi_i\phi$ and the index $L - 1$. Due to (22) there holds

$$\zeta_i S\chi_i\phi \in B^{1,L-1}_{\beta,C',d'}(\Gamma_i \cup \{t_i\} \cup \Gamma_{i+1}) \tag{23}$$

and due to the analyticity of the kernel of the Poincaré-Steklov operator aside the diagonal $x = y$ we also have

$$\zeta_j S\chi_i\phi \in B^{1,L-1}_{\beta,C',d'}(\Gamma \setminus \overline{\Gamma_i \cup \{t_i\} \cup \Gamma_{i+1}})\quad (j \neq i). \tag{24}$$

Putting (23) and (24) together for each of the corners $\{t_i\}$ we obtain

$$S\phi = \sum_{j=1}^{m}\sum_{i=1}^{m} \zeta_j S\chi_i\phi \in B^{1,L-1}_{\beta,C^\star,d^\star}(\Gamma) \tag{25}$$

where $C^\star$ and $d^\star$ are the largest numbers of the different C's and d's, respectively. Since the constants C' and d' do not depend on the parameter L and the partitions of unity $\{\zeta_i\}$ and $\{\chi_i\}$ corresponding to L, eq. (25) finally yields together with the representation (12)

$$S\phi \in B^{1}_{\beta,C^\star,d^\star}(\Gamma).$$

$\square$

We now investigate the inverse of the Poincaré-Steklov operator. First let us formulate the local regularity result which corresponds to the continuity given by Lemma 3.

Lemma 4. *Let $\rho = 3/2 - \min\{\frac{\pi}{2\pi-\omega}, \frac{\pi}{\omega}\}$. For $\phi \in H^1(\Gamma^\omega)$ such that $S\phi \in H^{j-1,1}_\beta(\Gamma^\omega)$ for $j \geq 2$ there holds $\phi \in H^{j,2}_\beta(\Gamma^\omega)$ for $\rho < \beta < 1$ and*

$$|\phi|_{H^{j,2}_\beta(\Gamma^\omega)} \leq C|S\phi|_{H^{j-1,1}_\beta(\Gamma^\omega)}.$$

The constant C does not depend on j.

Proof. Again we use the representation of functions on Γ^ω by their even and odd parts on Γ^- and Γ^+ and the induced matrix representation of the Poincaré-Steklov operator as in the proof of Lemma 3, cf. (15). With regard to (20) there holds

$$\hat\phi^e(\lambda) = \frac{\cosh \pi\lambda + \cosh(\pi-\omega)\lambda}{2\lambda\big(\sinh\pi\lambda - \sinh(\pi-\omega)\lambda\big)} \mathcal{M}(S_{ee}\phi^e)(\lambda - i),\quad \mathrm{Im}\,(\lambda) \in (-1,1).$$

Noting that $|f_j(\lambda)/\lambda| = |f_{j-1}(\lambda - i)|$ we obtain by Lemma 2 for $\beta \in (1/2,1)$

$$|\phi^e|^2_{H^{j,2}_\beta(\mathbb{R}_+)} \simeq \int_{\mathrm{Im}\,(\lambda)=3/2-\beta} |f_j(\lambda)|^2 |\hat\phi^e(\lambda)|^2\, d\lambda \tag{26}$$

$$= \int_{\mathrm{Im}\,(\lambda)=3/2-\beta} |f_j(\lambda)|^2 \left|\frac{\cosh\pi\lambda + \cosh(\pi-\omega)\lambda}{2\lambda(\sinh\pi\lambda - \sinh(\pi-\omega)\lambda)}\right|^2 |\mathcal{M}(S_{ee})(\lambda)|^2 |\hat\phi^e(\lambda)|^2\, d\lambda$$

$$= \frac{1}{4}\int_{\mathrm{Im}\,(\lambda)=3/2-\beta} |f_{j-1}(\lambda-i)|^2 \left|\frac{\cosh\pi\lambda + \cosh(\pi-\omega)\lambda}{\sinh\pi\lambda - \sinh(\pi-\omega)\lambda}\right|^2 |\mathcal{M}(S_{ee}\phi^e)(\lambda-i)|^2\, d\lambda$$

$$= \frac{1}{4}\int_{\mathrm{Im}\,(\lambda)=1/2-\beta} |f_{j-1}(\lambda)|^2 \left|\frac{\cosh\pi(\lambda+i) + \cosh(\pi-\omega)(\lambda+i)}{\sinh\pi(\lambda+i) - \sinh(\pi-\omega)(\lambda+i)}\right|^2 |\mathcal{M}(S_{ee}\phi^e)(\lambda)|^2\, d\lambda.$$

With $\lambda = x + iy$ there holds

$$\sinh \pi \lambda - \sinh(\pi - \omega)\lambda = 2 \sinh \frac{\omega}{2}\lambda \cosh[(\pi - \frac{\omega}{2})\lambda]$$
$$= 2 \left(\sinh \frac{\omega}{2}x \cos \frac{\omega}{2}y + i \cosh \frac{\omega}{2}x \sin \frac{\omega}{2}y \right)$$
$$\times \left(\cosh(\pi - \frac{\omega}{2})x \cos(\pi - \frac{\omega}{2})y + i \sinh(\pi - \frac{\omega}{2})x \sin(\pi - \frac{\omega}{2})y \right)$$
$$= 0$$

if and only if

$$x = 0 \quad \text{and} \quad y = \frac{k\pi}{\frac{\omega}{2}} = \frac{2k\pi}{\omega} \quad (k \text{ integer})$$

or

$$x = 0 \quad \text{and} \quad y = \frac{(k + \frac{1}{2})\pi}{\pi - \frac{\omega}{2}} = \frac{(2k+1)\pi}{2\pi - \omega} \quad (k \text{ integer}).$$

Therefore, $\sinh \pi(\lambda + i) - \sinh(\pi - \omega)(\lambda + i)$ does not vanish for

$$\text{Im}(\lambda) = 1/2 - \beta \in \left(-\frac{1}{2}, \min\{\frac{2\pi}{\omega} - 1, \frac{\pi}{2\pi - \omega} - 1, 0\} \right) = \left(-\frac{1}{2}, \min\{\frac{\pi}{2\pi - \omega} - 1, 0\} \right),$$

i.e., for $\beta > 3/2 - \pi/(2\pi - \omega)$. Since

$$\left| \frac{\cosh \pi(\lambda + i) + \cosh(\pi - \omega)(\lambda + i)}{\sinh \pi(\lambda + i) - \sinh(\pi - \omega)(\lambda + i)} \right|$$

is bounded for $\text{Re}(\lambda) \to \pm\infty$ provided the denominator does not vanish at the horizontal strip under consideration, we obtain by (26) and Lemma 2

$$|\phi^e|^2_{H^{j,2}_\beta(\mathbb{R}_+)}$$
$$\simeq \int_{\text{Im}(\lambda)=1/2-\beta} |f_{j-1}(\lambda)|^2 \left| \frac{\cosh \pi(\lambda + i) + \cosh(\pi - \omega)(\lambda + i)}{\sinh \pi(\lambda + i) - \sinh(\pi - \omega)(\lambda + i)} \right|^2 |\mathcal{M}(S_{ee}\phi^e)(\lambda)|^2 \, d\lambda$$
$$\leq C \int_{\text{Im}(\lambda)=1/2-\beta} |f_{j-1}(\lambda)|^2 |\mathcal{M}(S_{ee}\phi^e)(\lambda)|^2 \, d\lambda \simeq |S_{ee}\phi^e|^2_{H^{j-1,1}_\beta(\mathbb{R}_+)}$$

for $\beta > 3/2 - \pi/(2\pi - \omega)$. Analogously, we obtain for the odd part

$$\hat{\phi}^o(\lambda) = \frac{\cosh \pi \lambda - \cosh(\pi - \omega)\lambda}{2\lambda(\sinh \pi \lambda + \sinh(\pi - \omega)\lambda)} \mathcal{M}(S_{oo}\phi^o)(\lambda - i), \quad \text{Im}(\lambda) \in (-1, 1)$$

and therefore,

$$|\phi^o|^2_{H^{j,2}_\beta(\mathbb{R}_+)}$$
$$\simeq \int_{\text{Im}(\lambda)=1/2-\beta} |f_{j-1}(\lambda)|^2 \left| \frac{\cosh \pi(\lambda + i) - \cosh(\pi - \omega)(\lambda + i)}{\sinh \pi(\lambda + i) + \sinh(\pi - \omega)(\lambda + i)} \right|^2 |\mathcal{M}(S_{oo}\phi^o)(\lambda)|^2 \, d\lambda$$
$$\leq C \int_{\text{Im}(\lambda)=1/2-\beta} |f_{j-1}(\lambda)|^2 |\mathcal{M}(S_{oo}\phi^o)(\lambda)|^2 \, d\lambda \simeq |S_{oo}\phi^o|^2_{H^{j-1,1}_\beta(\mathbb{R}_+)}$$

for $\beta > 3/2 - \pi/\omega$.

Altogether, since $\phi \in H^1(\Gamma^\omega)$ by assumption, we proved that $\phi^e \in H_\beta^{j,2}(\mathbb{R}_+)$ and $\phi^o \in H_\beta^{j,2}(\mathbb{R}_+)$ for $j \geq 2$ and therefore we have $\phi \in H_\beta^{j,2}(\Gamma^\omega)$ and the proof of the lemma is finished. □

Now we use again the partitions of unity to prove the regularity of the Poincaré-Steklov operator on the whole polygon.

Theorem 2. *Let $\rho = 3/2 - \min\{\frac{\pi}{2\pi-\omega_j}, \frac{\pi}{\omega_j}; j = 1, \ldots, J\}$ and $\rho < \beta < 1$. Then there holds $\phi \in B_\beta^2(\Gamma)$ if $S\phi \in B_\beta^1(\Gamma)$ with $\int_\Gamma \phi ds = 0$, where ϕ is unique up to a constant.*

Proof. The proof is analogous to the proof of Theorem 1. The regularity with respect to the global Sobolev norms is given by Proposition 1. The boundedness with respect to the seminorms $|\cdot|_{H_\beta^{j,l}(\Gamma_k^j)}$ follows from Lemma 4. Here again, we have to use a partition of unity as in the proof of Theorem 1. □

References

[1] I. Babuška, B. Q. Guo, E. P. Stephan: On the exponential convergence of the h-p version for boundary element Galerkin methods on polygons, *Math. Methods Appl. Sci.*, **12** (1990), 413–427.

[2] T. F. Chan, T. P. Mathew: Domain decomposition algorithms, *Acta Numerica* (1994), 61–143.

[3] M. Costabel: Boundary integral operators on Lipschitz domains: Elementary results, *SIAM J. Math. Anal.*, **19** (1988), 613–626.

[4] M. Costabel, E. P. Stephan: A direct boundary integral equation method for transmission problems, *J. Math. Anal. Appl.*, **106** (1985), 367–413.

[5] M. Costabel, E. P. Stephan: Coupling of finite element and boundary element methods for an elasto-plastic interface problem, *SIAM J. Numer. Anal.*, **27** (1990), 1212–1226.

[6] B. Q. Guo, I. Babuška: The h-p version of the finite element method, Part 1: The basic approximation results, *Comp. Mech.*, **1** (1986), 21–41.

[7] B. Q. Guo, N. Heuer, E. P. Stephan: The h-p version of the boundary element method for transmission problems with piecewise analytic data, *SIAM J. Numer. Anal.*, **33** (1996), 789–808.

[8] N. Heuer, E. P. Stephan: Boundary integral operators in countably normed spaces, *Math. Nachr.*, **191** (1998), 123–151.

[9] G. C. Hsiao, W. L. Wendland: Domain decomposition in boundary element methods. In: Domain Decomposition Methods for Partial Differential Equations, R. Glowinski, Y. A. Kuznetsov, G. A. Meurant, J. Périaux, and O. B. Widlund, eds., SIAM, Philadelphia (1991), 41–49.

[10] V. G. Maz'ya: Boundary integral equations, *Encyclopeadia of Mathematical Sciences*, Springer–Verlag Berlin, **17** (1991), 127–223.

[11] A. H. Schatz, V. Thomée, W. L. Wendland: *Mathematical Theory of Finite and Boundary Element Methods*, Birkhäuser, Basel, 1990.

[12] O. Steinbach: Gebietszerlegungsmethoden mit Randintegralgleichungen und effiziente numerische Lösungsverfahren für gemischte Randwertprobleme. PhD–Thesis, Universität Stuttgart, 1996.

[13] W. L. Wendland: On asymptotic error estimates for the combined BEM and FEM. In: Innovative Numerical Methods in Engineering, R. P. Shaw et al., eds., Springer-Verlag Berlin (1986), 55–70.

Norbert Heuer, Institut für Wissenschaftliche Datenverarbeitung, Universität Bremen, Universitätsallee 29, 28359 Bremen, Germany;

Ernst P. Stephan, Institut für Angewandte Mathematik, Universität Hannover, Welfengarten 1, 30167 Hannover, Germany.

T. IVANOV, V. MAZ'YA and G. SCHMIDT

Boundary Layer Approximate Approximations for the Cubature of Potentials

1. Introduction

In this article we study a new method for the computation of multivariate integral operators with singular difference kernels over bounded domains

$$\mathcal{K}u(x) = \int_\Omega g(x-y)\,u(y)\,dy\,. \tag{1}$$

The accurate computation of such integrals and their derivatives is one of the main tasks, for example, in the solution of boundary value problems for partial differential equations with inhomogeneous right–hand side u by using boundary element methods.

Our approach is based on the concept of *approximate approximations* introduced in [3]. This concept offers new possibilities for high order approximations of integral and pseudodifferential operators of mathematical physics. This is demonstrated in [4] and [5], where also the cubature error for integrals over the whole space is studied. The concept of approximate approximations was applied recently in [2] to develop efficient cubature formulas for bounded Lipschitz domains. Here we consider a modification of this method to polyhedral domains leading to formulas of the same accuracy but of reduced complexity.

Our method approximates on the whole space $\mathbb{R}^n$ the function u by a quasi–interpolant which is a linear combination of shifted and scaled versions of a special basis function η. This function is sufficiently smooth and rapidly decaying in $\mathbb{R}^n$ and satisfies the moment condition

$$\int_{\mathbb{R}^n} \eta(x)\,dx = 1\,, \quad \int_{\mathbb{R}^n} x^\alpha \eta(x)\,dx = 0\,, \quad 0 < |\alpha| < N \tag{2}$$

for all multiindeces $\alpha = (\alpha_1,\ldots,\alpha_n)$, $|\alpha| = \alpha_1+\ldots+\alpha_n$. In the sequel m denotes the integer vector $m = (m_1,\ldots,m_n) \in \mathbb{Z}^n$.

In the case that the density u has compact support in Ω and belongs to the Sobolev space W^N, $N > n/2$, the quasi–interpolant

$$\mathcal{M}_h u(x) := \mathcal{D}^{-n/2} \sum_{hm \in \Omega} u(hm)\,\eta\Big(\frac{x-hm}{h\sqrt{\mathcal{D}}}\Big)\,. \tag{3}$$

approximates u in the L_p–norm in such a way that for any $\varepsilon > 0$ one can choose the parameter $\mathcal{D} > 0$ large enough that

$$\|\mathcal{M}_h u - u\|_{L_p(\mathbb{R}^n)} \le c\,(\sqrt{\mathcal{D}}h)^N \|\nabla_N u\|_{L_p(\Omega)} + \varepsilon \|u\|_{W_p^{N-1}(\Omega)}\,. \tag{4}$$

Thus the quasi–interpolant (3) behaves up to the so–called saturation error ε, which can be prescribed, like a high order approximation. Note that with respect to negative norms even the estimate

$$\|\mathcal{M}_h u - u\|_{H_p^{-t}(\mathbb{R}^n)} \leq c(\sqrt{\mathcal{D}}h)^N \|\nabla_N u\|_{L_p(\Omega)} + h^t \varepsilon \|u\|_{W_p^{N-1}(\Omega)}, \quad t > 0, \qquad (5)$$

is valid, here H_p^t, $t \in (-\infty, \infty)$, $p \in [1, \infty)$, denotes the scale of Bessel potential spaces. Since the results of numerical computations are of interest only with certain tolerance, this *approximate approximation* procedure can be applied in numerical methods if the saturation error ε is smaller than this tolerance. The approach allows to use new classes of basis functions adapted to the special numerical problem. If, for example, η is chosen such that the integral $\mathcal{K}\eta$ can be analytically determined, which is in general impossible for piecewise polynomials or other finite–element functions, then

$$\mathcal{K}_h u(x) := \mathcal{K}\mathcal{M}_h u(x) = h^n \sum_{hm \in \Omega} u(hm) \int_{\mathbb{R}^n} g\left(\sqrt{\mathcal{D}}h\left(\frac{x-hm}{h\sqrt{\mathcal{D}}} - y\right)\right) \eta(y)\, dy \qquad (6)$$

is a semi–analytic cubature formula for the integral operator. If $\mathcal{K}$ is a bounded operator from $H_p^{-t}(\mathbb{R}^n)$, $t \geq 0$, into a Banach space X then obviously the error estimate (5) is valid also for $\|\mathcal{K}u - \mathcal{K}_h u\|_X$.

The quasi–interpolation on uniformly distributed nodes $\{hm\} \subset \Omega$, $m \in \mathbb{Z}^n$, provides good approximations only if the function u is smooth in $\mathbb{R}^n$. If u does not vanish at the boundary $\partial\Omega$ the computation of the integral $\mathcal{K}u$ by using linear combinations of $\mathcal{K}\eta$ requires a new idea. Since the approximant composed of dilated shifts of η shall approximate in some suitable norm the function u extended by zero outside the domain Ω, mesh refinement towards the boundary is necessary. In [2] a method was introduced which uses a system of M boundary layers $\mathcal{Q}_k \subset \Omega$ and constructs at any layer special approximants for uniform isotropic meshes of size $h_k = \mu^k h$, $\mu^{-1} \in \mathbb{N}$. The generating function η is supposed to belong to the Schwartz class $\mathcal{S}(\mathbb{R}^n)$ with positive Fourier transform $\mathcal{F}\eta > 0$ and to satisfy (2). It is shown that for a bounded Lipschitz domain Ω and $u \in W_p^N(\Omega)$ there exists a quasi–interpolant of the form

$$\mathcal{B}_M u(x) = \mathcal{D}^{-n/2} \sum_{k=0}^{M} \sum_{h_k m \in \mathcal{Q}_k} c_{k,m}\, \eta\left(\frac{x - h_k m}{h_k \sqrt{\mathcal{D}}}\right), \qquad (7)$$

with coefficients $c_{k,m}$ depending on the values of $u(h_k m)$, which approximates u with

$$\|u - \mathcal{B}_M u\|_{L_p(\mathbb{R}^n)} \leq c_1 (\mathcal{D}h)^N \|\nabla_N u\|_{L_p(\Omega)} + c_2 (\mu^M h)^{1/p} \|u\|_{L_\infty(\Omega)} + \varepsilon \|u\|_{W_p^N(\Omega)}. \qquad (8)$$

Thus for $p < \infty$ and sufficiently large M the second term in estimate (8) can be made comparable to the first term. Thus an estimate similar to (4) is valid and $\mathcal{K}\mathcal{B}_M u$ is a high order cubature formula of $\mathcal{K}u$. The construction of $\mathcal{B}_M u$ will be given in the next section. In section 3 we specialize this method for polyhedral domains and generating functions having tensor product structure. At the boundary layers again approximants on uniform meshes are constructed, but the meshes will be refined only in the direction to the boundary. This anisotropic mesh refinement leads to a considerable reduction of data points and, which is most important, of the number of summands in $\mathcal{K}\mathcal{B}_M u$ for the cubature of the singular integral $\mathcal{K}u$. Finally, in section 4 we give an example how potentials of special anisotropic tensor product generating functions can be computed.

2. Isotropic boundary layer approximate approximation

In the following we suppose that the generating function η belongs to the Schwartz class $\mathcal{S}(\mathbb{R}^n)$ and satisfies the moment condition (2). The construction of $\mathcal{B}_M$ providing estimate (8) is based on an approximate refinement relation which can be formulated as follows.

Lemma 1. ([6]) *Suppose that $\eta \in \mathcal{S}(\mathbb{R}^n)$ has a positive Fourier transform $\mathcal{F}\eta > 0$ and that for given $\mu \in (0,1)$ the function $\tilde{\eta} \in \mathcal{S}(\mathbb{R}^n) = \mathcal{F}^{-1}(\mathcal{F}\eta(\cdot)/\mathcal{F}\eta(\mu\cdot))$ belongs also to $\mathcal{S}(\mathbb{R}^n)$. Then for any $\varepsilon > 0$ there exists $\mathcal{D} > 0$ such that*

$$\eta\left(\frac{x}{\sqrt{\mathcal{D}}}\right) = \mathcal{D}^{-n/2} \sum_{m \in \mathbb{Z}^n} \tilde{\eta}\left(\frac{\mu m}{\sqrt{\mathcal{D}}}\right) \eta\left(\frac{x - \mu m}{\sqrt{\mathcal{D}}\mu}\right) + R_{\eta,\mu,\mathcal{D}}(x), \qquad (9)$$

where $R_{\eta,\mu,\mathcal{D}} \in \mathcal{S}(\mathbb{R}^n)$ fulfills $|R_{\eta,\mu,\mathcal{D}}| < \varepsilon$.

Corollary 1. ([2]) *If $\mu^{-1} \in \mathbb{N}$ then (9) implies the factorization of the quasi–interpolation operator (3)*

$$\mathcal{M}_h = \mathcal{M}_{\mu h} \widetilde{\mathcal{M}}_h + \mathcal{R}_h,$$

with the quasi-interpolant

$$\widetilde{\mathcal{M}}_h f(x) := \mathcal{D}^{-n/2} \sum_{m \in \mathbb{Z}^n} f(hm) \, \tilde{\eta}\left(\frac{x - hm}{\sqrt{\mathcal{D}}h}\right), \qquad (10)$$

and the remainder

$$\mathcal{R}_h f(x) = \mathcal{D}^{-n/2} \sum_{m \in \mathbb{Z}^n} f(hm) \, R_{\eta,\mu,\mathcal{D}}\left(\frac{x - hm}{\sqrt{\mathcal{D}}h}\right) \quad \text{with} \quad \|\mathcal{R}_h f\|_{L_\infty} \leq \varepsilon \|\mathcal{R}_h f\|_{L_\infty}.$$

Note that $\tilde{\eta}$ satisfies also the moment condition (2) and therefore $\widetilde{\mathcal{M}}_h$ provides the same approximation properties as $\mathcal{M}_h$.

Let us fix some tolerance $\varepsilon > 0$. Then in accordance with (4) (see also [5]) one can choose $\mathcal{D}$ such that the saturation errors of $\mathcal{M}_h$ and $\widetilde{\mathcal{M}}_h$ are less than $\varepsilon/2$. Further, due to the decay of η and $\tilde{\eta}$ one can truncate the summation in (3) and (10).

Lemma 2. ([2]) *There exists $N_s > 0$ such that the quasi–interpolant*

$$\mathcal{M}_h^s u(x) := \mathcal{D}^{-n/2} \sum_{|x-hm| \leq N_s h} u(hm) \, \eta\left(\frac{x - hm}{h\sqrt{\mathcal{D}}}\right) \qquad (11)$$

approximates $u \in W_p^N(\Omega)$ in the subdomain $\Omega_{N_s h} = \{x \in \Omega : \mathrm{dist}(x, \partial\Omega) > N_s h\}$ with the error estimate

$$\|\mathcal{M}_h^s u - u\|_{L_p(\Omega_{N_s h})} \leq c \, (\sqrt{\mathcal{D}} h)^N \|\nabla_N u\|_{L_p(\Omega)} + \varepsilon \|u\|_{W_p^{N-1}(\Omega_{N_s h})}.$$

To ensure that the norm of the approximant outside Ω is small we introduce another parameter N_o. Consider the subset $\mathcal{Q}_0 = \{x \in \Omega : \mathrm{dist}(x, \partial\Omega) \geq N_o h\}$ and its characteristic function $\chi_0^\star$. Obviously it is possible to choose the number N_o so that

$$\|\mathcal{M}_h^s \chi_0^\star u\|_{L_p(\mathbb{R}^n \setminus \Omega)} < \varepsilon \|u\|_{L_\infty(\Omega)}.$$

Note that these parameters do not depend on the mesh width h. Now we are in position to define the boundary layers. Fix h and the scaling factor μ with $\mu^{-1} \in \mathbb{N}$. In the following we denote by $h_k = \mu^k h$ the sequence of step sizes and define the boundary layers

$$\mathcal{Q}_k = \{x \in \Omega : N_o h_k \leq \mathrm{dist}(x, \partial\Omega) \leq (N_o + \widetilde{N}_s) h_{k-1}\}.$$

Here $\widetilde{N}_s$ is the truncation parameter of $\widetilde{\mathcal{M}}_h$. Now we define the approximant as

$$\mathcal{B}_M u(x) = \mathcal{D}^{-n/2} \sum_{k=0}^{M} \sum_{h_k m \in \mathcal{Q}_k} c_{k,m}\, \eta\left(\frac{x - h_k m}{h_k \sqrt{\mathcal{D}}}\right), \qquad (12)$$

with the coefficients

$$c_{k,m} = \begin{cases} u(h_0 m), & k = 0 \\ u(h_k m) - (\widetilde{\mathcal{M}}^s_{\mu^{k-1} h} \chi^\star_{k-1} u)(h_k m), & k \geq 1 \end{cases} \qquad (13)$$

Here $\chi^\star_k$ is the characteristic function of the subset $\{x \in \Omega : \mathrm{dist}(x, \partial\Omega) > N_o h_k\}$. Note that

$$\begin{aligned}
(\widetilde{\mathcal{M}}^s_{\mu^{k-1} h} \chi^\star_{k-1} u)(h_k m) &= \mathcal{D}^{-n/2} \sum_{|h_k m - h_{k-1} j| \leq \widetilde{N}_s h_{k-1}} u(h_{k-1} j)\, \tilde\eta\left(\frac{h_k m - h_{k-1} j}{h_{k-1} \sqrt{\mathcal{D}}}\right) \\
&= \mathcal{D}^{-n/2} \sum_{|\mu m - j| \leq \widetilde{N}_s} u(h_{k-1} j)\, \tilde\eta\left(\frac{\mu m - j}{\sqrt{\mathcal{D}}}\right)
\end{aligned}$$

where additionally $\mathrm{dist}(h_{k-1} j, \partial\Omega) > N_o h_{k-1}$. Since the coefficients $\tilde\eta((\mu m - j)/\sqrt{\mathcal{D}})$ do not depend on the number of the boundary layer they can be precomputed and used for any $k \geq 1$.

Under the additional condition that N_o is chosen such that $N_o > (1 - \mu)^{-1} \widetilde{N}_s$ the following theorems have been proved in [2].

Theorem 1. *Suppose that Ω is a bounded domain with Lipschitz boundary and let $u \in W_p^N(\Omega)$ with $N > n/p$. Under the assumptions made above, the quasi-interpolant $\mathcal{B}_M$ approximates u with the error*

$$\|u - \mathcal{B}_M u\|_{L_p(\mathbb{R}^n)} \leq c_1 (\mathcal{D} h)^N \|\nabla_N u\|_{L_p(\Omega)} + c_2 (\mu^M h)^{1/p} \|u\|_{L_\infty(\Omega)} + \varepsilon \|u\|_{W_p^N(\Omega)}.$$

Theorem 2. *If Ω is a bounded domain with Lipschitz boundary, then for any $u \in W_p^N(\Omega)$ with $N > n/p$ and for any $\varepsilon > 0$ there exists $\mathcal{D} > 0$ and a boundary layer approximation $\mathcal{B}_M$ such that*

$$\|u - \mathcal{B}_M u\|_{H_p^{-s}(\mathbb{R}^n)} \leq (c_1 (\mathcal{D} h)^N + c_2 (\mu^M h)^{1/p+r} + \varepsilon\, h^s) \|u\|_{W_p^N(\Omega)},$$

where $0 < r < s/n$ and $r \leq (p-1)/p$.

These results follow from the factorization property of the quasi–interpolation operator given in Corollary 1. The factorization leads to the multiresolution decomposition of the quasi–interpolant

$$\mathcal{M}_{\mu^M h}\, \chi_M^\star = \mathcal{M}_h\, \chi_0^\star + \sum_{k=1}^{M} \mathcal{M}_{\mu^k h}(\chi_k^\star - \widetilde{\mathcal{M}}_{\mu^{k-1} h}\, \chi_{k-1}^\star) - \sum_{k=0}^{M-1} \mathcal{R}_{\mu^k h}\, \chi_k^\star\,, \qquad (14)$$

which holds for $\mu^{-1} \in \mathbb{N}$ and any sequence of bounded operators $\{\chi_k^\star\}$. The main idea in the construction of the operator $\mathcal{B}_M$ is to perform on the layer $\mathcal{Q}_k$ only k levels of this multiresolution decomposition. Formally this can be written in the following form: If χ_k denotes the characteristic function of the boundary layer $\mathcal{Q}_k$, then $\mathcal{B}_M$ is given as

$$\mathcal{B}_M := \mathcal{M}_{\mu h}^s\, \chi_0^\star + \sum_{k=1}^{M} \mathcal{M}_{\mu^k h}^s\, \chi_k\, (\chi_k^\star - \widetilde{\mathcal{M}}_{\mu^{k-1} h}^s\, \chi_{k-1}^\star)\,.$$

It turns out that for $x \in \Omega_{(N_o+N_s)h_k} \setminus \Omega_{(N_o+N_s)h_{k-1}}$ the quasi–interpolant $\mathcal{B}_M u$ approximates u modulo the saturation error with the order $O((\mathcal{D}\mu^k h)^N)$. Near the boundary of Ω, precisely in the subset $\Omega \setminus \Omega_{(N_o+N_s)h_M}$, the difference $\mathcal{B}_M u - u$ is not small, but its L_p-norm can be estimated by $c_2 (\mu^M h)^{1/p} \|u\|_{L_\infty(\Omega)}$ due to $\mathrm{meas}(\Omega \setminus \Omega_{(N_o+N_s)h_M}) = O((N_o + N_s)\mu^M h)$. Thus, by choosing $\mu^M = O(h^{Np-1})$, for given N and $p < \infty$, this error behaves like $O(h^N)$.

As mentioned in the introduction, the cubature formula $\mathcal{K}_h u$ for the integral operator

$$\mathcal{K} u(x) = \int_\Omega g(x-y) u(y) dy\,.$$

is obtained from the quasi–interpolant of the density u by setting

$$\mathcal{K}_h u(x) := \mathcal{K} \mathcal{B}_M u(x) = \sum_{k=0}^{M} h_k^n \sum_{h_k m \in \mathcal{Q}_k} c_{k,m} \int_{\mathbb{R}^n} g\Big(\sqrt{\mathcal{D}} h_k \Big(\frac{x - h_k m}{h_k \sqrt{\mathcal{D}}} - y\Big)\Big) \eta(y)\, dy \quad (15)$$

if η is chosen such that the integrals can be obtained analytically or by simple one-dimensional quadrature.

Let us consider a class of generating functions introduced in [4] which allows to obtain accurate cubature formulas for the inverse of elliptic partial differential operators connected with the Laplacian. Consider for $x \in \mathbb{R}^n$ and $M = 1, 2, \ldots$ the functions

$$\eta_{2M}(x) := \pi^{-n/2} L_{M-1}^{(n/2)}(|x|^2)\, e^{-|x|^2}\,, \quad \text{where } L_j^{(\gamma)}(y) = \frac{e^y y^{-\gamma}}{j!} \Big(\frac{d}{dy}\Big)^j \Big(e^{-y} y^{j+\gamma}\Big) \qquad (16)$$

are the generalized Laguerre polynomials. Since

$$\mathcal{F}\eta_{2M}(\lambda) = e^{-\pi^2 |\lambda|^2} \sum_{j=0}^{M-1} \frac{(\pi^2 |\lambda|^2)^j}{j!} \qquad (17)$$

the conditions of Lemma 1 as well as the moment condition (2) with $N = 2M$ are satisfied by η_{2M}. Using the additive representation

$$L_{M-1}^{(n/2)}(|x|^2) e^{-|x|^2} = \sum_{j=0}^{M-1} \frac{(-1)^j}{j! \, 4^j} \Delta^j e^{-|x|^2}, \tag{18}$$

the harmonic potential $\mathcal{H}_n$ of η_{2M} has the analytic form

$$\mathcal{H}_n \eta_{2M}(x) = \frac{\Gamma(\frac{n}{2} - 1)}{4\pi^{n/2}} \int_{\mathbb{R}^n} \frac{\eta_{2M}(y)}{|x-y|^{n-2}} dy$$

$$= \frac{1}{4|x|^{n-2} \pi^{n/2}} \int_0^{|x|^2} \tau^{n/2-2} e^{-\tau} d\tau + \pi^{-n/2} e^{-|x|^2} \sum_{j=0}^{M-2} \frac{L_j^{(n/2-1)}(|x|^2)}{4(j+1)}.$$

Thus the approximation of the harmonic potential $\mathcal{H}_n u$ is obtained from (15) after calculating $\mathcal{H}_n \eta_{2M}$ for the argument $(x - h_k m)/h_k \sqrt{\mathcal{D}}$.

Some further examples for the action of different potentials of mathematical physics on the generating functions η_{2M} in any space dimension, including the elastic, hydrodynamic and diffraction potentials, can be found in [4] and [5].

To estimate the accuracy of (15) we note that the interesting integral operators with singular kernel function are bounded mappings

$$\mathcal{K} : L_p(\Omega) \to W_p^m(\Omega_1), \tag{19}$$

with $\Omega, \Omega_1 \subset \mathbb{R}^n$; we write $\mathcal{K} \in \mathcal{L}(L_p(\Omega), W_p^m(\Omega_1))$. Note that the case $m = 0$ corresponds to singular integral operators, whereas the volume potentials associated with partial differential equations satisfy relation (19) with the order of the differential equations m. In any case the kernel $g(x - y)$ is singular at the diagonal $x = y$, so that the approximation of such multivariate integrals by usual cubature approaches is quite complicated.

Since in general supp $\mathcal{B}_M u = \mathbb{R}^n$ one has to estimate also integrals of the form

$$\int_{\mathbb{R}^n \setminus \Omega} g(x - y) \mathcal{B}_M u(y) \, dy, \quad x \in \Omega_1.$$

To this end we make the following assumption: For a ball B_R with radius R around the origin containing the domains Ω and Ω_1 the kernel function g satisfies for all multi-indexes $0 \leq |\alpha| \leq m$ the estimate

$$|\partial^\alpha g(x - y)| \leq r_\alpha(|y|), \quad \text{for } x \in \Omega_1, \, y \in \mathbb{R}^n \setminus B_R,$$

where r_α are functions of at most polynomial growth.

Theorem 3. ([2]) *Let $u \in W_p^N(\Omega)$ with $N > n/p$ and $\mathcal{K} \in \mathcal{L}(L_p(\Omega), W_p^m(\Omega_1))$. Under the assumptions made above for any $\varepsilon > 0$ there exists $\mathcal{D} > 0$ such that*

$$\|\mathcal{K}u - \mathcal{K}_h u\|_{W_p^m(\Omega_1)} \leq c_1(\mathcal{D}h)^N \|\nabla_N u\|_{L_p(\Omega)} + c_2(\mu^M h)^{1/p} \|u\|_{L_\infty(\Omega)} + \varepsilon \|u\|_{W_p^N(\Omega)}.$$

If additionally $\mathcal{K} \in \mathcal{L}((W_{p/(p-1)}^m(\Omega))', L_p(\Omega_1))$ then

$$\|\mathcal{K}u - \mathcal{K}_h u\|_{L_p(\Omega_1)} \leq (c_1(\mathcal{D}h)^N + c_2(\mu^M h)^{1/p+r})\|u\|_{W_p^N(\Omega)} + \varepsilon\, h^m\, \|u\|_{W_p^N(\Omega)},$$

where $0 < r < m/n$, $r \leq (p-1)/p$.

Remark 1. Theorem 3 shows that for $p < \infty$ and $\mathcal{D}$ large enough an N-th order approximation of the integral $\mathcal{K}u$ up to a prescribed accuracy can be obtained by fixing $\mu^{-1} \in \mathbb{N}$ and choosing M large enough, for example $M = O((pN-1)\log h/\log\mu)$. The computational costs are of course proportional to the number of grid points used for the construction of $\mathcal{B}_M$. The thickness of the layers $\mathcal{Q}_k$, $k = 1, \ldots, M$, is proportional to the size $\mu^k h$ of the mesh chosen there, hence the number of grid points at $\mathcal{Q}_k$ is $O((\mu^k h)^{-(n-1)})$. Therefore the computational costs for evaluating the sum over all nodes of the layers $\mathcal{Q}_k$, $k = 1, \ldots, M$, are of the order

$$O\left(\left(\frac{(\mu-\mu^M)h}{1-\mu}\right)^{-(n-1)}\right).$$

Thus the boundary layer approximation with isotropic meshes reduces the order of the computational costs by 1. In the next section we consider the use of anisotropic meshes for special domains leading to a further reduction of the complexity of the cubature. Note that a more detailed analysis performed in [1] indicate that an optimal choice for scaling subsequent meshes is $\mu = 1/3$ if the generating functions (16) are used.

Remark 2. Note that the previous construction and results are applicable if instead of the integer points $\{m \in \mathbb{Z}^n\}$ the set of lattice points $\{Am : m \in \mathbb{Z}^n\}$ with a nonsingular $n \times n$–matrix A is taken as basis grid. This case can be transformed to that considered above by introducing the coordinates $y = Ax$ and the new generating function $\eta_A := |\det A|^{-1} \eta(A\cdot)$, which satisfies all necessary conditions.

3. Anisotropic boundary layer approximate approximation

Consider a three–dimensional bounded polyhedral domain Ω. For approximating the integral operator acting on functions u over Ω we partition the domain in simpler parts and process them separately. Using a partition of unity the function u can be decomposed into the sum

$$u = u\left(\varphi_{int} + \underbrace{\sum \varphi_{c_k}}_{corners} + \underbrace{\sum \varphi_{e_k}}_{edges} + \underbrace{\sum \varphi_{f_k}}_{faces}\right) \qquad (20)$$

where the cut–off functions $\varphi \in C^\infty$, $0 \leq \varphi \leq 1$, are different from zero only on special parts of Ω. So $\varphi_{c_k} = 1$ at a neighbourhood of the k-th corner point and vanishes outside a larger neighbourhood. Correspondingly, $\varphi_{e_k} = 1$ at a neighbourhood of the interior of the k-th edge, $\varphi_{f_k} = 1$ at some interior part of the k-th face and $\varphi_{int} = 1$ on some interior part of Ω. The approximation of $u\varphi_{int}$ can be performed by the usual quasi–interpolant (3) with a suitable generating function, whereas for the approximation of the functions $u\,\varphi_{c_k}$ the operator $\mathcal{B}_M$ considered in the previous section will be used on the domain $\operatorname{supp} \varphi_{c_k}$.

Here we study the approximation of the functions $u\,\varphi_{e_k}$ and $u\,\varphi_{f_k}$ by using mesh refinement only in the direction normal to the boundary side.

To start with the approximation near the faces of Ω we consider a sufficiently smooth function u given in $\mathbb{R}^3_+ = \{x = (x', x_3) \in \mathbb{R}^3 : x_3 \geq 0\}$ with bounded support and $u(x', 0) \not\equiv 0$. More precisely, suppose that $G \subset \mathbb{R}^3_+$ is a bounded domain whose boundary contains a two-dimensional domain $\Gamma \subset \{(x', 0) : x' \in \mathbb{R}^2\}$. In the following $\mathring{W}^N_p(G)$ denotes the subspace of all functions from $W^N_p(G)$ which can be smoothly extended by zero through $\partial G \backslash \Gamma$.

We are interested in constructing approximants to $u \in \mathring{W}^N_p(G)$ with respect to the sequence of anisotropically distributed mesh points $\{(hm', \mu^k hm_3) : (m', m_3) \in \mathbb{Z}^3_+\}$. Therefore we consider generating functions of the tensor product form

$$\eta_3(x', x_3) = \eta_2(x')\,\eta_1(x_3) \tag{21}$$

where η_2 is a two- and η_1 a one-dimensional generating function of the Schwartz class with positive Fourier transform.

Lemma 3. *Suppose that both $\eta(x')$ and $\eta(x_3)$ satisfy the moment condition (2) for some given N and let $\mu \in (0, 1)$. There exists $N_s > 0$ such that the quasi-interpolant*

$$\mathcal{M}^s_{h,\mu} u(x) := \mathcal{D}^{-3/2} \sum_{|x'-hm'|+|x_3/\mu-hm_3|\leq N_s h} u(hm)\,\eta_3\!\left(\frac{x'-hm'}{h\sqrt{\mathcal{D}}}, \frac{x_3 - \mu h m_3}{\mu h \sqrt{\mathcal{D}}}\right) \tag{22}$$

approximates $u \in \mathring{W}^N_p(G)$, $N > n/p$, in the subdomain $G_{N_s \mu h} = \{x \in G : x_3 > N_s \mu h\}$ with the estimate

$$\|\mathcal{M}^s_{h,\mu} u - u\|_{L_p(G_{N_s h})} \leq c\,(\sqrt{\mathcal{D}} h)^N \|\nabla_N u\|_{L_p(G)} + \varepsilon \|u\|_{W^{N-1}_p(G_{N_s \mu h})}.$$

Thus, for any $\delta > 0$ one obtains approximations of order $O((\mathcal{D} h)^N)$ modulo the saturation error for all $x \in G$ with exception of the strip $0 < x_3 < \delta$, if μ is chosen sufficiently small. Therefore it suffices to refine the mesh only in x_3-direction, i.e. to use the refinement relation for the one-dimensional generating function η_1. If we denote by $\tilde\eta_1$ the corresponding function described in Lemma 1 then obviously the assertion of Corollary 1 remains true for $\mathcal{M}_{h,\mu}$ and

$$\widetilde{\mathcal{M}}_{h,\mu} f(x) := \mathcal{D}^{-3/2} \sum_{m \in \mathbb{Z}^3} f(hm)\,\eta_2\!\left(\frac{x'-hm'}{\sqrt{\mathcal{D}} h'}\right) \tilde\eta_1\!\left(\frac{x_3 - \mu h m_3}{\mu h \sqrt{\mathcal{D}}}\right),$$

i.e.
$$\mathcal{M}_{h,\mu} = \mathcal{M}_{h,\mu^2}\,\widetilde{\mathcal{M}}_{h,\mu} + \mathcal{R}_{h,\mu}, \qquad \|\mathcal{R}_{h,\mu}\| < \varepsilon.$$

Analogously to the isotropic case we may therefore define the face approximant as

$$\mathcal{B}_{M,f} u(x) = \mathcal{D}^{-3/2} \sum_{k=0}^M \sum_{(hm', \mu^k h m_3) \in \mathcal{Q}_k} c_{k,m}\,\eta_3\!\left(\frac{x'-hm'}{h\sqrt{\mathcal{D}}}, \frac{x_3 - \mu^k h m_3}{\mu^k h \sqrt{\mathcal{D}}}\right) \tag{23}$$

with the coefficients

$$c_{k,m} = \begin{cases} u(hm', hm_3), & k = 0 \\ u(hm', \mu^k hm_3) - (\widetilde{\mathcal{M}}^s_{h,\mu^{k-1}} \chi^\star_{k-1} u) \end{cases}$$

where now

$$\mathcal{Q}_k = \{x \in G : N_o h \mu^k \leq x_3 \leq (N_o + \widetilde{N}_s) h \mu^{k-1}\},$$

$\chi^\star_k(x) = 1$ for $x_3 > N_o \mu^k$ and otherwise zero, and the parameters N_o, $\widetilde{N}_s$ are obtained from the one–dimensional functions η_1 and $\tilde{\eta}_1$.

Obviously, for the approximant on the anisotropic mesh (23) the assertions of Theorems 1 and 2 remain true. In contrast to the approximation (12) using isotropic mesh refinement the pointwise approximation quality on the layer $(N_o + N_s)h\mu^k \leq x_3 \leq (N_o + N_s)h\mu^{k-1}$ is worse, $O((\mathcal{D}h)^N)$ compared with $O((\mathcal{D}\mu^k h)^N)$. However, this does not influence the estimate in integral norms and, which is more important, the approximation quality of the resulting cubature formula. But most importantly, anisotropic nmesh refinement leads to a considerable reduction of mesh points and therefore summands in (23), necessary to get N-th order approximation of potentials. Since any $\mathcal{Q}_k$ contains $O(h^{-2})$ mesh points the numerical costs to obtain the sum over these nodes are of the order $O(Mh^{-2})$.

Quite similarly the approximation of edge functions $u \, \varphi_{e_k}$ can be performed with the order $O((\mathcal{D}h)^N)$ using anisotropic mesh refinement. Consider the bounded domain G obtained after intersecting some wedge $\mathcal{W} = \{(x_1, x_2, x_3) \in \mathbb{R}^3 : 0 < x_2 < ax_1\}$, $0 < a \leq \infty$, with another domain. The common part of ∂G and the boundary of the wedge $\partial \mathcal{W}$ is denoted by Γ and again we denote by $\mathring{W}^N_p(G)$ the subspace of all functions from $W^N_p(G)$ which can be smoothly extended by zero through $\partial G \backslash \Gamma$.

Now we are interested in constructing approximants to $u \in \mathring{W}^N_p(G)$ with respect to the sequence of meshes which are refined in the two directions normal to Γ. As basic mesh points we choose the lattice $\{(hAm', hm_3) : m' \in \mathbb{Z}^2_{++}, m_3 \in \mathbb{Z}\}$, where A is a 2×2 matrix transforming the quarter plane $\mathbb{R}^2_{++}$ onto $\{x \in \mathcal{W} : x_3 = 0\}$ with $\det A = 1$.

In view of remark 2 the following assertion for the tensor product generating function (21) is valid.

Lemma 4. *Suppose that both $\eta(x')$ and $\eta(x_3)$ satisfy the moment condition (2) for some given $N > n/p$ and let $\mu \in (0,1)$. There exists $N_s > 0$ such that the quasi-interpolant*

$$\mathcal{M}^s_{h,\mu} u(x) := \mathcal{D}^{-3/2} \sum_{|x'/\mu - hAm'| + |x_3 - hm_3| \leq N_s h} u(hm) \, \eta_3 \left(\frac{x' - \mu hAm'}{\mu h \sqrt{\mathcal{D}}}, \frac{x_3 - hm_3}{h\sqrt{\mathcal{D}}} \right) \quad (24)$$

approximates $u \in \mathring{W}^N_p(G)$ in the subdomain $G_{N_s \mu h} = \{x \in G : \text{dist}(x, \Gamma) > N_s \mu h\}$ with the estimate

$$\|\mathcal{M}^s_{h,\mu} u - u\|_{L_p(G_{N_s h})} \leq c \, (\sqrt{\mathcal{D}} h)^N \|\nabla_N u\|_{L_p(G)} + \varepsilon \|u\|_{W^{N-1}_p(G_{N_s \mu h})} \, .$$

The mesh refinement towards Γ without loss of approximation order is possible due to the factorization

$$\mathcal{M}_{h,\mu} = \mathcal{M}_{h,\mu^2}\widetilde{\mathcal{M}}_{h,\mu} + \mathcal{R}_{h,\mu}, \quad \|\mathcal{R}_{h,\mu}\| < \varepsilon.$$

where now again $\mu^{-1} \in \mathbb{N}$,

$$\widetilde{\mathcal{M}}_{h,\mu}f(x) := \mathcal{D}^{-3/2}\sum_{m \in \mathbb{Z}^3} f(hm)\,\tilde{\eta}_2\left(\frac{x' - \mu h Am'}{\mu h'\sqrt{\mathcal{D}}}\right)\eta_1\left(\frac{x_3 - hm_3}{h\sqrt{\mathcal{D}}}\right),$$

and $\tilde{\eta}_2$ is the adjoint function of η_2 defined in Lemma 1. This follows immediately from the fact that the quasi–interpolant with $\eta_{2,A} := \eta_2(A\,\cdot)$ is based on the nodes $h\mathbb{Z}^2$ and that $\mathcal{F}\eta_{2,A} = \mathcal{F}\eta_2(A^{-1}\,\cdot)$.

Now the edge approximant to $u \in \mathring{W}_p^N(G)$ is defined as

$$\mathcal{B}_{M,e}u(x) = \mathcal{D}^{-3/2}\sum_{k=0}^{M}\sum_{(\mu^k hAm', hm_3) \in \mathcal{Q}_k} c_{k,m}\,\eta_3\left(\frac{x' - \mu^k hAm'}{\mu^k h\sqrt{\mathcal{D}}}, \frac{x_3 - hm_3}{h\sqrt{\mathcal{D}}}\right) \quad (25)$$

with the coefficients

$$c_{k,m} = \begin{cases} u(hAm', hm_3), & k=0 \\ u(\mu^k hAm', hm_3) - (\widetilde{\mathcal{M}}^s_{h,\mu^{k-1}}\chi^\star_{k-1}u)(\mu^k hAm', hm_3), & k \geq 1 \end{cases}$$

where

$$\mathcal{Q}_k = \{x \in G : N_o h\mu^k \leq \mathrm{dist}(x,\Gamma) \leq (N_o + \widetilde{N}_s)h\mu^{k-1}\},$$

$\chi_k^\star(x) = 1$ for $\mathrm{dist}(x,\Gamma) > N_o\mu^k$ and otherwise zero, and the parameters $N_o, \widetilde{N}_s$ are obtained from the two–dimensional functions η_2 and $\tilde{\eta}_2$. Again, for (25) the assertions of Theorems 1 and 2 remain true. Remark that the numerical costs to evaluate $\mathcal{B}_{M,e}u$ are of the order $O(M^2 h^{-2})$.

Now we are in position to define the approximant for the function u given in the polyhedral domain Ω. Turning to the decomposition (20) and using the different types of quasi–interpolants defined in (11), (12), (23), and (25) we introduce the sum of approximants

$$\widetilde{\mathcal{B}}_M u = \mathcal{M}_h^s(u\varphi_{int}) + \underbrace{\sum \mathcal{B}_M(u\varphi_{c_k})}_{corners} + \underbrace{\sum \mathcal{B}_{M,e}(u\varphi_{e_k})}_{edges} + \underbrace{\sum \mathcal{B}_{M,f}(u\varphi_{f_k})}_{faces}.$$

Theorem 4. *For any $\varepsilon > 0$ there exists $\mathcal{D} > 0$ and a boundary layer approximation $\widetilde{\mathcal{B}}_M$, which is anisotropic near faces and edges of the polyhedral domain $\Omega \in \mathbb{R}^3$, such that for any $u \in W_p^N(\Omega)$, $N > 3/p$,*

$$\|u - \widetilde{\mathcal{B}}_M u\|_{L_p(\mathbb{R}^3)} \leq c_1(\mathcal{D}h)^N \|\nabla_N u\|_{L_p(\Omega)} + c_2(\mu^M h)^{1/p}\|u\|_{L_\infty(\Omega)} + \varepsilon\|u\|_{W_p^N(\Omega)}$$

and

$$\|u - \widetilde{\mathcal{B}}_M u\|_{H_p^{-s}(\mathbb{R}^3)} \leq (c_1(\mathcal{D}h)^N + c_2(\mu^M h)^{1/p+r} + \varepsilon\,h^s)\|u\|_{W_p^N(\Omega)},$$

where $0 < r < s/3$ and $r \leq (p-1)/p$.

4. Potentials of tensor product generating functions

From the results of section 2 and Theorem 4 it is clear that the assertions of Theorem 3 are valid also for the approximation $\mathcal{K}\tilde{\mathcal{B}}_M$ of the integral operator $\mathcal{K}$ over a polyhedral domain $\Omega \in \mathbb{R}^3$. Therefore it remains to investigate the efficient computation of potentials of generating functions occuring in $\tilde{\mathcal{B}}_M u$. Here we consider the tensor product function η_3 with factors of the class η_{2M} defined in (16), i.e. we set

$$\eta_3(x) = \pi^{-3/2} L_{M-1}^{(1)}(|x'|^2) L_{M-1}^{(1/2)}(x_3^2) e^{-|x'|^2 - x_3^2} = \pi^{-3/2} \sum_{j,k=0}^{M-1} \frac{(-1)^{j+k}}{j!\,k!\,4^{j+k}} \Delta_{x'}^j \frac{\partial^{2k}}{\partial x_3^{2k}} e^{-|x|^2}$$

where $\Delta_{x'} = (\partial/\partial x_1)^2 + (\partial/\partial x_2)^2$. For potentials of $\eta_3(x', ax_3)$, $a > 0$, we have

$$\int_{\mathbb{R}^3} g(x-y)\, \eta_3(y', ay_3)\, dy = \pi^{-3/2} \sum_{j,k=0}^{M-1} \frac{(-1)^{j+k}}{j!\,k!\,4^{j+k}} \int_{\mathbb{R}^3} g(x-y) \Delta_{y'}^j \frac{\partial^{2k}}{\partial y_3^{2k}} e^{-|y'|^2 - a^2 y_3^2}\, dy$$

$$= \pi^{-3/2} \sum_{j,k=0}^{M-1} \frac{(-1)^{j+k}}{j!\,k!\,4^{j+k}} \Delta_{x'}^j \frac{\partial^{2k}}{\partial x_3^{2k}} \int_{\mathbb{R}^3} g(x-y) e^{-|y'|^2 - a^2 y_3^2}\, dy\,,$$

i.e. higher order cubature formulas are easily derived from the integrals

$$\int_{\mathbb{R}^3} g(x-y)\, e^{-|y'|^2 - a^2 y_3^2}\, dy \,. \tag{26}$$

For $a = 1$ these integrals are known for different potentials of mathematical physics. It remains the case $a \neq 1$, which covers both types (23) for face function and (25) for edge function approximation. For kernels g, often occuring in applications, the integral (26) can be reduced to a one–dimensional integral with a smooth and rapidly decaying integrand. Moreover, by simple differentiation of the integrand with respect to $|x'|$ and x_3 the potential of $\eta_3(x', ax_3)$ can be reduced to a similar one–dimensional integral, i.e. it can be computed with efficient quadrature methods.

Consider for example the solution of the Poisson equation

$$\Delta f = -e^{-|x'|^2 - a^2 x_3^2}$$

tending to 0 if $|x| \to \infty$, i.e. f is the Newton potential of $e^{-|y'|^2 - a^2 y_3^2}$. Taking the two–dimensional Fourier transform with respect to x' leads to the differential equation

$$\frac{\partial^2 \hat{f}}{\partial x_3^2} - 4\pi^2 |\xi'|^2 \hat{f} = -\pi\, e^{-\pi^2 |\xi'|^2 - a^2 x_3^2}$$

for $\hat{f}(\xi', x_3) = (\mathcal{F}_{x' \to \xi'} f)(\xi', x_3)$. Its solution is given by

$$\hat{f}(\xi', x_3) = \frac{e^{-\pi^2 |\xi'|^2}}{4\,|\xi'|} \int_{-\infty}^{\infty} e^{-2\pi |\xi'|\, |x_3 - t|}\, e^{-a^2 t^2}\, dt$$

$$= \frac{e^{-\pi^2 |\xi'|^2 - a^2 x_3^2}}{4a\,|\xi'|} \left(W\!\left(i\Big(\frac{\pi|\xi'|}{a} + a x_3\Big)\right) + W\!\left(i\Big(\frac{\pi|\xi'|}{a} - a x_3\Big)\right) \right),$$

with the Faddejewa function

$$W(z) := e^{-z^2} \int_{-iz}^{\infty} e^{-t^2}\, dt = \frac{\sqrt{\pi}e^{-z^2}}{2}\, \text{erfc}(-iz)\,.$$

Thus $\hat{f}(\xi', x_3) = \hat{f}(|\xi'|, x_3)$ and, consequently, the inverse Fourier transform for radial functions leads to the one-dimensional integral

$$f(x) = f(|x'|, x_3) = \frac{e^{-a^2 x_3^2}}{2} \int_0^\infty e^{-a^2 t^2}\, J_0(2a|x'|t)\Big(W\big(i(t+ax_3)\big) + W\big(i(t-ax_3)\big)\Big)\, dt\,,$$

where J_0 is the Bessel function. Denoting by

$$\mathcal{W}(t, x_3) = e^{-a^2 x_3^2}\Big(W\big(i(t+ax_3)\big) + W\big(i(t-ax_3)\big)\Big)$$

and using $\Delta_{x'} J_0(a|x'|) = -a^2 J_0(a|x'|)$, obviously

$$\frac{1}{4\pi}\int_{\mathbb{R}^3} \frac{\eta_3(y', ay_3)}{|x-y|}\, dy = \frac{1}{2\pi^{3/2}} \int_0^\infty J_0(2a|x'|t) \sum_{j=0}^{M-1} \frac{(at)^{2j}}{j!} e^{-a^2 t^2} \sum_{k=0}^{M-1} \frac{(-1)^k}{k!\, 4^k} \frac{\partial^{2k} \mathcal{W}(t, x_3)}{\partial x_3^{2k}}\, dt\,,$$

i.e., the Newton potential of the anisotropic function $\eta_3(x', ax_3)$ is a one-dimensional integral with smooth and rapidly decaying integrand, and therefore very efficiently computable with some high order quadrature method.

References

[1] T. Ivanov: *Boundary Layer Approximate Approximations and Cubature of Potentials in Domains.* PhD thesis, Linköping University, 1997.

[2] T. Ivanov, V. Maz'ya, G. Schmidt: *Boundary Layer Approximate Approximations and Cubature of Potentials in Domains.* WIAS Preprint No. 402, (1998).

[3] V. Maz'ya: *A New Approximation Method and its Applications to the Calculation of Volume Potentials. Boundary Point Method.* In:*3. DFG-Kolloqium des DFG-Forschungsschwerpunktes "Randelementmethoden"*, 1991.

[4] V. Maz'ya: Approximate approximations. In: *The Mathematics of Finite Elements and Applications. Highlights.* (J. R. Whiteman, ed.), Chichester, Wiley & Sons (1994), 77–104.

[5] V. Maz'ya, G. Schmidt: "Approximate Approximations" and the cubature of potentials. *Rend. Mat. Acc. Lincei*, s. 9, **6** (1995), 161–184.

[6] V. Maz'ya, G. Schmidt: Approximate wavelets and the approximation of pseudodifferential operators. To appear in: *Applied and Computational Harmonic Analysis*.

T. Ivanov, V. Maz'ya: Linköping University, Institute of Technology,
Dept. of Mathematics,
S–581 83 Linköping, Sweden
Email: ivanov@math.liu.se and vlmaz@math.liu.se

G. Schmidt: Weierstraß–Institut für Angewandte Analysis und Stochastik Berlin
Mohrenstraße 39, D–10117 Berlin, Germany
Email: schmidt@wias-berlin.de

D. MAUERSBERGER and I.H. SLOAN

A Simplified Approach to the Semi-discrete Galerkin Method for the Single-layer Equation for a Plate

1. Introduction

This paper presents a simple analysis of a class of semi-discrete Galerkin methods for the three-dimensional single-layer equation on a polygonal plate,

$$Vu(x) := \frac{1}{4\pi} \int_\Gamma \frac{u(y)}{|x-y|} \, dy = f(x), \quad x \in \Gamma, \tag{1}$$

where $\Gamma \subset \mathbb{R}^2$ is a polygon, and $|x - y|$ is the Euclidean distance between x and y. The solution $u(x)$ has the physical interpretation of the electrostatic charge density at $x \in \Gamma$, if f is a prescribed potential on Γ, considered as a charged thin plate in $\mathbb{R}^3$.

The paper follows recent work by R.D Grigorieff and I. Sloan [7] (which in turn followed analyses of plane problems with Lipschitz boundaries in [12] and [1]), in which a boundary element Galerkin method with discontinuous piecewise-constant rectangular elements was applied to the problem above, and then further approximated by replacing the exact integral over each element in the inner-product integral for the Galerkin method by a composite quadrature rule. The quadrature rule for the kth element was assumed to be the M^2-fold copy of a basic rule q_k, obtained by applying an appropriately scaled version of q_k to M^2 smaller rectangles scaled by a factor $1/M$. The essence of the analysis was to show that the semi-discrete approximation exists and is unique if M is sufficiently large, and has an error estimate that mirrors that for the Galerkin method.

The present paper goes beyond the paper [7] in three main ways. First, it uses a quasi-uniform sequence of shape-regular triangulations, with piecewise-constant triangular elements, instead of the rectangular elements of [7]. In this respect it follows the thesis [9]. Second, the technically difficult arguments of [7] are here replaced by relatively simple arguments. Third, the restriction in [7] that quadrature points are not allowed on an element boundary is here removed.

For simplicity, we here consider, as in [7], only piecewise-constant elements. The basic rule is assumed to be exact for linear functions. Thus the semi-discrete approximation is shown to be well defined for $M \geq M_0$, where M_0 is independent of the mesh diameter h, and to converge with optimal order to the exact solution in suitable norms as $h \to 0$. Higher-order discontinuous elements are considered in [9], but in the present context do not result in a higher order of convergence.

Fully discrete Galerkin methods have been discussed in [10] and [5]; see also [6] in this volume.

The authors acknowledge the support of the Australian Research Council.

2. The setting and the approximation scheme

The plate Γ is considered to be a bounded, polygonal region of the plane $\mathbb{R}^2$. The plate is partitioned by a family of triangulations

$$\mathcal{T}_h = \{\triangle_k : k = 1, 2, ..., N\},$$

where $\triangle_k \cap \triangle_l = \emptyset$ for $k \neq l$, and $\bigcup_k \overline{\triangle}_k = \Gamma$. Because we consider only discontinuous elements, non-standard triangulations are allowed; in particular, a vertex of one triangle may be an interior point of the edge of another. However, we shall always require shape regularity: we assume that there exists a constant $\rho > 0$, the same for all triangulations $\mathcal{T}_h$, such that

$$\rho h_k^2 \leq |\triangle_k|, \quad k \geq 1.$$

Here $|\triangle_k|$ is the area of triangle $\triangle_k$, and h_k is its diameter, i.e. the largest edge. Also needed for the final results is quasi-uniformity of the mesh, i.e. we assume the existence of $\sigma > 0$, the same for all $\mathcal{T}_h$, such that

$$h_k \geq \sigma h, \quad \forall k, \tag{2}$$

where we set $h := \max_k h_k$. However, for an important part of the argument, to be discussed in the next section, a local form of quasi-uniformity is sufficient.

Let S_h denote the set of piecewise-constant functions on the triangulation $\mathcal{T}_h$. The Galerkin method is: find $u_h^G \in S_h$, such that

$$(V u_h^G, v_h) = (f, v_h) \quad \forall v_h \in S_h. \tag{3}$$

Here $(\cdot, \cdot)$ denotes the L_2 inner product,

$$(f, v) = \int_\Gamma f(x) v(x) \, dx,$$

and dx is area measure.

The corresponding semi-discrete approximation is: find $u_h \in S_h$, such that

$$(V u_h, v_h)_M = (f, v_h) \quad \forall v_h \in S_h. \tag{4}$$

Here $(\cdot, \cdot)_M$ is a discrete version of the inner-product integral, obtained by replacing the exact integral by a composite quadrature rule Q_M,

$$(f, v)_M = Q_M(fv),$$

where Q_M denotes an M^2-fold composite rule defined this way: first, each element $\triangle_k$ is divided into M^2 congruent copies, obtained by scaling $\triangle_k$ by a factor of $1/M$, and then rotating and translating M^2 such small triangles appropriately to form $\triangle_k$; and then applying to each small triangle an appropriately transformed version of a certain fixed quadrature rule q. This fixed quadrature rule is conveniently defined for a standard reference triangle $\triangle$ with boundaries $x = 0$, $y = 0$ and $x + y = 1$. For example, q might be the rule which places a single point at the centroid,

$$q(f) = \frac{1}{2} f\left(\frac{1}{3}, \frac{1}{3}\right),$$

or the rule that uses the three vertices weighted equally,

$$q(f) = \frac{1}{6}\left(f(0,0) + f(1,0) + f(0,1)\right).$$

In general we will write the rule q as

$$q(f) = \frac{1}{2} \sum_{i=1}^{I} w_i f(t_i),$$

where $\sum_{i=1}^{I} w_i = 1$. The factor $1/2$ is the area of the reference triangle, making the rule q exact for constants. For simplicity, only positive weights w_i are allowed. The corresponding rule for the triangle $\triangle_k$ will be written as

$$q_k(f) = |\triangle_k| \sum_{i=1}^{I} w_i f(t_{k,i}),$$

where $t_{k,i}$ is the image of t_i under the affine mapping of $\triangle$ to $\triangle_k$.

REMARK: The definition of the semi-discrete approximation in (4) above differs slightly from that in [7], in that a modification is introduced in [7] which leads to a symmetric matrix upon the introduction of a basis for S_h. If the basis of S_h is denoted by $\{\phi_1, ..., \phi_N\}$, the construction in [7] uses the formula $(V\phi_i, \phi_j)_M$ (which corresponds to (4)) only for the elements of the matrix for which $i \leq j$, and then enforces symmetry of the matrix to complete the definition. From the point of view of the analysis the difference between the two approaches is of little real importance.

The main result of this paper is stated in Theorem 1 below. In the theorem, as in [7], $\tilde{H}^t(\Gamma)$ denotes the completion of $C_0^\infty(\Gamma)$ in the norm of $H^t(\mathbb{R}^2)$, and $\|\cdot\|_{H^s(\Gamma)}$ denotes the dual norm to $\|\cdot\|_{\tilde{H}^{-s}(\Gamma)}$, i.e.

$$\|g\|_{H^s(\Gamma)} = \sup_{v \in C_0^\infty(\Gamma)} \frac{|(g,v)|}{\|v\|_{\tilde{H}^{-s}(\Gamma)}}.$$

Theorem 1. *Assume that the family of triangulations $\mathcal{T}_h$ is quasiuniform, and let $f \in H^{s+1}(\Gamma)$, where $-1/2 \leq s < 0$. Assume also that the rule q integrates exactly all linear functions. Then there exists M_0, independent of h, such that if $M \geq M_0$ then $u_h \in S_h$ exists and is unique, and for $-1 < t \leq s$ there exists c such that*

$$\|u_h - u\|_{\tilde{H}^t(\Gamma)} \leq ch^{s-t} \|u\|_{\tilde{H}^s(\Gamma)} \leq ch^{s-t} \|f\|_{H^{s+1}(\Gamma)}.$$

In the next section we begin the proof, and give in Lemma 2 the key part of the proof, under a less restrictive regularity assumption on the mesh. The proof will be completed in Section 4. Throughout the paper c is a generic constant, which may take different values in different places.

3. The heart of the matter

To begin the theoretical analysis, it is convenient to write the Galerkin approximation (3) as

$$P_h V u_h^G = P_h f, \quad u_h^G \in S_h, \tag{5}$$

where P_h denotes the L_2-orthogonal projection onto S_h. Note that $P_h f$ can be written concretely as

$$P_h f = \sum_k \frac{\chi_k}{|\triangle_k|} \int_\Gamma f \chi_k = \sum_k \frac{\chi_k}{|\triangle_k|} \int_{\triangle_k} f,$$

where χ_k is the characteristic function of the element $\triangle_k$, which obviously satisfies

$$(\chi_k, \chi_l) = \delta_{k,l} |\triangle_k|. \tag{6}$$

In a similar way we can write the semi-discrete approximation as

$$V_h u_h = P_h f, \quad u_h \in S_h. \tag{7}$$

Here $V_h : S_h \to S_h$ is defined by

$$V_h v_h = \sum_k \frac{\chi_k}{|\triangle_k|} Q_M\big((Vv_h)\chi_k\big) = \sum_k \frac{\chi_k}{|\triangle_k|} q_{k,M}(Vv_h),$$

where $q_{k,M}$ denotes the M^2-fold copy of the basic quadrature rule q applied to the element $\triangle_k$.

As in [7] we define

$$E_{kl} := \int_\Gamma (V\chi_l)\chi_k - Q_M\big((V\chi_l)\chi_k\big) = \int_{\triangle_k} V\chi_l - q_{k,M}(V\chi_l).$$

Note that in general $E_{kl} \neq E_{lk}$, because of the lack of symmetry in the second term. In the following lemma the norm on both sides is the L_2-norm, $\| v \| = (v,v)^{1/2}$.

Lemma 1. *For $v_h \in S_h$,*

$$\| (P_h V - V_h) v_h \| \leq \left(\max_k \frac{1}{|\triangle_k|} \sum_l |E_{kl}| \right)^{1/2} \left(\max_l \frac{1}{|\triangle_l|} \sum_k |E_{kl}| \right)^{1/2} \| v_h \|.$$

Proof. Writing $v_h = \sum_l c_l \chi_l$, we obtain the explicit representation

$$(P_h V - V_h) v_h = \sum_k \frac{\chi_k}{|\triangle_k|} \sum_l E_{kl} c_l.$$

We use (6) and the Cauchy-Schwarz inequality to derive

$$\begin{aligned}
\| (P_h V - V_h)v_h \|^2 &= ((P_h V - V_h)v_h, (P_h V - V_h)v_h) \\
&= \sum_k \frac{1}{|\triangle_k|} \left(\sum_l E_{kl} c_l \right)^2 \\
&\leq \sum_k \frac{1}{|\triangle_k|} \left(\sum_l |E_{kl}| \right) \left(\sum_l |E_{kl}| c_l^2 \right) \\
&\leq \left(\max_k \frac{1}{|\triangle_k|} \sum_l |E_{kl}| \right) \sum_l \sum_k \frac{1}{|\triangle_l|} |E_{kl}| |\triangle_l| c_l^2 \\
&\leq \left(\max_k \frac{1}{|\triangle_k|} \sum_l |E_{kl}| \right) \left(\max_l \frac{1}{|\triangle_l|} \sum_k |E_{kl}| \right) \sum_l |\triangle_l| c_l^2.
\end{aligned}$$

Since $\| v_h \|^2 = \sum_l |\triangle_l| c_l^2$, the claimed bound is proved. □

The essence of the proof of Theorem 1 (and the principal place where originality shows itself in this paper) lies in Lemma 2 below. It requires not full quasi-uniformity, but only the local regularity property expressed in the following assumption:

Definition 1. *[Local regularity assumption] A family of triangulations $\{\mathcal{T}_h\}$ satisfies the local regularity assumption if there exists $\kappa \geq 1$, with κ the same for all $\mathcal{T}_h$, such that*

$$h_k \leq \kappa \max \left(h_l, \operatorname{dist}(\triangle_k, \triangle_l) \right) \quad \forall k, l.$$

Here $\operatorname{dist}(\triangle_k, \triangle_l)$ is the Euclidean distance between $\triangle_k$ and $\triangle_l$,

$$\operatorname{dist}(\triangle_k, \triangle_l) := \inf \{|s - t| : t \in \triangle_k, s \in \triangle_l\}.$$

Note that the local regularity assumption holds for geometrically graded meshes.

Lemma 2. *Assume that the local regularity assumption holds for some constant $\kappa \geq 1$. Then*

$$\| (P_h V - V_h)v_h \| \leq c \frac{h}{M} \| v_h \| \quad \forall v_h \in S_h.$$

Proof. The result will follow from Lemma 1 if we can show both

$$\max_k \frac{1}{|\triangle_k|} \sum_l |E_{kl}| \leq c \frac{h}{M}, \tag{8}$$

and

$$\max_l \frac{1}{|\triangle_l|} \sum_k |E_{kl}| \leq c \frac{h}{M}. \tag{9}$$

It is sufficient to prove (8) and (9) for $M = 1$, since the results for general M can then be obtained by replacing h_k by h_k/M and h_l by h_l/M throughout the argument. (Note that E_{kl} is then representable as a double sum, each sum having M^2 terms, corresponding to the partitioning of each of $\triangle_k$ and $\triangle_l$ into M^2 pieces.)

Starting with the second result, we therefore set $M = 1$, and seek to prove

$$\max_l \frac{1}{|\triangle_l|} \sum_k |E_{kl}| \leq ch.$$

For fixed l the definition of E_{kl} together with $\sum_i w_i = 1$ gives

$$\begin{aligned} E_{kl} &= \int_{\triangle_k} (V\chi_l)(t)\,dt - |\triangle_k| \sum_{i=1}^{I} w_i (V\chi_l)(t_{k,i}) \\ &= \sum_{i=1}^{I} w_i \int_{\triangle_k} \left[(V\chi_l)(t) - (V\chi_l)(t_{k,i})\right] dt \\ &= \frac{1}{4\pi} \sum_{i=1}^{I} w_i \int_{\triangle_k}\int_{\triangle_l} \left[\frac{1}{|s-t|} - \frac{1}{|s-t_{k,i}|}\right] ds\,dt \\ &= \frac{1}{4\pi} \sum_{i=1}^{I} w_i \int_{\triangle_l} \left(\int_{\triangle_k}\left[\frac{1}{|s-t|} - \frac{1}{|s-t_{k,i}|}\right] dt\right) ds. \end{aligned} \quad (10)$$

It is convenient to classify the terms E_{kl} according to the distance of $\triangle_k$ from $\triangle_l$. For fixed l, the 'close' terms, denoted by $E_{kl}^{(c)}$, are defined to be those for which

$$\mathrm{dist}(\triangle_k, \triangle_l) < h_l;$$

note in particular that E_{ll} is a close term, and that so is E_{kl} if $\overline{\triangle}_k$ and $\overline{\triangle}_l$ have one or more points in common. The 'intermediate' terms, denoted by $E_{kl}^{(i)}$, are those for which

$$h_l \leq \mathrm{dist}(\triangle_k, \triangle_l) < h,$$

where $h := \max_k h_k$. And finally the 'far' terms E_{kl}, denoted by $E_{kl}^{(f)}$, are those for which

$$h \leq \mathrm{dist}(\triangle_k, \triangle_l).$$

For the close terms $E_{kl}^{(c)}$ we use the bound

$$\left|\frac{1}{|s-t|} - \frac{1}{|s-t_{k,i}|}\right| = \frac{\left||s-t_{k,i}| - |s-t|\right|}{|s-t||s-t_{k,i}|} \leq \frac{h_k}{|s-t||s-t_{k,i}|}, \quad (11)$$

for $t, t_{k,i} \in \triangle_k$ and $s \in \triangle_l$. Thus we obtain

$$|E_{kl}^{(c)}| \leq \frac{1}{4\pi} h_k \sum_{i=1}^{I} w_i \int_{\triangle_l} \left(\int_{\triangle_k} \frac{1}{|s-t|} dt \right) \frac{1}{|s-t_{k,i}|} ds.$$

For the innermost integral we use the result (see Appendix) that for any domain $\Omega \subset \mathbb{R}^2$ with area $|\Omega|$,

$$\int_\Omega \frac{1}{|s-t|} dt \leq 2(\pi|\Omega|)^{1/2} \quad \forall s \in \mathbb{R}^2. \tag{12}$$

This and the shape-regularity assumption, $h_k \leq \rho^{-1/2}|\triangle_k|^{1/2}$, imply

$$|E_{kl}^{(c)}| \leq c|\triangle_k| \sum_{i=1}^{I} w_i \int_{\triangle_l} \frac{1}{|s-t_{k,i}|} ds \tag{13}$$

$$\leq c|\triangle_k| h_l, \tag{14}$$

where we used (12) again. On summing over all the close terms for fixed l, we obtain

$$\sum_k |E_{kl}^{(c)}| \leq ch \sum_{\text{dist}(\triangle_k, \triangle_l) \leq h_l} |\triangle_k|.$$

For a given l the close elements $\triangle_k$ are all contained within a circle centered at an arbitrary point in $\triangle_l$ and of radius $2h_l + h_k$, where (since $\text{dist}(\triangle_k, \triangle_l) < h_l$)

$$2h_l + h_k \leq 2h_l + \kappa h_l = (2+\kappa) h_l,$$

from which it follows that

$$\sum_{\text{dist}(\triangle_k, \triangle_l) < h_l} |\triangle_k| \leq \pi(2+\kappa)^2 h_l^2 \leq c|\triangle_l|,$$

where again we use the shape-regularity property. Thus the sum over the close terms satisfies the required bound

$$\sum_k |E_{kl}^{(c)}| \leq ch|\triangle_l|.$$

For the intermediate terms we again apply the bound (11), but this time we use, since $\text{dist}(\triangle_k, \triangle_l) \geq h_l$,

$$|s - t_{k,i}| \geq \text{dist}(\triangle_k, \triangle_l) \geq \frac{1}{\kappa} h_k, \quad s \in \triangle_l. \tag{15}$$

So the bound (11) together with (10) gives for the intermediate terms

$$|E_{kl}^{(i)}| \leq \frac{\kappa}{4\pi} \int_{\triangle_l} \left(\int_{\triangle_k} \frac{1}{|s-t|} dt \right) ds. \tag{16}$$

Summing over the intermediate terms, we obtain

$$\sum_k |E_{kl}^{(i)}| \leq \frac{\kappa}{4\pi} \int_{\Delta_l} \left(\int_{\Delta_l^{(i)}} \frac{1}{|s-t|} dt \right) ds,$$

where $\Delta_l^{(i)}$ is the union of all the elements which are at an intermediate distance with respect to Δ_l. It is easy to see that all such elements are contained in a circle of radius at most $h_l + h + h_k \leq 3h$ centered anywhere in Δ_l. Applying the result (12) again, we obtain for the intermediate sum

$$\sum_k |E_{kl}^{(i)}| \leq ch \int_{\Delta_l} ds = ch|\Delta_l|.$$

Finally, we consider the far terms $E_{kl}^{(f)}$. The Taylor expansion with respect to an arbitrarily chosen point $\tau \in \Delta_k$ gives the following representation ([8], lemma A.3):

$$\frac{1}{|s-t|} = \frac{1}{|s-\tau|} + \frac{(t-\tau)\cdot(s-\tau)}{|s-\tau|^3} + \mathcal{O}\left(\frac{h_k^2}{|s-\tau|^3}\right), \quad t \in \Delta_k,\ s \in \Delta_l,$$

which implies

$$\frac{1}{|s-t|} - \frac{1}{|s-t_{k,i}|} = \frac{(t-t_{k,i})\cdot(s-\tau)}{|s-\tau|^3} + \mathcal{O}\left(\frac{h_k^2}{|s-t|^3}\right). \tag{17}$$

On substituting this expression into (10), and using the fact that the quadrature sum is exact for all linear functions, and hence

$$|\Delta_k| \sum_{i=1}^{I} w_i t_{k,i} = \int_{\Delta_k} t\, dt,$$

we see that the first term on the right of (17) contributes a zero result. Thus we derive

$$E_{kl}^{(f)} \leq ch^2 \int_{\Delta_l} \left(\int_{\Delta_k} \frac{1}{|s-t|^3} dt \right) ds, \tag{18}$$

and by summing over all far elements,

$$\sum_k |E_{kl}^{(f)}| \leq ch^2 \int_{\Delta_l} \left(\int_{\Delta_l^{(f)}} \frac{1}{|s-t|^3} dt \right) ds,$$

where $\Delta_l^{(f)}$ denotes the union of all elements Δ_k whose distance from Δ_l is greater than h. A bound on the inner integral is easily obtained by changing the variable to $t' := t - s$ and exploiting the fact that $|t'| = |t-s| \geq h$ for $t \in \Delta_l^{(f)}$ and $s \in \Delta_l$. Thus we obtain, with $r := |t'|$,

$$\int_{\Delta_l^{(f)}} \frac{1}{|s-t|^3} dt \leq \int_h^\infty \frac{1}{r^3} 2\pi r\, dr = \frac{2\pi}{h},$$

and hence
$$\sum_k |E_{kl}^{(f)}| \le ch^2|\triangle_l|h^{-1} = ch|\triangle_l|.$$

The above three bounds together imply
$$\sum_k |E_{kl}| \le ch|\triangle_l| \qquad \forall l,$$

completing the proof of (9). □

To prove (8) we again set $M = 1$, and seek to prove
$$\max_k \frac{1}{|\triangle_k|} \sum_l |E_{kl}| \le ch.$$

For fixed k, it is convenient this time to define the close terms $E_{kl}^{(c)}$ as those for which
$$\text{dist}(\triangle_k, \triangle_l) < h,$$

and the far terms $E_{kl}^{(f)}$ as those for which
$$h \le \text{dist}(\triangle_k, \triangle_l).$$

For the close terms, we use the bound (13), from which it follows that
$$\sum_l |E_{kl}^{(c)}| \le c|\triangle_k| \sum_{i=1}^I w_i \int_{\triangle_k^{(c)}} \frac{1}{|s - t_{k,i}|} ds$$

where $\triangle_k^{(c)}$ denotes the union of all elements $\triangle_l$ whose distance from $\triangle_k$ is less than h. Thus $\triangle_k^{(c)}$ is contained in a circle of radius $3h$ with centre in $\triangle_k$. Now we use the result (12) to prove the required bound
$$\sum_l |E_{kl}^{(c)}| \le ch|\triangle_k|.$$

Finally, for the far terms $E_{kl}^{(f)}$ we again use (18), and by interchanging k and l in the subsequent argument find
$$\sum_l |E_{kl}^{(f)}| \le ch|\triangle_k|,$$

completing the proof of (8). □

4. Completing the proof of Theorem 1

First we note that, as shown in [11] (in Theorem 4.1 and an example),

$$V : \tilde{H}^\sigma(\Gamma) \to H^{\sigma+1}(\Gamma), \quad -1 \leq \sigma < 0$$

is continuous and bijective and has a continuous inverse. Thus under the assumptions in the theorem $u \in \tilde{H}^s(\Gamma)$, and

$$\|u\|_{\tilde{H}^s(\Gamma)} \leq c\|f\|_{H^{s+1}(\Gamma)}, \quad \text{for} \quad -\frac{1}{2} \leq s < 0.$$

Following [7], the proof now builds upon well-known properties of the Galerkin method. In particular, from [2], [3] we have

$$\|u_h^G - u\|_{\tilde{H}^t(\Gamma)} \leq C h^{s-t} \|u\|_{\tilde{H}^s(\Gamma)} \tag{19}$$

if $-\frac{1}{2} \leq s < 0$, $-1 < t < s$. Setting $t = s = \sigma$, it follows from (19) and the triangle inequality that

$$\|u_h^G\|_{\tilde{H}^\sigma(\Gamma)} \leq c\|u\|_{\tilde{H}^\sigma(\Gamma)} \leq c\|f\|_{H^{\sigma+1}(\Gamma)}, \tag{20}$$

if $-\frac{1}{2} \leq \sigma < 0$ and $f \in H^{\sigma+1}(\Gamma)$. From this and (5) it follows, for arbitrary $v_h \in S_h$, that

$$|(v_h, f)| = |(v_h, P_h V u_h^G)| = |(P_h V v_h, u_h^G)| \leq \|P_h V v_h\| \|u_h^G\|$$
$$\leq c h^\sigma \|P_h V v_h\| \|u_h^G\|_{\tilde{H}^\sigma(\Gamma)} \leq c h^\sigma \|P_h V v_h\| \|f\|_{H^{\sigma+1}(\Gamma)},$$

where we used the inverse inequality

$$\|w_h\|_{\tilde{H}^s(\Gamma)} \leq c h^{t-s} \|w_h\|_{\tilde{H}^t(\Gamma)} \quad \text{if} \quad w_h \in S_h, \ -1 \leq t \leq s \leq 0, \tag{21}$$

which follows from the quasi-uniformity assumption (2). The duality of the $H^{-t}(\Gamma)$ and $\tilde{H}^t(\Gamma)$ norms then gives

$$\|v_h\|_{\tilde{H}^{-1-\sigma}(\Gamma)} \leq c h^\sigma \|P_h V v_h\|, \tag{22}$$

as a convenient expression of the stability of the Galerkin method.

This stability property now extends to the semi-discrete approximation method: by virtue of (22) and Lemma 2, for arbitrary $v_h \in S_h$ we find

$$\|v_h\|_{\tilde{H}^{-1-\sigma}(\Gamma)} \leq c h^\sigma \|V_h v_h\| + c h^\sigma \|(P_h V - V_h) v_h\|$$
$$\leq c h^\sigma \|V_h v_h\| + c \frac{h^{\sigma+1}}{M} \|v_h\|$$
$$\leq c h^\sigma \|V_h v_h\| + \frac{c_1}{M} \|v_h\|_{\tilde{H}^{-1-\sigma}(\Gamma)},$$

where again we used the inverse inequality (21). On taking $M_0 := 2c_1$ (which we note is independent of h), we see that for $M \geq M_0$

$$\|v_h\|_{\tilde{H}^{-1-\sigma}(\Gamma)} \leq ch^\sigma \|V_h v_h\| \quad \forall v_h \in S_h, \tag{23}$$

if $-\frac{1}{2} \leq \sigma < 0$.

For $M \geq M_0$ the uniqueness and hence existence of the solution $u_h \in S_h$ of the semi-discrete scheme (4) is assured by (23).

Initially we restrict $-1 < t \leq \frac{1}{2}$. Then by choosing $\sigma = -1 - t$, it follows from (23), (7), (21), (20) and Lemma 2 that

$$\begin{aligned}
\|u_h - u_h^G\|_{\tilde{H}^t(\Gamma)} &\leq ch^{-1-t}\|V_h(u_h - u_h^G)\| = ch^{-1-t}\|P_h f - V_h u_h^G\| \\
&= ch^{-1-t}\|(P_h V - V_h)u_h^G\| \leq ch^{-t}\|u_h^G\| \\
&\leq ch^{s-t}\|u_h^G\|_{\tilde{H}^s(\Gamma)} \leq ch^{s-t}\|u\|_{\tilde{H}^s(\Gamma)},
\end{aligned}$$

if $-\frac{1}{2} \leq s < 0$. The result can now be extended to the full range $-1 < t < s$ by another application of the inverse inequality, and the proof completed with the triangle inequality together with (19). □

5. Numerical example

In this example (extracted from [9]), the domain Γ is the square $[0, 2]^2$, and the triangulation $\mathcal{T}_h$ is obtained by dividing the square into $2N^2$ congruent right-angled isosceles triangles, with edges parallel to the x and y axes and $x + y = 0$. The quadrature rule q is the centroid rule,

$$qg = \frac{1}{2}g\left(\frac{1}{3}, \frac{1}{3}\right).$$

The right-hand side function in (1) is given by $f(x) = 1$, and the quantity computed is the capacitance.

The errors in the capacitance are given in Table 1, with the exact value taken to be 0.7355 (see [10] and [4]). Also shown are the apparent orders of convergence with respect to h. As expected, these approach 1 as N increases.

The column labelled $M = 1$ corresponds to the method of collocation at the element centroids. The subsequent columns show the results obtained with $M = 2$ and 4. In this particular problem it is evident that the additional stabilisation associated with increasing M is not needed, in that the numerical performance of the collocation method is already satisfactory. But of course we have as yet no proof that this is always true.

N	$M = 1$		$M = 2$		$M = 4$	
2	0.0811		0.0571		0.0477	
4	0.0434	0.90	0.0311	0.88	0.0262	0.86
8	0.0226	0.94	0.0163	0.93	0.0139	0.92
16	0.0115	0.96	0.0083	0.96	0.0071	0.96

Table 1: Errors in capacitance, and apparent sets of convergence

Appendix

In Section 3 we used the following result.

Lemma 3. *Let Ω denote a domain in $\mathbb{R}^2$. Then*

$$\int_\Omega \frac{1}{|s-t|} dt \leq 2\sqrt{\pi|\Omega|} \quad \text{for all} \quad s \in \mathbb{R}^2.$$

Note that the bound is sharp if the domain Ω is a circle and s is its centre.

Proof. Let R denote the circle with centre s and radius $a = \sqrt{|\Omega|/\pi}$. In this case R and Ω have the same area, from which follows $|R\backslash\Omega| = |\Omega\backslash R|$. We also use $|s-t| \geq a$ and $|s-t| \leq a$ for $s \in \Omega\backslash R$ and $s \in R\backslash\Omega$ respectively, so obtaining

$$\int_\Omega \frac{dt}{|s-t|} = \int_{\Omega\cap R} \frac{dt}{|s-t|} + \int_{\Omega\backslash R} \frac{dt}{|s-t|} \leq \int_{R\cap\Omega} \frac{dt}{|s-t|} + \frac{|\Omega\backslash R|}{a}$$

$$\leq \int_{R\cap\Omega} \frac{dt}{|s-t|} + \int_{R\backslash\Omega} \frac{dt}{|s-t|} = \int_R \frac{dt}{|s-t|} = 2\pi a = 2\sqrt{\pi|\Omega|}.$$

$\square$

References

[1] Ainsworth, M., Grigorieff, R.D., Sloan, I.H.: Semi-discrete Galerkin approximation of the single-layer equation by general splines, *Numer. Math.* **79**, (1998), 157-174.

[2] Costabel, M.: Boundary integral operators on Lipschitz domains: elementary results, *SIAM J. Math. Anal.*, **19** (1988), 613–626.

[3] Costabel, M., Stephan, E.P.,: Duality estimates for the numerical solution of integral equations, *Numer. Math.* **54** (1988), 339–353.

[4] Ervin, V., Stephan, E.P., Abou El-Seoud, S.: An improved boundary element method for the charge density of a thin electrified plate in $\mathbb{R}^3$, *Math. Meth. Appl. Sci.* **13** (1990), 291–303.

[5] Graham, I.G., Hackbusch, W., Sauter, S.A.: *Discrete boundary element methods on general meshes in $3D$*, University of Bath Preprint 97/19, 1997.

[6] Graham I.G., Hackbusch, W., Sauter, S.A.: Hybrid Galerkin boundary elements on degenerate meshes, this volume.

[7] Grigorieff, R.D., Sloan I.H.: Galerkin approximation with quadrature for the screen problem in $\mathbb{R}^3$, *J. Integral Equations Appl.*, to appear.

[8] Hackbusch, W., Nowak, Z.P.: On the fast matrix multiplication in the boundary element method by panel clustering, *Numer. Math.* **54** (1989), 463-491.

[9] Mauersberger, D.: *Galerkin Verfahren mit Quadratur für das Einfachschichtpotential einer Platte in* $\mathbb{R}^3$, Diplom thesis, Technical University of Berlin, (1997).

[10] Penzel, F.: Error estimates for a discretized Galerkin method for a boundary integral equation in two dimensions, *Numer. Math. for PDE* **8** (1992), 405–421.

[11] Schneider, R.: Reduction of order for pseudodifferential operators on Lipschitz domains, *Commun. Partial Diff. Eqns.* **16** (1991), 1263–1286.

[12] Sloan, I.H., Atkinson, K.E.: Semi-discrete Galerkin approximation for a single-layer equation on Lipschitz curves, *J. Int. Eqns. and Applics.* **9** (1997), 279–292.

School of Mathematics, University of New South Wales
2052 Sydney, Australia
Email: I.Sloan@unsw.edu.au

V. MAZ'YA and G. SCHMIDT

Construction of Basis Functions for High Order Approximate Approximations

1. Introduction

The paper is devoted to the construction of simple approximation formulas with the rate $\mathcal{O}(h^N)$ for arbitrary $N \in \mathbb{N}$ and certain range of step size h relevant in numerical computations. The use of approximation methods approaching a given function up to some in numerical computations negligible saturation error, but in general not converging as $h \to 0$, was introduced in [3]. The concept of *approximate approximations* reflects the fact that for computers "h tending to 0" is impossible, since floating point arithmetic and memory resources always give lower bounds for step sizes. On the other hand, the numerical results for any relevant practical application are needed only within some tolerance, prescribed for example by the accuracy of input data. Therefore, numerical computations do not need approximations for arbitrarily small h, but rather high order approximations in some range of h accessible to the computing system.

In [3] and [4] some approximation procedures of this kind, which possess additionally many interesting features, were described. Because the requirement of convergence is given up, the choice of approximating functions becomes much more flexible. This can be used to adapt the functions better to the underlying numerical problem or to achieve more computational simplicity which is very important, especially in the multidimensional case. For example, consider in $\mathbb{R}^n$ the quasi–interpolant

$$\mathcal{M}_h u(\mathbf{x}) := \mathcal{D}^{-n/2} \sum_{\mathbf{m} \in \mathbb{Z}^n} u(h\mathbf{m})\, \eta\left(\frac{\mathbf{x} - h\mathbf{m}}{\sqrt{\mathcal{D}}h}\right). \qquad (1)$$

with the step width h and a fixed parameter $\mathcal{D} > 0$. Here the generating function η is assumed to be M–times differentiable with M the smallest integer greater than $n/2$ and to satisfy together with all derivatives $\partial^\beta \eta := \partial_{x_1}^{\beta_1} \ldots \partial_{x_n}^{\beta_n}$, $[\beta] := \beta_1 + \cdots + \beta_n \leq M$, the decay condition

$$|\partial^\beta \eta(\mathbf{x})| \leq A_{K,\beta}\,(1 + |\mathbf{x}|^2)^{-K/2}\,, \quad \mathbf{x} \in \mathbb{R}^n\,, \qquad (2)$$

for some number $K > N + n$ and constants $A_{K,\beta}$. As usual $|\mathbf{x}| := |\mathbf{x}|_2 = \langle \mathbf{x}, \mathbf{x} \rangle^{-1/2}$ denotes the Euclidean norm in $\mathbb{R}^n$. The simple formula (1) provides the following pointwise approximation result.

Theorem 1. ([5]) *Suppose that the generating function η is subject to the decay* (2) *and satisfies the moment condition*

$$\mathcal{F}\eta(0) = 1\,, \ \partial^\alpha \mathcal{F}\eta(0) = 0\,, \quad \text{with } \mathcal{F}\eta(\lambda) = \int_{\mathbb{R}^n} \eta(\mathbf{x})\, e^{-2\pi i \langle \mathbf{x}, \lambda \rangle}\, d\mathbf{x}\,, \qquad (3)$$

for all multiindices $\alpha = (\alpha_1, \ldots, \alpha_n)$ with $1 \leq [\alpha] < N$. If the function $u \in L_\infty(\mathbb{R}^n)$ is N-times continuously differentiable in a closed ball around $\mathbf{x} \in \mathbb{R}^n$ with radius $\kappa > 0$, $u \in C^N(B(\mathbf{x}, \kappa))$, then

$$|\mathcal{M}_h u(\mathbf{x}) - u(\mathbf{x})| \leq c(\sqrt{\mathcal{D}}h)^N \sum_{[\alpha]=N} \frac{\|\partial^\alpha u\|_{C(B(\mathbf{x},\kappa))}}{\alpha!} + \sum_{[\alpha]=0}^{N-1} \varepsilon_{\alpha,\mathcal{D}} \left(\frac{\sqrt{\mathcal{D}}h}{2\pi}\right)^{[\alpha]} \frac{|\partial^\alpha u(\mathbf{x})|}{\alpha!},$$

where $\varepsilon_{\alpha,\mathcal{D}} = \sum_{\nu \in \mathbb{Z}^n \setminus \{0\}} |\partial^\alpha \mathcal{F}\eta(\sqrt{\mathcal{D}}\nu)|$ and the constant c depends only on η and κ. Moreover, for any $\varepsilon > 0$ there exists $\mathcal{D} > 0$ such that $|\varepsilon_{\alpha,\mathcal{D}}| < \varepsilon$.

Since $\mathcal{D} > 0$ is fixed and, in general, $\varepsilon_{0,\mathcal{D}} \neq 0$ the sum $\mathcal{M}_h u(\mathbf{x})$ does not converge to $u(\mathbf{x})$ as $h \to 0$. However, the quasi–interpolant (1) approximates a smooth function u with the order $\mathcal{O}((\sqrt{\mathcal{D}}h)^N)$ up to the saturation error ε which can be made smaller than a prescribed tolerance by choosing the parameter $\mathcal{D}$ large enough. It is important that for a large class of generating functions η this saturation error is comparable to negligible errors already for relatively small $\mathcal{D}$. As h has to be chosen such that $\sqrt{\mathcal{D}}h < 1$, the application of (1) in numerical methods with those generating functions is very attractive. It is possible to obtain simple formulas which approximate with high order up to some prescribed accuracy, given by the practical problem to be solved. Giving up the convergence of approximate approximations is not feasible in numerical methods, but allows to use new classes of approximating functions specially adapted to the given problem. Especially we note the possibility to use generating functions giving explicit formulas for the values of various integral and pseudodifferential operators of mathematical physics applied to them. This approach results in new methods for solving differential and integral equation problems (some examples are contained e.g. in [4] and [2]).

Let us present some examples of useful generating functions. The table below lists some univariate functions, the last row contains a lower bound $\mathcal{D}_{min}$ such that for $\mathcal{D} \geq \mathcal{D}_{min}$ the main term of the saturation error $\varepsilon_0(\eta, \mathcal{D}) < 10^{-8}$.

$\eta(x)$	$\mathcal{F}\eta(\lambda)$	$\mathcal{D}_{min}$
$\pi^{-1/2} e^{-x^2}$	$e^{-\pi^2 \lambda^2}$	2.0
$\dfrac{1}{\pi} \operatorname{sech} x$	$\operatorname{sech} \pi^2 \lambda$	4.1
$\dfrac{2}{\pi^2} x \operatorname{cosech} x$	$\operatorname{sech}^2 \pi^2 \lambda$	1.1
$\dfrac{2}{\pi(1+x^2)^2}$	$(1 + 2\pi|\lambda|) e^{-2\pi|\lambda|}$	12.0
$\sqrt{\dfrac{e}{\pi}} e^{-x^2} \cos \sqrt{2}x$	$e^{-\pi^2 \lambda^2} \cosh \sqrt{2}\pi\lambda$	2.6

The first four functions provide the estimate of Theorem 1 with $N = 2$, whereas for the last function $N = 4$.

For arbitrary space dimension n one can take tensor products of the above listed functions. Some further examples which are radial and correspond to $N = 2$ are given in the following table.

$\eta(\mathbf{x})$	$\mathcal{F}\eta(\lambda)$
$\pi^{-n/2} e^{-|\mathbf{x}|^2}$	$e^{-\pi^2|\lambda|^2}$
$\dfrac{4}{3\pi^{n+1/2}} \dfrac{\Gamma(\frac{n+5}{2})}{(1+|\mathbf{x}|^2)^{(n+5)/2}}$	$\left(1 + 2\pi|\lambda| + \dfrac{4}{3}\pi^2|\lambda|^2\right) e^{-2\pi|\lambda|}$
$\dfrac{\Gamma(k+1+\frac{n}{2})}{\pi^{n/2}\Gamma(k+1)} (1-|\mathbf{x}|^2)^k \chi(\mathbf{x})$	$\Gamma(k+1+\frac{n}{2}) \dfrac{J_{k+n/2}(2\pi|\lambda|)}{(\pi|\lambda|)^{k+n/2}}$
$(-1)^k \dfrac{\pi^{(n+1)/2}}{\Gamma(k+\frac{n+1}{2})} \dfrac{\partial^k}{\partial\tau^k} \dfrac{1}{\sqrt{\tau}} e^{-2\pi\sqrt{\tau}|\mathbf{x}|}\bigg|_{\tau=1}$	$(1+|\lambda|^2)^{-k-\frac{n+1}{2}}$

Here χ denotes the characteristic function of the unit ball $B(0,1)$ and J_ν the Bessel function of the first kind.

Estimates similar to Theorem 1 are also valid for integral norms (cf. [6]). It is interesting that the saturation error tends to zero with respect to weak norms. We formulate an estimate for the quasi–interpolation in the negative Sobolev norm.

Theorem 2. ([6]) *Suppose that η is as in Theorem 1 and u belongs to the Sobolev space $W_2^s(\mathbb{R}^n)$, $n/2 < s \in \mathbb{N}$. Then for any $t > 0$ and $\varepsilon > 0$ there exists $\mathcal{D} > 0$ such that*

$$\|\mathcal{M}_h u - u\|_{W_2^{-t}} \leq c \left(\sqrt{\mathcal{D}} h\right)^N \|\nabla_N u\|_{L_p} + h^t \varepsilon \|u\|_{W_p^{s-1}} . \tag{4}$$

A direct application of the approximate approximation formula (1) is the construction of new cubature formulas for volume potentials of partial differential equations with constant coefficients, i.e. for the convolution with the fundamental solution g

$$\mathcal{K}u(\mathbf{x}) = \int_{\mathbb{R}^n} g(\mathbf{x} - \mathbf{y}) u(\mathbf{y}) \, d\mathbf{y} . \tag{5}$$

It is well known that the problem of accurate computation of such Newton potentials and their derivatives arises for example in the solution of boundary value problems for partial differential equations with an inhomogeneous right–hand side u in combination with boundary element methods. Due to the singularity of g usual cubature methods are very time consuming. For this particular problem the greater flexibility offered by the concept of approximate approximations can be used to choose appropriate functions η with the property that $\mathcal{K}\eta$ can be analytically determined. The examples mentioned above possess this property for many important integral operators. If additionally η satisfies the conditions of Theorem 1 then approximating the density u by the quasi–interpolant (1) results in a high order semi–analytic cubature formula

$$\mathcal{K}_h u(\mathbf{x}) := \mathcal{K}\mathcal{M}_h u(\mathbf{x}) = h^n \sum_{\mathbf{m}\in\mathbb{Z}^n} u(h\mathbf{m}) \int_{\mathbb{R}^n} g\left(\sqrt{\mathcal{D}} h \left(\dfrac{\mathbf{x} - h\mathbf{m}}{\sqrt{\mathcal{D}} h} - \mathbf{y}\right)\right) \eta(\mathbf{y}) \, d\mathbf{y} \tag{6}$$

of the volume potential (5). It turns out that due to the smoothing property of the integral operators the difference between $\mathcal{K}u$ and $\mathcal{K}_h u$ can be estimated in integral norms or even pointwise similar to (4) (cf. [6]). Moreover, by simple differentiation of (6) one obtains immediately approximations of the corresponding derivatives of (5).

Thus, if it is possible to construct from a given basis function new generating functions η_N satisfying the moment condition (3) for large N in such a way that $\mathcal{K}\eta_N$ can be effectively determined, then one gets immediately cubature formulas of high approximation rates. Note that a simple technique to improve the approximation order of a quasi–interpolant by replacing the basis function by a linear combination of its translates has been used already in many papers (cf. [9], [1], [7]). Here we discuss several other construction methods which use in particular analytic methods to obtain generating functions η_N for large N. Thus if an analytic expression of $\mathcal{K}\eta$ is known then this remains true also for $\mathcal{K}\eta_N$. Therefore the approximation method under consideration generates semi–analytic cubature formulas for important integral operators of mathematical physics possessing the rate $\mathcal{O}(h^N)$ up to some prescribed accuracy.

2. Constructions of generating functions for arbitrary N

2.1. A general formula

Theorem 3. *Suppose that η satisfies (2), is $(N-1)$-times continuously differentiable and that*

$$\int_{\mathbb{R}^n} \eta(\mathbf{t})\, d\mathbf{t} \neq 0\,, \quad \int_{\mathbb{R}^n} |\mathbf{t}|^{N-1} |\partial^\alpha \eta(\mathbf{t})|\, d\mathbf{t} < \infty,\ 0 \leq |\alpha| \leq N-1.$$

Then the function

$$\eta_N(\mathbf{x}) = \sum_{[\alpha]=0}^{N-1} \frac{\partial^\alpha (\mathcal{F}\eta)^{-1}(0)}{\alpha!\, (2\pi i)^{[\alpha]}} \partial^\alpha \eta(\mathbf{x}) \quad \text{with } \partial^\alpha(\mathcal{F}\eta)^{-1}(0) := \partial^\alpha \left(\frac{1}{\mathcal{F}\eta(\lambda)}\right)\bigg|_{\lambda=0} \quad (7)$$

satisfies the moment condition (3).

Proof. With the N-th order Taylor polynomial of $1/\mathcal{F}\eta(\lambda)$

$$P_N(\lambda) = \sum_{[\alpha]=0}^{N-1} \frac{\partial^\alpha (\mathcal{F}\eta)^{-1}(0)}{\alpha!} \lambda^\alpha\,.$$

obviously the equations

$$\partial^\beta (P_N(\lambda)\mathcal{F}\eta(\lambda))\big|_{\lambda=0} = \partial^\beta \left(\frac{1}{\mathcal{F}\eta(\lambda)} \mathcal{F}\eta(\lambda)\right)\bigg|_{\lambda=0} = \delta_{[\beta]0}$$

are valid for all $[\beta] \leq N-1$. But

$$P_N(\lambda)\mathcal{F}\eta(\lambda) = \mathcal{F}\left(\sum_{|\alpha|=0}^{N-1} \frac{\partial^\alpha (\mathcal{F}\eta)^{-1}(0)}{\alpha!\, 2\pi i^{[\alpha]}} \partial^\alpha \eta(\mathbf{x})\right)(\lambda)\,.$$

Since the Fourier transform of η_N is given by

$$\mathcal{F}\eta_N(\lambda) = \mathcal{F}\eta(\lambda) \sum_{|\alpha|=0}^{N-1} \frac{\partial^\alpha (\mathcal{F}\eta)^{-1}(0)}{\alpha!} \lambda^\alpha$$

the saturation error of the quasi–interpolant with the generating function η_N is controlled by the values $\partial^\alpha \mathcal{F}\eta(\sqrt{\mathcal{D}}\nu)$ and $\partial^\alpha (\mathcal{F}\eta)^{-1}(0)$.

If the basis function η is symmetric with respect to the coordinate planes $x_i = 0$,

$$\eta(x_1, \ldots, x_i, \ldots, x_n) = \eta(x_1, \ldots, -x_i, \ldots, x_n), \ i = 1, \ldots, n. \tag{8}$$

then automatically $\partial^\alpha \mathcal{F}\eta(0) = 0$ for any multiindex $\alpha = (\alpha_1, \ldots, \alpha_n)$ containing at least one odd α_i and formula (7) simplifies to

$$\eta_N(\mathbf{x}) = \sum_{[\alpha]=0}^{M-1} \frac{\partial^{2\alpha}(\mathcal{F}\eta)^{-1}(0)}{(2\alpha)!\,(-4\pi^2)^{[\alpha]}} \,\partial^{2\alpha}\eta(\mathbf{x}), \quad N = 2M.$$

For radial basis functions, i.e. $\eta(\mathbf{x}) = \psi(r^2/2)$, $r = |\mathbf{x}|$, this approach leads to

Theorem 4. *Suppose that η satisfies the conditions of Theorem 3 and is additionally a radial function. Then*

$$\eta_{2M}(\mathbf{x}) = \Gamma\left(\frac{n}{2}\right) \sum_{j=0}^{M-1} \frac{\Delta^j (\mathcal{F}\eta)^{-1}(0)}{j!\,(4\pi)^{2j}\,\Gamma\left(j+\frac{n}{2}\right)} (-\Delta)^j \eta(\mathbf{x}) \tag{9}$$

is subject to the moment condition (3) with $N = 2M$ and has the Fourier transform

$$\mathcal{F}\eta_{2M}(\lambda) = \mathcal{F}\eta(\lambda)\,\Gamma\left(\frac{n}{2}\right) \sum_{j=0}^{M-1} \frac{\Delta^j (\mathcal{F}\eta)^{-1}(0)}{j!\,4^j\,\Gamma\left(j+\frac{n}{2}\right)} |\lambda|^{2j}.$$

An interesting feature of formula (9) is its additive structure. The order of a given quasi–interpolant can be increased by adding a new formula (1) with the next term of (9) as generating function.

As an example we consider the function $\eta(\mathbf{x}) = \exp(-\langle A^{-1}\mathbf{x}, \mathbf{x}\rangle)$, where A is a nonsingular symmetric $n{\times}n$-matrix with constant complex entries satisfying $\mathrm{Re}\,A \geq 0$. The application of the above formula leads to the generating function

$$\eta_{2M}(\mathbf{x}) = \frac{1}{\pi^{n/2}\sqrt{\det A}} L_{M-1}^{(n/2)}(\langle A^{-1}\mathbf{x},\mathbf{x}\rangle)\, e^{-\langle A^{-1}\mathbf{x},\mathbf{x}\rangle},$$

with the generalized Laguerre polynomial $L_{M-1}^{(n/2)}$ and

$$\mathcal{F}\eta_{2M}(\lambda) = e^{-\pi^2 \langle A\lambda,\lambda\rangle} \sum_{j=0}^{M-1} \frac{(\pi^2 \langle A\lambda,\lambda\rangle)^j}{j!}.$$

Note that (9) is useful for obtaining cubature formulas for the inverse of the operator $\Delta + k^2$, $k \in \mathbb{C}$. If for some appropriate radial function η an analytic expression of $(\Delta + k^2)^{-1}\eta$ is known, then a high order semi–analytic cubature formula for this potential can be obtained by simple analytic operations.

2.2. Using different $\mathcal{D}$

In this subsection we show that $\mathcal{D}$ is not only a scaling parameter which controls the saturation error. It is also useful to obtain other generating functions. We suppose that η satisfies (8). For some M-tuple $\mathbb{D} = (\mathcal{D}_1, \ldots, \mathcal{D}_M)$, $\mathcal{D}_j > 0$, $\mathcal{D}_j \neq \mathcal{D}_l$, $j \neq l$, we consider the function

$$\tilde{\eta}_\mathbb{D}(\mathbf{x}) := \sum_{j=1}^M a_j \mathcal{D}_j^{-n/2} \eta\left(\frac{\mathbf{x}}{\sqrt{\mathcal{D}_j}}\right).$$

possessing the moments

$$\int_{\mathbb{R}^n} \mathbf{x}^\alpha \tilde{\eta}_\mathbb{D}(\mathbf{x})\, d\mathbf{x} = \sum_{j=1}^M a_j \mathcal{D}_j^{-n/2} \int_{\mathbb{R}^n} \mathbf{x}^\alpha \eta\left(\frac{\mathbf{x}}{\sqrt{\mathcal{D}_j}}\right) d\mathbf{x} = \sum_{j=1}^M a_j \mathcal{D}_j^{[\alpha]/2} \int_{\mathbb{R}^n} \mathbf{x}^\alpha \eta(\mathbf{x})\, d\mathbf{x}.$$

Since $\partial^\alpha \mathcal{F}\tilde{\eta}_\mathbb{D}(0) = 0$ for odd $[\alpha]$ we see that $\tilde{\eta}_\mathbb{D}$ satisfies the moment condition (3) with $N = 2M$ if the numbers a_j solve the linear system

$$\sum_{j=1}^M a_j = 1, \quad \sum_{j=1}^M a_j \mathcal{D}_j^k = 0, \quad k = 1, \ldots, M-1.$$

Theorem 5. *Suppose that the function η with nonzero moment satisfies (2) with $K > 2M$ and (8). Then for any M-tuple of mutually different positive numbers $\mathcal{D}_j$ the generating function*

$$\tilde{\eta}_\mathbb{D}(\mathbf{x}) := \sum_{j=1}^M \prod_{\substack{l=1 \\ l \neq j}}^M \frac{\mathcal{D}_l \mathcal{D}_j^{-n/2}}{\mathcal{D}_l - \mathcal{D}_j} \eta\left(\frac{\mathbf{x}}{\sqrt{\mathcal{D}_j}}\right) \tag{10}$$

satisfies the moment condition (3) with $N = 2M$.

A practical realization of this construction is the following: Suppose that for given values $\{u(h\mathbf{m}), \mathbf{m} \in \mathbb{Z}^n\}$ and different $\mathcal{D}_j$ we have quasi–interpolants

$$\mathcal{M}_{j,h} u(\mathbf{x}) = \mathcal{D}_j^{-n/2} \sum_{\mathbf{m} \in \mathbb{Z}^n} u(h\mathbf{m})\, \eta\left(\frac{\mathbf{x} - h\mathbf{m}}{\sqrt{\mathcal{D}_j} h}\right), \quad j = 1, \ldots, M,$$

providing order $\mathcal{O}(h^2)$. Then the linear combination $\sum_{j=1}^M \prod_{\substack{l=1 \\ l \neq j}}^M \frac{\mathcal{D}_l}{\mathcal{D}_l - \mathcal{D}_j} \mathcal{M}_{j,h} u(\mathbf{x})$ gives an approximate approximation to u of the order $\mathcal{O}(h^{2M})$.

2.3. Derivatives with respect to $\mathcal{D}$

This construction is derived from formula (10) if the parameters $\mathcal{D}_j$ tend to some positive limit. By using the well-known relation

$$\sum_{j=1}^{M} \prod_{\substack{l=1 \\ l \neq j}}^{M} \frac{x_l}{x_l - x_j} f(x_j) \longrightarrow \sum_{l=0}^{M-1} \frac{(-z)^l}{l!} f^{(l)}(z) \quad \text{as } x_j \to z, \, j=1,\ldots,M,$$

one obtains for $\mathcal{D}_j \to \tau > 0$, $j = 1,\ldots,M$,

$$\sum_{j=1}^{M} \prod_{\substack{l=1 \\ l \neq j}}^{M} \frac{\mathcal{D}_l}{\mathcal{D}_l - \mathcal{D}_j} \mathcal{D}_j^{-n/2} \eta\left(\frac{\mathbf{x}}{\sqrt{\mathcal{D}_j}}\right) \longrightarrow \sum_{j=0}^{M-1} \frac{(-\tau)^j}{j!} \frac{d^j}{d\tau^j}\left(\tau^{-n/2} \eta\left(\frac{\mathbf{x}}{\sqrt{\tau}}\right)\right).$$

To show that the moment condition holds we note that

$$\sum_{j=0}^{M-1} \frac{(-\tau)^j}{j!} \frac{d^j}{d\tau^j}\left(\tau^{-n/2} \eta\left(\frac{\mathbf{x}}{\sqrt{\tau}}\right)\right) = \frac{(-1)^{M-1} \tau^M}{(M-1)!} \left(\frac{d}{d\tau}\right)^{M-1} \left(\tau^{-1-n/2} \eta\left(\frac{\mathbf{x}}{\sqrt{\tau}}\right)\right).$$

Hence, for any multiindex α, $[\alpha] < 2M$ and $\tau > 0$

$$\tau^M \left(\frac{d}{d\tau}\right)^{M-1} \left(\tau^{-1-n/2} \int_{\mathbb{R}^n} \eta\left(\frac{\mathbf{x}}{\sqrt{\tau}}\right) \mathbf{x}^\alpha \, d\mathbf{x}\right) = \int_{\mathbb{R}^n} \eta(\mathbf{x}) \mathbf{x}^\alpha \, d\mathbf{x} \, \tau^M \left(\frac{d}{d\tau}\right)^{M-1} \tau^{[\alpha]/2-1} = \delta_{[\alpha]0},$$

where we use (8) if $[\alpha]$ is an odd number.

Theorem 6. *Suppose that η is $(M-1)$-times continuously differentiable and satisfies the conditions of Theorem 5. Then the generating function*

$$\tilde{\eta}_M(\mathbf{x}) := \frac{(-1)^{M-1}}{(M-1)!} \left(\frac{d}{d\tau}\right)^{M-1} \left(\tau^{-1-n/2} \eta\left(\frac{\mathbf{x}}{\sqrt{\tau}}\right)\right)\bigg|_{\tau=1} \quad (11)$$

satisfies (3) with $N = 2M$. The Fourier transform of $\tilde{\eta}_M(\mathbf{x})$ is given by

$$\mathcal{F}\tilde{\eta}_M(\lambda) = \frac{(-1)^{M-1}}{(M-1)!} \left(\frac{d}{d\tau}\right)^{M-1} \left(\tau^{-1} \mathcal{F}\eta(\sqrt{\tau}\lambda)\right)\bigg|_{\tau=1}.$$

Formula (11) is useful for obtaining high order cubature formulas for integral operators $\mathcal{K}$ if an analytic formula of $\mathcal{K}\eta$ for a suitable basis function η is available. Then differentiation with respect to the parameter τ leads to high order semi-analytic cubature formulas for this integral operator.

2.4. Mellin convolution integrals

Here we suppose as before that η satisfies (2), (8) and that additionally

$$\left\|\sum_{\nu\in\mathbb{Z}^n\setminus\{0\}} |\partial^\alpha \mathcal{F}\eta(\cdot\sqrt{\mathcal{D}}\nu)|\right\|_{L_\infty(1,\infty)} \longrightarrow 0 \quad \text{as } \mathcal{D}\to\infty, \quad 0\le [\alpha] < N. \tag{12}$$

For a given function f with $\tau^{N/2} f(\tau) \in L_1(1,\infty)$ we consider the generating function

$$\hat{\eta}(\mathbf{x}) = \int_1^\infty f(\tau)\,\eta\!\left(\frac{\mathbf{x}}{\sqrt{\tau}}\right) \frac{d\tau}{\tau^{n/2}}, \tag{13}$$

The moments of $\hat{\eta}$ are given by

$$\int_{\mathbb{R}^n} \mathbf{x}^\alpha d\mathbf{x} \int_1^\infty f(\tau)\,\eta\!\left(\frac{\mathbf{x}}{\sqrt{\tau}}\right) \frac{d\tau}{\tau^{n/2}} = \int_1^\infty f(\tau)\tau^{[\alpha]/2}\, d\tau \int_{\mathbb{R}^n} \mathbf{x}^\alpha \eta(\mathbf{x})\, d\mathbf{x}.$$

Thus, if f is chosen such that

$$\int_1^\infty f(\tau)\, d\tau = 1/\mathcal{F}\eta(0),\quad \int_1^\infty f(\tau)\tau^k\, d\tau = 0, \quad k=1,\ldots,M-1, \tag{14}$$

then in view of (8) $\hat\eta$ satisfies the moment condition (3) for $|\alpha| < 2M$. Furthermore

$$\int_{\mathbb{R}^n} \mathbf{x}^\alpha \hat\eta(\mathbf{x}) e^{-2\pi i\langle \mathbf{x},\lambda\rangle} d\mathbf{x} = \int_1^\infty f(\tau) \tau^{[\alpha]/2} \int_{\mathbb{R}^n} \mathbf{x}^\alpha \eta(\mathbf{x}) e^{-2\pi i\sqrt{\tau}\langle \mathbf{x},\lambda\rangle}\, d\mathbf{x}\, d\tau$$

$$= \int_1^\infty f(\tau) \tau^{[\alpha]/2} \mathcal{F}(\mathbf{x}^\alpha \hat\eta(\mathbf{x}))(\sqrt{\tau}\lambda)\, d\tau ,$$

such that

$$\sum_{\nu\in\mathbb{Z}^n\setminus\{0\}} |\partial^\alpha \mathcal{F}\hat\eta(\sqrt{\mathcal{D}}\nu)| \le \int_1^\infty |f(\tau)|\tau^{[\alpha]/2} \sum_{\nu\in\mathbb{Z}^n\setminus\{0\}} |\partial^\alpha \mathcal{F}\eta(\sqrt{\mathcal{D}\tau}\nu)|\, d\tau.$$

Theorem 7. *Suppose that η satisfies (2), (8), (12), and $\mathcal{F}\eta(0)\ne 0$. Then the generating function $\hat\eta$ defined in (13) with f subject to (14) satisfies the moment condition (3) with $N=2M$.*

Note that (13) leads to

$$\mathcal{D}^{-n/2}\hat\eta\!\left(\frac{\mathbf{x}}{\sqrt{\mathcal{D}}}\right) = \int_1^\infty f(\tau)\,\eta\!\left(\frac{\mathbf{x}}{\sqrt{\mathcal{D}\tau}}\right) \frac{d\tau}{(\mathcal{D}\tau)^{n/2}}$$

$$= \int_\mathcal{D}^\infty \frac{\tau}{\mathcal{D}} f\!\left(\frac{\tau}{\mathcal{D}}\right) \tau^{-n/2}\eta\!\left(\frac{\mathbf{x}}{\sqrt{\tau}}\right)\frac{d\tau}{\tau} = \int_0^\infty k\!\left(\frac{\tau}{\mathcal{D}}\right)\tau^{-n/2}\eta\!\left(\frac{\mathbf{x}}{\sqrt{\tau}}\right)\frac{d\tau}{\tau},$$

hence (13) represents a Mellin convolution with the kernel function

$$k(\tau) := \begin{cases} \tau f(\tau) & , \quad 1 \leq \tau, \\ 0 & , \quad 0 \leq \tau < 1. \end{cases} \qquad (15)$$

Let us mention that in general it is not necessary that f is a function. Obviously all assertions remain valid if a new generating function is defined by

$$\hat{\eta}(\mathbf{x}) = \int_1^\infty \eta\left(\frac{\mathbf{x}}{\sqrt{\tau}}\right) \frac{d\mu(\tau)}{\tau^{n/2}}, \qquad (16)$$

where the measure μ satisfies the conditions

$$\int_1^\infty d\mu(\tau) = 1/\mathcal{F}\eta(0), \quad \int_1^\infty \tau^k \, d\mu(\tau) = 0 \quad, \quad k = 1, \ldots, M-1.$$

2.5. Linear combinations of translates

Finally we consider the case of the quasi–interpolation formula

$$\mathcal{D}^{-n/2} \sum_{\mathbf{m} \in \mathbb{Z}^n} u(h\mathbf{m}) \, \tilde{\eta}\left(\frac{\mathbf{x} - h\mathbf{m}}{\sqrt{\mathcal{D}} h}\right), \qquad (17)$$

where for given $\mathcal{D}$ the generating function $\tilde{\eta}(\mathbf{x}/\sqrt{\mathcal{D}})$ is a linear combination of a finite number of translates

$$\tilde{\eta}\left(\frac{\mathbf{x}}{\sqrt{\mathcal{D}}}\right) = \sum_{\mathbf{k} \in \Lambda} \gamma_{\mathbf{k}} \, \eta\left(\frac{\mathbf{x} - \mathbf{k}}{\sqrt{\mathcal{D}}}\right), \quad \Lambda \subset \mathbb{Z}^n, \, |\Lambda| < \infty. \qquad (18)$$

If η is subjected to (2) and the set Λ and the coefficients $\gamma_{\mathbf{k}}$ are chosen such that $\tilde{\eta}$ satisfies the moment condition for large N then from (18)

$$\mathcal{D}^{-n/2} \sum_{\mathbf{m} \in \mathbb{Z}^n} u(h\mathbf{m}) \, \tilde{\eta}\left(\frac{\mathbf{x} - h\mathbf{m}}{\sqrt{\mathcal{D}} h}\right) = \mathcal{D}^{-n/2} \sum_{\mathbf{m} \in \mathbb{Z}^n} \left(\sum_{\mathbf{k} \in \Lambda} \gamma_{\mathbf{k}} \, u(h(\mathbf{m} - \mathbf{k}))\right) \eta\left(\frac{\mathbf{x} - h\mathbf{m}}{\sqrt{\mathcal{D}} h}\right), \qquad (19)$$

thus we obtain an approximate approximation formula of order $\mathcal{O}(h^N)$ using the given generating function η and some linear combination of the values of u at the mesh points $h\mathbf{m} - h\Lambda$. Note that for basis functions satisfying the Strang–Fix condition (cf. [9], [7]) this method of improving the approximation order is well–known.

Since

$$\tilde{\eta}(\mathbf{x}) = \sum_{\mathbf{k} \in \Lambda} \gamma_{\mathbf{k}} \, \eta\left(\mathbf{x} - \frac{\mathbf{k}}{\sqrt{\mathcal{D}}}\right) \quad \text{implies} \quad \mathcal{F}\tilde{\eta}(\lambda) = \mathcal{F}\eta(\lambda) \sum_{\mathbf{k} \in \Lambda} \gamma_{\mathbf{k}} \, e^{-2\pi i \langle \mathbf{k}, \lambda \rangle / \sqrt{\mathcal{D}}},$$

any trigonometric polynomial P_Λ of period $\sqrt{\mathcal{D}}$ satisfying

$$\partial^\alpha P_\Lambda(0) = \partial^\alpha \sum_{\mathbf{k}\in\Lambda} \gamma_{\mathbf{k}} e^{-2\pi i \langle \mathbf{k},\lambda\rangle/\sqrt{\mathcal{D}}}\Big|_{\lambda=0} = \partial^\alpha (\mathcal{F}\eta)^{-1}(0), \quad \forall \alpha, [\alpha] < N,$$

can be taken to determine $\tilde{\eta}$. So the coefficients $\gamma_{\mathbf{k}}$ have to solve the equations

$$\sum_{\mathbf{k}\in\Lambda} \gamma_{\mathbf{k}} \mathbf{k}^\alpha = \left(-\frac{\sqrt{\mathcal{D}}}{2\pi i}\right)^{|\alpha|} \partial^\alpha (\mathcal{F}\eta)^{-1}(0), \quad [\alpha] < N. \tag{20}$$

It is well known (cf.[8]) that for any N there exist finite sets $\Lambda_N \subset \mathbb{Z}^n$ with the cardinality $|\Lambda_N| = \dfrac{(N+n-1)!}{(N-1)!\, n!}$ such that (20) is uniquely solvable. Furthermore, the periodicity of $P_\Lambda(\lambda)$ implies

$$\partial^\alpha \mathcal{F}\tilde{\eta}(\sqrt{\mathcal{D}}\nu) = \sum_{\beta \leq \alpha} \frac{\alpha!}{\beta!\,(\alpha-\beta)!} \partial^\beta \mathcal{F}\eta(\sqrt{\mathcal{D}}\nu)\, \partial^{\alpha-\beta} P_\Lambda(0).$$

Theorem 8. *Suppose that η satisfies (2) and $\mathcal{F}\eta(0) \neq 0$. Then for any positive $N \leq K$ there exists a finite set $\Lambda_N \subset \mathbb{Z}^n$ and coefficients $\gamma_{\mathbf{k}}$, $\mathbf{k} \in \Lambda_N$, such that*

$$\tilde{\eta}(\mathbf{x}) = \sum_{\mathbf{k}\in\Lambda_N} \gamma_{\mathbf{k}}\, \eta\!\left(\mathbf{x} - \frac{\mathbf{k}}{\sqrt{\mathcal{D}}}\right)$$

satisfies the moment condition (3). Moreover,

$$\partial^\alpha \mathcal{F}\tilde{\eta}(\sqrt{\mathcal{D}}\nu) = \sum_{\beta \leq \alpha} \frac{\alpha!}{\beta!\,(\alpha-\beta)!} \partial^\beta \mathcal{F}\eta(\sqrt{\mathcal{D}}\nu)\, \partial^{\alpha-\beta}(\mathcal{F}\eta)^{-1}(0), \quad [\alpha] < N,\ \nu \in \mathbb{Z}^n.$$

This result is closely connected with the problem of the best approximation by elements of the space

$$S_h := \mathrm{span}\left\{\eta\!\left(\frac{\cdot - h\mathbf{m}}{\sqrt{\mathcal{D}}h}\right),\ \mathbf{m} \in \mathbb{Z}^n\right\}.$$

We sketch only a simple consequence of Theorems 1 and 8.

Lemma 1. *Suppose that η belongs to the Schwartz space $\mathcal{S}(\mathbb{R}^n)$ with $\mathcal{F}\eta(0) \neq 0$. Then for any $N \leq K$ and any $\varepsilon > 0$ there exist $\mathcal{D} > 0$ and a constant c_N such that for all $u \in C^N(\mathbb{R}^n) \cap L_\infty(\mathbb{R}^n)$*

$$\inf_{v_h \in S_h} \|u - v_h\|_{L_\infty(\mathbb{R}^n)} \leq c_N(\sqrt{\mathcal{D}}h)^N \sum_{[\alpha]=N} \frac{\|\partial^\alpha u\|_{L_\infty(\mathbb{R}^n)}}{\alpha!} + \varepsilon \sum_{[\alpha]=0}^{N-1} (\sqrt{\mathcal{D}}h)^{[\alpha]}\, \frac{|\partial^\alpha u(\mathbf{x})|}{\alpha!},$$

Note that the construction of these approximants in the case of symmetric generating functions η and $N = 2M$ can be based on the trigonometric polynomial

$$P_M(\lambda) = \sum_{[\beta]<M} a_\beta \prod_{j=1}^n \cos\frac{2\pi}{\sqrt{\mathcal{D}}}\beta_j\lambda_j, \quad \beta = (\beta_1,\ldots,\beta_n) \in \mathbb{Z}^n_{\geq 0}. \quad (21)$$

Then $\tilde\eta = \mathcal{F}^{-1}(P_M \mathcal{F}\eta)$ satisfies (8), therefore (20) leads to the system of linear equations

$$\sum_{[\beta]<M} a_\beta \beta^{2\alpha} = \left(-\frac{\mathcal{D}}{4\pi^2}\right)^{[\alpha]} \partial^{2\alpha}(\mathcal{F}\eta)^{-1}(0), \quad [\alpha] < M. \quad (22)$$

Its unique solution $\{a_\beta\}$ provides the generating function

$$\tilde\eta(\mathbf{x}) = \sum_{[\beta]<M} 2^{-\kappa(\beta)} a_\beta \sum_{\{|\mathbf{k}|\}=\beta} \eta\left(\mathbf{x} - \frac{\mathbf{k}}{\sqrt{\mathcal{D}}}\right) \quad (23)$$

satisfying the moment condition (3) with $N = 2M$. Here $\{|\mathbf{k}|\}$, $\mathbf{k} \in \mathbb{Z}^n$, denotes the vector $(|k_1|,\ldots,|k_n|)$ and $\kappa(\beta)$ is the number of nonzero components of β. This construction leads to the approximation formula

$$\mathcal{D}^{-n/2} \sum_{\mathbf{m}\in\mathbb{Z}^n} \eta\left(\frac{\mathbf{x}-h\mathbf{m}}{\sqrt{\mathcal{D}}h}\right)\left(\sum_{[\beta]<M} 2^{-\kappa(\beta)} a_\beta \sum_{\{|\mathbf{k}|\}=\beta} u(h(\mathbf{m}-\mathbf{k}))\right)$$

of order $\mathcal{O}(h^{2M})$ based on the function values at the mesh points $h(\mathbf{m}-\mathbf{k})$ with minimal distance to $h\mathbf{m}$. A detailed study of these approximations with the Gaussian as generating function is contained in [5].

Note that these formulas are useful to obtain approximations of different order to integral operators in parallel. If for example for a suitable symmetric basis function η the values of the integral operator

$$a_\mathbf{m} := \mathcal{K}\eta\left(\frac{\cdot - h\mathbf{m}}{\sqrt{\mathcal{D}}h}\right)(0) = \int_{\mathbb{R}^n} k\left(\sqrt{\mathcal{D}}h\left(\frac{\mathbf{m}}{\sqrt{\mathcal{D}}}-\mathbf{y}\right)\right)\eta(\mathbf{y})\,d\mathbf{y}$$

are precomputed and stored then the sums

$$\mathcal{D}^{-n/2} \sum_{\mathbf{m}\in\mathbb{Z}^n} a_{\mathbf{j}-\mathbf{m}}\left(\sum_{[\beta]<M} 2^{-\kappa(\beta)} a_\beta \sum_{\{|\mathbf{k}|\}=\beta} u(h(\mathbf{m}-\mathbf{k}))\right)$$

provide approximations of $\mathcal{K}u$ at the mesh point $h\mathbf{j}$ of order $\mathcal{O}(h^{2M})$.

References

[1] C. de Boor: The polynomials in the linear space of integer translates of a compactly supported function. *Constr. Approx.* **3** (1987), 199-208.

[2] V. Karlin, V. Maz'ya: Time-marching algorithms for non-local evolution equations based upon "approximate approximations". *SIAM J. Sc. Comp.* **18** (1997), 736–752.

[3] V. Maz'ya: A New Approximation Method and its Applications to the Calculation of Volume Potentials. Boundary Point Method. In: *3. DFG-Kolloqium des DFG-Forschungsschwerpunktes "Randelementmethoden"*, (1991).

[4] V. Maz'ya: Approximate approximations. In: *The Mathematics of Finite Elements and Applications. Highlights.* (J. R. Whiteman, ed.), Wiley & Sons Chichester (1994), 77–104.

[5] V. Maz'ya, G. Schmidt: On Approximate Approximations using Gaussian Kernels. *IMA J. Num. Anal.* **16** (1996), 13-29.

[6] V. Maz'ya, G. Schmidt: "Approximate Approximations" and the cubature of potentials. *Rend. Mat. Acc. Lincei*, s. 9, **6** (1995), 161–184.

[7] R. Schaback, Z. Wu: Construction techniques for highly accurate quasi–interpolation operators. *J. of Approx. Theory* **91** (1997), 320–331.

[8] S. L.Sobolev: *Introduction to the Theory of Cubature Formulas.* Nauka, Moscow, (1974) (in Russian).

[9] G.Strang, G.Fix: A Fourier analysis of the finite element variational method. In: *Constructive Aspects of Functional Analysis, C.I.M.E. II Ciclo 1971*, (G. Geymonat, ed.), (1973) 793–840.

Linköping University, Institute of Technology, Dept. of Mathematics
S–581 83, Linköping, Sweden

V. MAZ'YA and A. SOLOVIEV

L_p-Theory of Direct Boundary Integral Equations on a Contour with Peak

1. Introduction

In this paper we apply the so called direct variant of the method of boundary integral equations to solve the Dirichlet problem

$$\Delta u = 0 \text{ in } \Omega, \quad u = \varphi \text{ on } \Gamma \setminus \{O\} \tag{1}$$

and the Neumann problem

$$\Delta u = 0 \text{ in } \Omega, \quad \partial u/\partial n = \psi \text{ on } \Gamma \setminus \{O\} \tag{2}$$

with boundary data ψ satisfying $\int_\Gamma \psi ds = 0$, where Ω is a plane simply connected domain having compact closure and the boundary Γ with peak $z = 0$. Here and elsewhere we assume that the normal n is directed outwards.

The classical method for solving boundary value problems is their reduction to boundary integral equations by using potentials. In the case of the Dirichlet and Neumann problems for the Laplace equation the solutions of these problems are represented in the form of simple and double layer potentials whose densities satisfy boundary integral equations. However, there exists another way of reduction when solutions of integral equations are represented explicity by the solutions of the boundary value problems (1) and (2). In this case the integral equations can be obtained from the integral representation for a harmonic function:

$$u(z) = \frac{1}{2\pi}\left(V\frac{\partial u}{\partial n}\right)(z) - \frac{1}{2\pi}(Wu)(z), \quad z \in \Omega,$$

where V is the simple layer potential:

$$(V\sigma)(z) = \int_\Gamma \sigma(q) \log \frac{|z|}{|z-q|} ds_q$$

and W is the double layer potential:

$$(W\sigma)(z) = \int_\Gamma \sigma(q) \frac{\partial}{\partial n} \log \frac{1}{|z-q|} ds_q.$$

By making use of the continuity of the simple layer potential and the limit relation for the double layer potential we obtain

$$\pi u(z) = \left(V\frac{\partial u}{\partial n}\right)(z) - (Wu)(z), \quad z \in \Gamma \setminus \{O\} \tag{3}$$

where $(Wu)(z)$ is the direct value of Wu at the point z on $\Gamma \setminus \{O\}$. By substituting the known values φ of u on $\Gamma \setminus \{O\}$ to (1) we obtain that $\partial u/\partial n$ satisfies the integral equation of the first kind

$$V\gamma = \pi\varphi + W\varphi \text{ on } \Gamma \setminus \{O\}.$$

The normal derivative on $\Gamma \setminus \{O\}$ of the solution u of (2) is defined by the boundary data ψ. From (3) it follows that u on $\Gamma \setminus \{O\}$ is a solution of the integral equation

$$\pi \sigma + W\sigma = V\psi.$$

Using the results of [1]-[3] we study the equations obtained by the direct reduction of problems (2) and (3) for a domain with a peak to integral equations. For every integral equation we point out a pair of function spaces and prove the solvability of the equation in one of these spaces with a right-hand side from another. We also describe solutions of the corresponding homogeneous equation.

Let Ω be a plane simply connected domain with boundary Γ which has a peak at the origin O. We assume that $\Gamma \setminus \{O\}$ belongs to the class C^2. We say that O is an outward (inward) peak if Ω (the exterior domain Ω') is given near O by the inequalities $\kappa_-(x) < y < \kappa_+(x)$, $0 < x < \delta$, where

$$x^{-\mu-1}\kappa_\pm(x) \in C^2[0,\delta], \quad \lim_{x \to +0} x^{-\mu-1}\kappa_\pm(x) = \alpha_\pm$$

with $\mu > 0$ and $\alpha_+ > \alpha_-$. By $\Gamma_\pm$ we denote the arcs $\{(x, \kappa_\pm(x)) : x \in [0,\delta]\}$. Points on Γ_+ and Γ_- with equal abscissas will be denoted by q_+ and q_-.

We say that φ belongs to $\mathcal{L}_{p,\beta}(\Gamma)$ if $|q|^\beta \varphi \in L_p(\Gamma)$. The norm in this space is given by

$$\|\varphi\|_{\mathcal{L}_{p,\beta}(\Gamma)} = \| |q|^\beta \varphi \|_{L_p(\Gamma)}.$$

Let $\mathcal{L}^1_{p,\beta}(\Gamma)$ be the space of absolutely continuous functions on $\Gamma \setminus \{O\}$ with the finite norm

$$\|\varphi\|_{\mathcal{L}^1_{p,\beta}(\Gamma)} = \|(\partial/\partial s)\varphi\|_{\mathcal{L}_{p,\beta}(\Gamma)} + \|\varphi\|_{\mathcal{L}_{p,\beta-1}(\Gamma)}.$$

We introduce the space $\mathfrak{N}_{p,\beta}(\Gamma)$ of absolutely continuous functions φ on $\Gamma \setminus \{O\}$ with the finite norm

$$\|\varphi\|_{\mathfrak{N}_{p,\beta}(\Gamma)} = \left(\int_{\Gamma_+ \cup \Gamma_-} |\varphi(q_+) - \varphi(q_-)|^p |q|^{p(\beta-\mu)} ds_q \right)^{1/p} + \|\varphi\|_{\mathcal{L}^1_{p,\beta+1}(\Gamma)}.$$

By $\mathfrak{N}^{-1}_{p,\beta}(\Gamma)$ we denote the space of functions on $\Gamma \setminus \{O\}$ represented in the form $\varphi = (d/ds)\psi$, where $\psi \in \mathfrak{N}_{p,\beta}(\Gamma)$ and $\psi(z_0) = 0$ with a fixed point $z_0 \in \Gamma \setminus \{O\}$. A norm on $\mathfrak{N}^{-1}_{p,\beta}(\Gamma)$ is defined by

$$\|\varphi\|_{\mathfrak{N}^{-1}_{p,\beta}(\Gamma)} = \|\psi\|_{\mathfrak{N}_{p,\beta}(\Gamma)}.$$

Furthermore we introduce the space $\mathfrak{N}^{(+)}_{p,\beta}(\Gamma)$ of absolutely continuous functions φ on $\Gamma \setminus \{O\}$ with the finite norms

$$\|\varphi\|_{\mathfrak{N}^{(+)}_{p,\beta}(\Gamma)} = \left(\int_{\Gamma_+ \cup \Gamma_-} |\varphi(q_+) + \varphi(q_-)|^p |q|^{p(\beta-\mu)} ds_q \right)^{1/p} + \|\varphi\|_{\mathcal{L}^1_{p,\beta+1}(\Gamma)}.$$

Let $\mathfrak{P}(\Gamma)$ denote the space of restrictions to $\Gamma \setminus \{O\}$ of real functions of the form $p(z) = \sum_{k=0}^{m} t^{(k)} \operatorname{Re} z^k$, where $m = [\mu - \beta - p^{-1} + 2^{-1}]$. We endow $\mathfrak{P}(\Gamma)$ with the norm

$$\| p \|_{\mathfrak{P}(\Gamma)} = \sum_{k=0}^{m} | t^{(k)} |.$$

The space $\mathfrak{M}_{p,\beta}(\Gamma)$ is defined as the direct sum of $\mathfrak{N}_{p,\beta}^{(+)}(\Gamma)$ and $\mathfrak{P}(\Gamma)$.

Now we can describe our results. We assume that $0 < \beta + p^{-1} < \min\{\mu, 1\}$.

Let Ω have an outward peak. We introduce the double layer potential in Ω' by setting

$$(W^{ext}\sigma)(z) = \int_{\Gamma} \sigma(q) \left(\frac{\partial}{\partial n_q} \log \frac{1}{|z-q|} + 1 \right) ds_q, \quad z \in \Omega'.$$

The value of this potential at the point $z \in \Gamma \setminus \{O\}$ will be also denoted by $(W^{ext}\sigma)(z)$. In Theorem 1 we prove that the integral equation

$$\pi \sigma + W^{ext}\sigma = V\psi \tag{4}$$

with the function $\psi \in \mathfrak{N}_{p,\beta}^{-1}(\Gamma)$ on the right-hand side has a unique solution γ in $\mathfrak{N}_{p,\beta}(\Gamma)$.

As is shown in Theorem 3 the integral equation of the first kind

$$V\gamma = \pi\varphi + W\varphi \tag{5}$$

with the function $\varphi \in \mathfrak{N}_{p,\beta}(\Gamma)$ on the right-hand side has a unique solution σ in $\mathfrak{N}_{p,\beta}^{-1}(\Gamma)$.

Let Ω have an inward peak. In Theorem 2 we show that the integral equation

$$\pi\sigma + W^{ext}\sigma = V\psi', \tag{6}$$

with the function $\psi \in \mathcal{L}_{p,\beta+1}^{1}(\Gamma)$ is solvable in $\mathcal{L}_{p,\beta+1}^{1}(\Gamma)$ and that the solution σ satisfies $\int_{\Gamma} \sigma ds = 0$. In Theorem 4 we prove that the integral equation

$$V\gamma = \pi\varphi + W\varphi \tag{7}$$

with the function $\varphi \in \mathcal{L}_{p,\beta+1}^{1}(\Gamma)$ has a solution $\gamma \in \mathcal{L}_{p,\beta+1}(\Gamma)$ satisfying $\int_{\Gamma} \gamma ds = 0$.

In Theorems 2 and 4 we show that the homogeneous equations (6) and (7) have only trivial solutions for $0 < \beta + p^{-1} < 1/2$ and a one-dimensional space of solutions for $1/2 < \beta + p^{-1} < 1$.

We shall make use of the following statements proved in [1] - [3].

Theorem A (see [1]). *Let Ω have either an outward or inward peak and $0 < \beta + p^{-1} < \min\{\mu, 1\}$, $\beta + p^{-1} \neq 1/2$. Then the operator*

$$\mathcal{L}_{p,\beta+1}(\Gamma) \times \mathbf{R} \ni (\sigma, t) \overset{\nu}{\longmapsto} V\sigma + t \in \mathfrak{N}_{p,\beta}(\Gamma)$$

is surjective and
  a) $\ker \mathcal{V} = \{0\}$ for $0 < \beta + p^{-1} < 1/2$;
  b) $\dim \ker \mathcal{V} = 1$ for $1/2 < \beta + p^{-1} < 1$.

If Ω has an outward peak and $1/2 < \beta + p^{-1} < 1$ then

$$\ker \mathcal{V} = \left\{ \frac{t}{\pi} \frac{\partial}{\partial n} \operatorname{Im} \frac{1}{\gamma^{(out)}}, t \operatorname{Im} \frac{1}{\gamma^{(out)}(\infty)} \right\},$$

where $t \in \mathbf{R}$ and $\gamma^{(out)}$ is the conformal mapping of Ω' onto $\mathbf{R}_+^2 = \{z : \operatorname{Im} z > 0\}$ subject to $\gamma^{(out)}(0) = 0$ and $\gamma^{(out)}(\infty) = i$.

If Ω has an inward peak and $1/2 < \beta + p^{-1} < 1$ then

$$\ker \mathcal{V} = \left\{ \frac{t}{\pi} \frac{\partial}{\partial n} \operatorname{Im} \frac{1}{\gamma^{(in)}}, 0 \right\},$$

where $t \in \mathbf{R}$ and $\gamma^{(in)}$ is the conformal mapping Ω onto $\mathbf{R}_+^2$ subject to the conditions $\gamma^{(in)}(0) = 0$ and $\gamma^{(in)}(z_0) = i$ with a fixed point $z_0 \in \Omega$.

Theorem B (see [2]). *Let Ω have an inward peak and let $0 < \beta + p^{-1} < \min\{\mu, 1\}$, $\beta + p^{-1} \neq 1/2$. Then the operator*

$$\mathcal{L}_{p,\beta+1}^1(\Gamma) \ni \sigma \xmapsto{\mathcal{W}} (\pi I + W^{ext})\sigma \in \mathfrak{N}_{p,\beta}(\Gamma)$$

is surjective and
  a) $\operatorname{Ker} \mathcal{W} = \{0\}$ for $0 < \beta + p^{-1} < 1/2$,
  b) $\dim \operatorname{Ker} \mathcal{W} = 1$ for $1/2 < \beta + p^{-1} < 1$ and

$$\operatorname{Ker} \mathcal{W} = \left\{ t \operatorname{Re} \frac{1}{\gamma_0} \right\},$$

where $t \in \mathbf{R}$ and γ_0 is the conformal mapping of Ω onto $\mathbf{R}_+^2$ subject to

$$\gamma_0(0) = 0, \quad \int_\Gamma \operatorname{Re} \frac{1}{\gamma_0} ds = 0 \text{ and } \operatorname{Im} \gamma_0(z_0) = 1,$$

where z_0 is a fixed point in Ω.

We introduce the functions $\mathcal{I}_k^{ext}$, $k = 1, 2, \ldots$ by setting

$$\mathcal{I}_k^{ext}(z) = \operatorname{Im} \left(\frac{z z_0}{z_0 - z} \right)^{k-1/2}, \quad z \in \Omega',$$

where z_0 is a fixed point in Ω.

Theorem C (see [3]). *Let Ω have an outward peak and let $0 < \beta + p^{-1} < \min\{\mu, 1\}$, $\mu - \beta - p^{-1} + 2^{-1} \notin \mathbf{N}$. Then the operator*

$$\mathcal{L}_{p,\beta+1}^1(\Gamma) \times \mathbf{R}^m \ni (\sigma, t) \xmapsto{\mathcal{W}_\mu} \pi\sigma + W^{ext}\sigma + \sum_{k=1}^m t^{(k)} \mathcal{I}_k^{ext} \in \mathfrak{M}_{p,\beta}(\Gamma)$$

is bijective.

2. The Dirichlet and Neumann problems for domains with peaks

2.1. Let $G \subset \mathbf{R}^2$ be a domain with C^2-boundary such that the set $\{(\tau,\nu) \in G : \tau \leq 0\}$ has compact closure and $\{(\tau,\nu) \in G : \tau > 0\} = \{(\tau,\nu) : \tau > 0, |\nu| < 1\}$.

As usual by $C_0^\infty(G)$ we mean the space of infinitely differentiable functions with compact supports in G. We denote by $W_p^k(G)$, $k = 0,1,2$, $p \in (1,\infty)$, the Sobolev space of functions in $L_p(G)$ with derivatives up to order k in $L_p(G)$. By $\mathring{W}_p^k(G)$ the completion of $C_0^\infty(G)$ in $W_p^k(G)$ will be denoted. Let $W_p^{k-1/p}(\partial G)$ be the space of traces on ∂G of functions in $W_p^k(G)$. We introduce also the space $W_p^{-1}(G)$ of distributions on G with the finite norm

$$\|\varphi\|_{W_p^{-1}(G)} = \inf \sum_{k=0}^{2} \|\varphi_k\|_{L_p(G)},$$

where the infimum is taken over all representations $\varphi = \varphi_0 + (\partial/\partial \tau)\varphi_1 + (\partial/\partial \nu)\varphi_2$ with $\varphi_j \in L_p(G)$, $j = 0,1,2$.

Let $\alpha \in \mathbf{R}$. We say that $\varphi \in W_{p,\alpha}^k(G)$ if $(1+\tau^2)^{\alpha/2}\varphi \in W_p^k(G)$, $k = -1, 0, 1\ldots$, and define the norm

$$\|\varphi\|_{W_{p,\alpha}^k(G)} = \|(1+\tau^2)^{\alpha/2}\varphi\|_{W_p^k(G)}.$$

The spaces $W_{p,\alpha}^k(\partial G)$ and $W_{p,\alpha}^{k-1/p}(\partial G)$ are introduced in the same way. By $L_{p,\alpha}(\partial G)$ we denote the space of functions with the finite norm

$$\|\varphi\|_{L_{p,\alpha}(\partial G)} = \|(1+\tau^2)^{\alpha/2}\varphi\|_{L_p(\partial G)}.$$

We shall make use of the same definitions for the strip $\Pi = \{(\tau,\nu) : \tau \in \mathbf{R}, |\nu| < 1\}$.

The following lemma is contained in more general results of the article [4].

Lemma 1. *The operator*

$$W_{p,\alpha}^k(G) \ni u \to \{\Delta u, u|_{\partial G}\} \in W_{p,\alpha}^{k-2}(G) \times W_{p,\alpha}^{k-1/p}(\partial G)$$

realizes an isomorphism for every real α and $k = 1, 2$. The same is true for $k = 1, 2$ if G is replaced by the strip Π.

2.2. In this section we consider the Dirichlet problem in Ω with outward peak. The following proposition is an improvement of a statement from [1].

Proposition 1. *Let φ belong to space $\mathfrak{N}_{p,\beta}(\Gamma)$, where $0 < \beta + p^{-1} < \min\{\mu, 1\}$. Then there exists a harmonic extension u of φ onto Ω with normal derivative in $\mathfrak{N}_{p,\beta}^{-1}(\Gamma)$ satisfying*

$$\|\partial u/\partial n\|_{\mathfrak{N}_{p,\beta}^{-1}(\Gamma)} \leq \|\varphi\|_{\mathfrak{N}_{p,\beta}(\Gamma)}. \tag{8}$$

Proof. It is sufficient to obtain (8) under the assumption $\varphi = 0$ in a neighbourhood of the peak.

We start with the case when $\varphi \in \mathfrak{N}_{p,\beta}(\Gamma)$ equals zero on $\Gamma \cap \{|q| < \delta/2\}$. Let θ be a conformal mapping of the unit disk D onto Ω. We introduce the harmonic extension F of the continuous function $\varphi \circ \theta$ onto D. The normal derivative $\partial F/\partial n$ on ∂D has the form

$$\frac{\partial F}{\partial n}(\zeta) = \frac{1}{2\pi}\int_{-\pi}^{\pi}\frac{d}{dt}\varphi(\theta(e^{it}))\cot\frac{s-t}{2}dt, \ \zeta = e^{is}.$$

Hence $u = F \circ \theta^{-1}$ satisfies

$$\|\,\partial u/\partial n\,\|_{\mathcal{L}_{p,\gamma+1}(\Gamma)} \leq c_1 \ \|\,\partial F/\partial n\,\|_{L_p(\partial D)}$$
$$\leq c_2 \ \|\,(\varphi\circ\theta)'\,\|_{L_p(\partial D)} \leq c_3 \ \|\,\varphi\,\|_{\mathfrak{N}_{p,\gamma}(\Gamma)}$$

for every real γ. Since $\int_\Gamma (\partial u/\partial n)ds = 0$, $\partial u/\partial n$ is represented in the form $(\partial/\partial s)v$, where the function v satisfies

$$\|\,v\,\|_{\mathcal{L}_{p,\beta-\mu}(\Gamma)} + \|\,v'\,\|_{\mathcal{L}_{p,\beta+1}(\Gamma)} \leq c \ \|\,\varphi\,\|_{\mathfrak{N}_{p,\beta}(\Gamma)}.$$

It remains to prove (8) for $\varphi \in \mathfrak{N}_{p,\beta}(\Gamma)$ equals zero outside $\Gamma_+ \cup \Gamma_-$. Let G be the domain defined in 2.1. By $z = \omega(\tau + i\nu)$ we denote a conformal mapping of G onto Ω such that $\omega(\infty) = 0$ and $\omega(\partial G \cap \{(x,y) : x > 1\}) \supset \Gamma_+ \cup \Gamma_-$. We introduce the functions $\Phi_\pm$ on $\mathbf{R}$ by

$$\Phi_\pm(\tau) = \varphi(\omega(\tau \pm i)) \text{ for } \tau \geq 0 \text{ and } \Phi_\pm(\tau) = 0 \text{ for } \tau < 0.$$

We shall make use of the following inequality proved in [1]:

$$c_1 \ \|\,\varphi\,\|_{\mathfrak{N}_{p,\beta}(\Gamma)} \leq \|\,\Phi_+ - \Phi_-\,\|_{L_{p,1-\alpha}(\mathbf{R})}$$
$$+ \sum_\pm \left(\|\,\Phi_\pm\,\|_{L_{p,-\alpha}(\mathbf{R})} + \|\,\Phi'_\pm\,\|_{L_{p,1-\alpha}(\mathbf{R})}\right) \leq c_2 \ \|\,\varphi\,\|_{\mathfrak{N}_{p,\beta}(\Gamma)} \qquad (9)$$

with α defined by $\beta + p^{-1} = \mu(\alpha - p^{-1})$.

Let Π denote the same strip as in 2.1. We introduce the bounded harmonic function $\Phi^{(+)}$ on Π taking the equal values $(\Phi_+(\tau) + \Phi_-(\tau))/2$ at the points $(\tau, 1)$ and $(\tau, -1)$ of $\partial\Pi$. The Fourier transform of $\Phi^{(+)}(\tau, \nu)$ with respect to τ is given by

$$c\left(\widehat{\Phi_+}(\xi) + \widehat{\Phi_-}(\xi)\right)\cosh(\nu\xi)(\cosh\xi)^{-1},$$

where $\widehat{\Phi_\pm}$ denote the Fourier transform of $\Phi_\pm$. Therefore the Fourier transform of $(\partial/\partial n)\Phi^{(+)}(\tau, \pm 1)$ is equal to

$$ci\left(\widehat{\Phi'_+}(\xi) + \widehat{\Phi'_-}(\xi)\right)\tanh\xi.$$

Hence we have

$$\frac{\partial \Phi^{(+)}}{\partial n}(\tau, \pm 1) = c\int_\mathbf{R}\frac{d}{dt}(\Phi_+(t) + \Phi_-(t))\left(\sinh\left(\frac{\pi}{2}(\tau - t)\right)\right)^{-1}dt.$$

By Lemma 1
$$\| \partial \Phi^{(+)}/\partial n \|_{L_{p,1-\alpha}(\partial \Pi)} \leq c \| (\Phi_+ + \Phi_-)' \|_{L_{p,1-\alpha}(\partial \Pi)}.$$

We rewrite the Fourier transform of $(\partial/\partial n)\Phi^{(+)}(\tau,\nu)$ with respect to τ in the form
$$c\, i\xi \left(\widehat{\Phi'_+}(\xi) + \widehat{\Phi'_-}(\xi) \right) \frac{\sinh \xi}{\xi \cosh \xi}.$$

Hence and by the multiplier theorem on weighted L_p-spaces (see [5]) it follows that $(\partial/\partial n)\Phi^{(+)}(\tau,\pm 1)$ is represented as $(\partial/\partial s)Y^{(-)}(\tau,\pm 1)$, where $Y^{(-)}(\tau,+1) = -Y^{(-)}(\tau,-1)$ and
$$\| Y^{(-)} \|_{L_{p,1-\alpha}(\partial\Pi)} + \| (\partial/\partial \tau) Y^{(-)} \|_{L_{p,1-\alpha}(\partial \Pi)}$$
$$\leq c \| (\partial/\partial \tau)(\Phi_+ + \Phi_-) \|_{L_{p,1-\alpha}(\partial\Pi)}.$$

Let $\Phi^{(-)}$ be the bounded harmonic function on Π taking the opposite values $(\Phi_+(\tau) - \Phi_-(\tau))/2$ and $(\Phi_-(\tau) - \Phi_+(\tau))/2$ at the points $(\tau, 1)$ and $(\tau, -1)$ of $\partial \Pi$. The Fourier transform of $\Phi^{(-)}(\tau,\nu)$ with respect to τ has the form
$$c \left(\widehat{\Phi_+}(\xi) - \widehat{\Phi_-}(\xi) \right) \sinh(\nu \xi)\, (\sinh \xi)^{-1}.$$

Therefore the Fourier transform of $(\partial/\partial n)\Phi^{(-)}(\tau,\pm 1)$ equals
$$\pm c\, \xi \left(\widehat{\Phi_+}(\xi) - \widehat{\Phi_-}(\xi) \right) \cosh \xi\, (\sinh \xi)^{-1}$$
$$= \pm c \left\{ \left(\widehat{\Phi_+}(\xi) - \widehat{\Phi_-}(\xi) \right) \frac{\xi}{\sinh \xi} + i \left(\widehat{\frac{d}{dt}\Phi_+}(\xi) - \widehat{\frac{d}{dt}\Phi_-}(\xi) \right) \tanh \xi \right\}.$$

and hence
$$\frac{\partial \Phi^{(-)}}{\partial n}(\tau, \pm 1) = \pm c_1 \int_{\mathbf{R}} (\Phi_+(t) - \Phi_-(t)) \left(\cosh \frac{\pi}{2}(\tau - t) \right)^{-2} dt$$
$$\pm c_2 \int_{\mathbf{R}} \left(\frac{d\Phi_+}{dt}(t) - \frac{d\Phi_-}{dt}(t) \right) (\sinh \pi(\tau - t))^{-1} dt.$$

By Lemma 1
$$\| \partial \Phi^{(-)}/\partial n \|_{L_{p,1-\alpha}(\partial \Pi)} \leq c \| \Phi_+ - \Phi_- \|_{W^1_{p,1-\alpha}(\partial \Pi)}.$$

It is clear that $(\partial/\partial n)\Phi^{(-)}$ is written as $(\partial/\partial s)Y^{(+)}(\tau,\pm 1)$, where $Y^{(+)}(\tau,+1) = Y^{(+)}(\tau,-1)$ and according to the multiplier theorem on weighted L_p-spaces (see [5])
$$\| Y^{(+)} \|_{L_{p,-\alpha}(\partial \Pi)} + \| (\partial/\partial \tau) Y^{(+)} \|_{L_{p,1-\alpha}(\partial \Pi)} \leq c \| (\Phi_+ - \Phi_-) \|_{W^1_{p,1-\alpha}(\partial \Pi)}.$$

Let $\chi \in C^\infty(\mathbf{R})$ be equal to 1 for $t > 1$ and vanish for $t < 0$ and let $\Psi = \Delta(\chi(\Phi^{(-)} + \Phi^{(+)}))$. Using Lemma 1 we have

$$\|\Psi\|_{L_p(\Pi)} \leq c\left(\|\Phi^{(-)}\|_{W_{p,-\alpha}^{1-1/p}(\partial\Pi)} + \|\Phi^{(+)}\|_{W_{p,-\alpha}^{1-1/p}(\partial\Pi)}\right).$$

By the inclusion $W_p^1(\partial\Pi) \subset W_p^{1-1/p}(\partial\Pi)$ the right-hand side has the majorant

$$c\left(\|\Phi^{(-)}\|_{W_{p,1-\alpha}^1(\partial\Pi)} + \|\Phi^{(+)}\|_{L_{p,-\alpha}(\partial\Pi)} + \|d\Phi^{(+)}/dt\|_{L_{p,1-\alpha}(\partial\Pi)}\right).$$

Applying Lemma 1 with $k = 2$ we obtain that the Dirichlet problem

$$\Delta Z = -\Psi \text{ in } G, \quad Z = 0 \text{ on } \partial G$$

has a solution satisfying

$$\|\partial Z/\partial n\|_{L_{p,\gamma}(\partial G)} \leq c_1 \|Z\|_{W_{p,\gamma}^2(G)} \leq c_2 \|\Psi\|_{L_p(G)}$$

for every real γ. Since $\int_{\partial G}(\partial Z/\partial n)ds = 0$, the function $\partial Z/\partial n$ is represented in the form $\partial X/\partial s$, where X satisfies

$$\|X\|_{\mathcal{L}_{p,1-\alpha}(\Gamma)} + \|X'\|_{\mathcal{L}_{p,1-\alpha}(\Gamma)} \leq c \|\Phi_+ - \Phi_-\|_{L_{p,1-\alpha}(\mathbf{R})}$$

$$+ \sum_{\pm}\left(\|\Phi_\pm\|_{L_{p,-\alpha}(\mathbf{R})} + \|\Phi_\pm'\|_{L_{p,1-\alpha}(\mathbf{R})}\right).$$

Set $F = Z + \chi(\Phi^{(-)} + \Phi^{(+)})$. By virtue of (9) the function $u = F \circ \omega^{-1}$ is the required harmonic extension of φ onto Ω. $\square$

Since constants belong to $\mathfrak{N}_{p,\beta}(\Gamma)$, the following proposition is obtained from Cauchy-Riemann conditions.

Proposition 2. *Let ψ belong to $\mathfrak{N}_{p,\beta}^{-1}(\Gamma)$, where $0 < \beta + p^{-1} < 1$. Then the Neumann problem in Ω with boundary data ψ has a solution u in $\mathfrak{N}_{p,\beta}(\Gamma)$ satisfying*

$$\|u\|_{\mathfrak{N}_{p,\beta}(\Gamma)} \leq c \|\psi\|_{\mathfrak{N}_{p,\beta}^{-1}(\Gamma)}.$$

3. Integral equation of the Neumann problem

Theorem 1. *Let Ω have an outward peak, and let $0 < \beta + p^{-1} < \min\{\mu, 1\}$ and $\mu - \beta - p^{-1} + 2^{-1} \notin \mathbf{N}$. Then the integral equation*

$$\pi\sigma + W^{ext}\sigma = V\psi \tag{10}$$

has a unique solution σ in $\mathfrak{N}_{p,\beta}(\Gamma)$ for every $\psi \in \mathfrak{N}_{p,\beta}^{-1}(\Gamma)$ satisfying $\int_\Gamma \sigma\, ds = 0$.

Proof. Let ψ belong to $C_0^\infty(\Gamma\setminus\{O\})$. By h we denote a solution of Neumann problem in Ω with boundary data ψ. From the integral representation of the harmonic function h in Ω and the limit relation for the double layer potential we obtain

$$h(z) + \frac{1}{\pi}\int_\Gamma h(q)\frac{\partial}{\partial n_q}\log\frac{1}{|z-q|}ds_q = \frac{1}{\pi}\int_\Gamma \log\frac{|z|}{|z-q|}\psi(q)ds_q \tag{11}$$

We choose h so that $\int_\Gamma h(q)ds_q = 0$. According to Proposition 2 h belongs to $\mathfrak{N}_{p,\beta}(\Gamma)$. From (11) it follows that $\sigma = h$ is a solution of (10).

Now let ψ be an arbitrary function in $\mathfrak{N}_{p,\beta}^{-1}(\Gamma)$. For $\beta > -p^{-1}$ there exists a sequence $\{\psi_n\}_{n\geq 1}$ of smooth function on $\Gamma \setminus \{O\}$ vanishing near the peak and approaching ψ in $\mathfrak{N}_{p,\beta}^{-1}(\Gamma)$. By σ_n we denote a solution of (10) with ψ_n on the right-hand side which exists and is unique by Theorem C. Since the operator:

$$\mathcal{L}_{p,\beta+1}(\Gamma) \ni \psi \longmapsto V\psi \in \mathfrak{N}_{p,\beta}(\Gamma)$$

is continuous (see Theorem A) it follows that $\{V\psi_n\}$ converges in $\mathcal{L}^1_{p,\beta+1}(\Gamma)$ to a limit $V\psi$. According to Proposition 2 the sequence $\{\sigma_n\}$ converges in $\mathcal{L}^1_{p,\beta+1}(\Gamma)$ to a limit σ. Since the operator:

$$\mathcal{L}^1_{p,\beta+1}(\Gamma) \ni \sigma \xrightarrow{W} (\pi I + W^{ext})\sigma \in \mathfrak{M}_{p,\beta}(\Gamma) \subset \mathcal{L}^1_{p,\beta+1}(\Gamma)$$

is continuous (see Theorem C) we obtain by taking the limit that σ is a solution of (10).

The kernel of W in $\mathcal{L}^1_{p,\beta+1}(\Gamma)$ is trivial. Therefore equation (10) is uniquely solvable in $\mathfrak{N}_{p,\beta}(\Gamma)$. Hence and by Proposition 1 it follows that the range of the inverse operator: $\mathfrak{N}_{p,\beta}^{-1}(\Gamma) \ni \psi \longmapsto \sigma \in \mathfrak{N}_{p,\beta}(\Gamma)$ coincides with $\{\sigma \in \mathfrak{N}_{p,\beta}(\Gamma) : \int_\Gamma \sigma\, ds = 0\}$. $\square$

Remark 1. For $\psi \in \mathfrak{N}_{p,\beta}^{-1}(\Gamma)$ on the right-hand side of (10) $V\psi$ belongs on $\mathfrak{N}_{p,\beta}(\Gamma) \cap \mathfrak{M}_{p,\beta}(\Gamma)$ and $V\psi$ satisfies to the "orthogonality" conditions

$$\int_\Gamma \frac{\partial}{\partial s}(V\psi)\mathrm{Re}\,\frac{1}{\zeta^{2k-1}}ds = 0, \quad k = 1,\dots,m.$$

Proof. According to Proposition 3 [3] under the conditions of Theorem 1 there exists a harmonic extension of $\varphi \in \mathfrak{N}^{(+)}(\Gamma)$ onto Ω' such that the conjugate function g satisfying $g(\infty) = 0$ has the representation

$$\sum_{k=1}^{m} c^{(k)}(\varphi)\mathrm{Re}\, z^{k-1/2} + g^{\#}(z),$$

where $c^{(k)}(\varphi)$ are linear continuous functionals in $\mathfrak{N}^{(+)}(\Gamma)$ and $g^{\#}$ satisfies

$$\| g^{\#} \|_{\mathfrak{N}_{p,\beta}(\Gamma)} \leq c \, \| \varphi \|_{\mathfrak{N}^{(+)}_{p,\beta}(\Gamma)}$$

with c independent of φ. By ζ we denote the conformal mapping of Ω' onto $\mathbf{R}^2_+$ such that

$$\zeta(0) = 0, \quad \mathrm{Re}\,\zeta(\infty) = 0 \quad \text{and} \quad \mathrm{Re}\,(1/\zeta(z)) = \pm x^{-1/2} + O(1).$$

We apply the Green formula to the functions g and $\mathrm{Re}\,\zeta^{1-2k}$ in $\Omega' \cap \{|z| < \varepsilon\}$. Passing to the limit as $\varepsilon \to 0$ we obtain

$$c^{(k)}(\varphi) = c_k \int_\Gamma \frac{\partial g}{\partial n}\mathrm{Re}\,\frac{1}{\zeta^{2k-1}}ds = c_k \int_\Gamma \varphi'_s\mathrm{Re}\,\frac{1}{\zeta^{2k-1}}ds.$$

Let $(\sigma, t) \in \mathcal{L}^1_{p,\beta+1}(\Gamma) \times \mathbf{R}^m$ be the solution to the equation

$$(\pi I + W^{ext})\sigma + \sum_{k=1}^m t^{(k)} \mathcal{I}_k = \varphi$$

with the right-hand side $\varphi \in \mathfrak{M}_{p,\beta}(\Gamma)$. In Theorem 2 [3] we proved that the components $t^{(k)}$, $k = 1, \ldots, m$, are equal to $c^{(k)}(\varphi)$ respectively.
Now from Theorems C and 1 it follows that $V\psi \in \mathfrak{N}_{p,\beta}(\Gamma) \cap \mathfrak{M}_{p,\beta}(\Gamma)$ and $c^{(k)}(V\psi) = 0$, $k = 1, \ldots, m$. □

Theorem 2. *Let Ω have an inward peak, and let $0 < \beta + p^{-1} < \min\{\mu, 1\}$ and $\beta + p^{-1} \neq 1/2$. Then the boundary integral equation*

$$\pi\sigma + W^{ext}\sigma = V\psi' \tag{12}$$

has a solution $\sigma \in \mathcal{L}^1_{p,\beta+1}(\Gamma)$ for every $\psi \in \mathcal{L}^1_{p,\beta+1}(\Gamma)$ satisfying $\int_\Gamma \sigma \, ds = 0$. The homogeneous equation (12) has only trivial solutions for $0 < \beta + p^{-1} < 1/2$ and a one-dimensional space of solutions for $1/2 < \beta + p^{-1} < 1$ given by

$$\left\{ t \operatorname{Re} \frac{1}{\gamma_0} \right\},$$

where $t \in \mathbf{R}$ and γ_0 is the conformal mapping of Ω onto $\mathbf{R}^2_+$ subject to

$$\gamma_0(0) = 0, \quad \int_\Gamma \operatorname{Re} \frac{1}{\gamma_0} ds = 0 \quad \text{and} \quad \operatorname{Im} \gamma_0(z_0) = 1$$

with a fixed point $z_0 \in \Omega$.

Proof. Let $\psi \in \mathcal{L}^1_{p,\beta+1}(\Gamma)$. Then $V\psi' \in \mathfrak{N}_{p,\beta}(\Gamma)$ (see Theorem A). According to Theorem B equation (12) is solvable in $\mathcal{L}^1_{p,\beta+1}(\Gamma)$. Since the harmonic extension of $V\psi'$ vanishes at infinity, we have $\int_\Gamma \sigma \, ds = 0$. By virtue of Theorem B the set of solutions to the homogeneous equation (12) in $\mathcal{L}^1_{p,\beta+1}(\Gamma)$ is one-dimensional for $1/2 < \beta + p^{-1} < 1$ and trivial for $0 < \beta + p^{-1} < 1/2$. The set of solutions to the homogeneous equation (12) is described in Theorem B. □

4. Integral equation of the Dirichlet problem

Theorem 3. *Let Ω have an outward peak and let $0 < \beta + p^{-1} < \min\{\mu, 1\}$ and $\mu - \beta - p^{-1} + 2^{-1} \notin N$. Then the boundary integral equation*

$$V\gamma = \pi\varphi + W\varphi \tag{13}$$

has a unique solution $\gamma \in \mathfrak{N}^{-1}_{p,\beta}(\Gamma)$ for every $\varphi \in \mathfrak{N}_{p,\beta}(\Gamma)$.

Proof. Let $\varphi \in C_0^\infty(\Gamma \setminus \{O\})$. By u we denote the bounded harmonic extension of φ onto Ω. By the integral representation of the harmonic function u on Ω and by the limit relation for the simple layer potential we obtain for $z \in \Gamma \setminus \{O\}$

$$\int_\Gamma \log \frac{|z|}{|z-q|} \frac{\partial u}{\partial n}(q) ds_q = \pi \varphi(z) + \int_\Gamma \varphi(q) \frac{\partial}{\partial n_q} \log \frac{1}{|z-q|} ds_q. \tag{14}$$

According to Proposition 1 the function $\partial u/\partial n$ belongs to $\mathfrak{N}_{p,\beta}^{-1}(\Gamma)$. From (14) it follows that $\gamma = (\partial/\partial n)u$ satisfies to (13).

Now let φ be an arbitrary function in $\mathfrak{N}_{p,\beta}(\Gamma)$. There exists a sequence $\{\varphi_n\}_{n\geq 1}$ of smooth functions on $\Gamma \setminus \{O\}$ vanishing near the peak and converging to φ in $\mathfrak{N}_{p,\beta}^{-1}(\Gamma)$. By γ_n we denote the solution of (13) with φ_n instead of φ in the right-hand side. Since the operator $(\pi I + W) : \mathcal{L}_{p,\beta+1}^1(\Gamma) \to \mathfrak{M}_{p,\beta}(\Gamma)$ is continuous (see Theorem C) we obtain by taking the limit that $\{\pi \varphi_n + W \varphi_n\}_{n\geq 1}$ converges in $\mathcal{L}_{p,\beta+1}^1(\Gamma)$. By virtue of Proposition 1 the sequence $\{\partial u_n/\partial n\}$, where u is the bounded extension of φ_n onto Ω, converges in $\mathcal{L}_{p,\beta+1}(\Gamma)$. Since the operator:

$$\mathcal{L}_{p,\beta+1}(\Gamma) \ni \gamma \xmapsto{V} V\gamma \in \mathfrak{N}_{p,\beta}(\Gamma)$$

is continuous (see Theorem A), then by taking the limit in the equation $V\gamma_n = \pi \varphi_n + W\varphi_n$ we obtain that γ is a solution of (13). For Ω with outward peak $\operatorname{Ker} V$ is trivial by virtue of Theorem A. Therefore the solution of (13) just obtained is unique. □

Theorem 4. *Let Ω have an inward peak, and let $0 < \beta + p^{-1} < \min\{\mu, 1\}$ and $\beta + p^{-1} \neq 1/2$. Then the integral equation*

$$V\gamma = \pi\varphi + W\varphi \tag{15}$$

has a solution $\gamma \in \mathcal{L}_{p,\beta+1}(\Gamma)$ for every $\varphi \in \mathcal{L}_{p,\beta+1}^1(\Gamma)$. This solution is unique for $0 < \beta + p^{-1} < 1/2$ and the homogeneous equation (15) has a one-dimensional space of solutions for $1/2 < \beta + p^{-1} < 1$ given by

$$\left\{ t \frac{\partial}{\partial n} \operatorname{Im} \frac{1}{\gamma^{(in)}} \right\},$$

where $t \in \mathbf{R}$ and $\gamma^{(in)}$ is the conformal mapping Ω onto $\mathbf{R}_+^2$ subject to the conditions $\gamma^{(in)}(0) = 0$ and $\gamma^{(in)}(z_0) = i$ with a fixed point $z_0 \in \Omega$.

Proof. Let $\varphi \in \mathcal{L}_{p,\beta+1}^1(\Gamma)$. According to Theorem B $(\pi I + W)\varphi$ belongs to $\mathfrak{N}_{p,\beta}(\Gamma)$ and its harmonic extension onto Ω'

$$(W\sigma)(z) = \int_\Gamma \sigma(q) \frac{\partial}{\partial n_q} \log \frac{1}{|z-q|} ds_q, \quad z \in \Omega'.$$

vanishes at infinity. The range of the operator

$$\mathcal{L}_{p,\beta+1}(\Gamma) \ni \gamma \longmapsto V\gamma \in \mathfrak{N}_{p,\beta}(\Gamma)$$

consists of the elements of $\mathfrak{N}_{p,\beta}(\Gamma)$ whose harmonic extensions to Ω' vanish at infinity (see Theorem A). By virtue of Theorem B (15) has a solution in $\mathcal{L}_{p,\beta+1}(\Gamma)$. According to Theorem A the homogeneous equation (15) in $\mathcal{L}_{p,\beta+1}(\Gamma)$ has only zero solution for $0 < \beta + p^{-1} < 1/2$ and the one-dimensional space of solutions for $1/2 < \beta + p^{-1} < 1$. The set of solutions to the homogeneous equation (15) is described in Theorem A. □

References

[1] V. Maz'ya, A. Soloviev: L_p-theory of a boundary integral equation on a cuspidal contour. *Applicable Analysis* **65** (1997), 289-305.

[2] V. Maz'ya, A. Soloviev: L_p-theory of boundary integral equations on a contour with outward peak. *Integral Equations and Operator Theory* (to appear).

[3] V. Maz'ya, A. Soloviev: L_p-theory of boundary integral equations on a contour with inward peak. *Zeitschrift für Analysis und ihre Anwendungen* (to appear).

[4] V.G. Maz'ya, B.A. Plamenevskiĭ: Estimates in L_p and in Hölder classes and the Miranda-Agmon maximum principle for solutions of elliptic boundary value problems in domains with singular points on the boundary. *Amer. Math. Soc. Transl. (2)*, **123** (1984), 1-56.

[5] D.S. Kurtz, R.L. Wheeden: Results on weighted norm inequlities for multipliers. *Trans. Amer. Math. Soc.* (1979), **255**, 343-362.

V. Maz'ya: Linköping University, Institute of Technology, Dept. of Mathematics,
S-581 83, Linköping, Sweden
Email: vlmaz@math.liu.se

A. Soloviev: Departement of Mathematics, Chelyabinsk State University,
Chelyabinsk, 454136 Russia
Email: alsol@csu.ac.ru

D. MEDKOVÁ and J. KRÁL

Essential Norms of the Integral Operator Corresponding to the Neumann Problem for the Laplace Equation

If G is an open set in the Euclidean space with a smooth boundary, $u \in \mathcal{C}^1(\text{cl } G)$ is a harmonic function on G and

$$\frac{\partial u}{\partial n} = g \text{ on } \partial G$$

where g is a continuous function on the boundary of G and n is the exterior unit normal of G, then for $\varphi \in \mathcal{D}$ (= the space of all compactly supported infinitely differentiable functions in R^m) we have

$$\int_{\partial G} \varphi g \, d\mathcal{H}_{m-1} = \int_G \nabla \varphi \cdot \nabla u \, d\mathcal{H}_m.$$

Here $\mathcal{H}_k$ is the k–dimensional Hausdorff measure normalized so that $\mathcal{H}_k$ is the Lebesgue measure in R^k. This formula motivates our definition of the solution of the Neumann problem for the Laplace equation.

Suppose that $G \subset R^m (m \geq 2)$ is an open set with a nonempty compact boundary ∂G coinciding with the boundary of the closure of G. If h is a harmonic function on G such that

$$\int_H |\nabla h| \, d\mathcal{H}_m < \infty$$

for all bounded open subsets H of G we define the weak normal derivative $N^G h$ of h as a distribution

$$\langle N^G h, \varphi \rangle = \int_G \nabla \varphi \cdot \nabla h \, d\mathcal{H}_m$$

for $\varphi \in \mathcal{D}$.

Let μ be a finite real measure on the boundary of G. We formulate *the Neumann problem for the Laplace equation with a boundary condition μ* as follows:

Find a harmonic function u on G such that the gradient of u is integrable over all bounded open subsets of G and $N^G u = \mu$.

It is usual to look for a solution of the Neumann problem in the form of a single layer potential

$$\mathcal{U}\nu(x) = \int_{R^m} h_x(y) \, d\nu(y),$$

where $\nu \in \mathcal{C}'(\partial G)$(= the Banach space of all finite real Borel measures with support in ∂G and with the total variation as a norm),

$$h_x(y) = \begin{cases} (m-2)^{-1} A^{-1} |x-y|^{2-m}, & m > 2, \\ A^{-1} \log |x-y|^{-1}, & m = 2, \end{cases}$$

This work was supported by GAČR Grant No. 201/96/0431

A is the area of the unit sphere in R^m. The single layer potential $\mathcal{U}\nu$ is a harmonic function in G for which the weak normal derivative $N^G\mathcal{U}\nu$ has a sense. One has $N^G\mathcal{U}\nu \in \mathcal{C}'(\partial G)$ for each $\nu \in \mathcal{C}'(\partial G)$ if and only if so called cyclic variation of G is bounded ([2],[3], [14],[15],[21]). Denote

$$\partial_e G = R^m \setminus \{x \in R^m; d_G(x) = 0 \vee d_{R^m \setminus G}(x) = 0\},$$

where

$$d_M(x) = \lim_{r \to 0+} \frac{\mathcal{H}_m(M \cap \mathcal{U}(x;r))}{\mathcal{H}_m(\mathcal{U}(x;r))}$$

is the density of M at x and $\mathcal{U}(x;r)$ is the open ball with the centre x and the diameter r. Then $v^G(x)$, the cyclic variation of G at the point x, is given by

$$v^G(x) = \frac{1}{A} \int_{\partial \mathcal{U}(0;1)} \sum_{y \in \partial_e G \cap \{x+t\theta; 0<t\}} 1 \, d\, \mathcal{H}_{m-1}(\theta).$$

An example of a set with a bounded cyclic variation is a set with a piecewise-$C^{1+\alpha}$ boundary. However, the class of sets with a bounded cyclic variation is very rich and the typical set with a bounded cyclic variation has a boundary of positive Lebesgue measure ([17]). If the cyclic variation of G is bounded then $N^G\mathcal{U} : \nu \mapsto N^G\mathcal{U}\nu$ is a bounded linear operator on $\mathcal{C}'(\partial G)$ ([2], [3],[14],[15],[21]).

Suppose that σ is a nonnegative finite measure with the support equal to the boundary of G. We would like to solve the Neumann problem for the Laplace equation with the boundary condition $g\sigma$, where $g \in L^1(\sigma)$ (= the space of all σ-integrable functions on ∂G). (If we choose the surface measure on the boundary of G as σ this problem corresponds to the weak formulation of the classical Neumann problem for the Laplace equation with the boundary condition g.) J.Král proved ([16]) that for each $f \in L^1(\sigma)$ there is $g_f \in L^1(\sigma)$ such that $N^G\mathcal{U}(f\sigma) = g_f\sigma$ if and only if the cyclic variation of G is bounded and there is $h_G \in L^1(\sigma)$ such that

$$\mathcal{H}_{m-1}(M \cap \partial_e G) = \int_M h_G \, d\sigma$$

for each Borel set M. Suppose that this condition is fulfilled. Denote for $f \in L^1(\sigma)$, $x \in \partial G$

$$Tf(x) = d_G(x) + h_G(x) \int_{\partial_e G} f(y) \frac{n^G(x) \cdot (y-x)}{A|y-x|^m} \, d\sigma(y)$$

if the expression on the right–hand side makes sense. Here A is the surface of the unit sphere and $n^G(x)$ is the exterior unit normal of G at x in Federer's sense, i.e. a unit vector such that the symmetric difference of G and the half-space $\{y \in R^m; (y-x) \cdot n^G(x) < 0\}$ has m-dimensional density zero at x. If there is no interior normal of G at x in this sense, we denote by $n^G(x)$ the zero vector in R^m. (If there is the exterior unit normal of G at x in the classical sense then this normal is the exterior unit normal of G at x in Federer's sense,too.)

The operator T is a bounded linear operator on $L^1(\sigma)$ and

$$N^G(\mathcal{U}(f\sigma)) = (Tf)\sigma$$

for each $f \in L^1(\sigma)$. So, if we look for a solution of the Neumann problem with the boundary condition $g\sigma$ in the form of a single layer potential $\mathcal{U}(f\sigma)$, we get the equation
$$Tf = g.$$
If X is a complex Banach space and S is a bounded linear operator on X, the operator S is called Fredholm if the kernel of S is finite–dimensional and the range of S has a finite codimension. Let us denote
$$r_{ess}(S) = \inf\{r \geq 0; \lambda \in \mathbb{C}, |\lambda| > r \Rightarrow S - \lambda I \text{ is a Fredholm operator}\}$$
the essential spectral radius of S. Here I denotes the identity operator and $\mathbb{C}$ is the complex plane. The condition $r_{ess}(T - \frac{1}{2}I) < \frac{1}{2}$ ensures that we can take a solution of the Neumann problem in the form of a single layer potential ([22], [23]). It is interesting that this condition does not depend on the choice of σ. In fact $r_{ess}(T - \frac{1}{2}I) = r_{ess}(N^G\mathcal{U} - \frac{1}{2}I)$. This relation is an easy consequence of the following

Lemma 1. *([13]) Let X be a Banach space, Y be a closed subspace of X, S be a bounded linear operator on X such that $S(Y) \subset Y$. If we denote by S/Y the restriction of S onto Y then*
$$r_{ess}(S/Y) \leq r_{ess}S.$$

Since $\{f\sigma; f \in L^1(\sigma)\}$ is a closed subspace of $\mathcal{C}'(\partial G)$ and $N^G\mathcal{U}(f\sigma) = (Tf)\sigma$ for each $f \in L^1(\sigma)$, we have $r_{ess}(T - \frac{1}{2}I) \leq r_{ess}(N^G\mathcal{U} - \frac{1}{2}I)$. The opposite inequality is a consequence of the fact that $N^G\mathcal{U}$ is the adjoint operator of the restriction of the adjoint operator of T onto $\mathcal{C}(\partial G)$.

Theorem 1. *([22],[23]) Let $r_{ess}(T - \frac{1}{2}I) < \frac{1}{2}$, $g \in L^1(\sigma)$. Then there is a solution of the Neumann problem for the Laplace equation with the boundary condition $g\sigma$ if and only if*
$$\int_{\partial H} g \, d\sigma = 0$$
for each bounded component H of G. We can take the solution in the form $\mathcal{U}(f\sigma)$, where
$$f = g + \sum_{n=0}^{\infty}(2T - I)^n(2T)g,$$
$$||(I - 2T)^n(2T)g|| \leq Cq^n||g||,$$
$q \in (0,1), C > 0$ *do not depend on g.*

Let us concentrate on the condition $r_{ess}(T - \frac{1}{2}I) < \frac{1}{2}$. The classical supposition is that G has a Ljapunov boundary (i.e. of class $C^{1+\alpha}$). In this case the operator $(T - \frac{1}{2}I)$ is compact. Unfortunately, if $(T - \frac{1}{2}I)$ is a compact operator then G has a boundary of class C^1 ([15]). To overcome this difficulty J. Radon substituted this condition by the weaker condition that the essential norm of the operator $(T - \frac{1}{2}I)$ is smaller than $\frac{1}{2}$. If X is a Banach space equipped with the norm p, S is a bounded linear operator on X then the essential norm of the operator S
$$p_{ess}(S) = \inf\{p(S - K); K \text{ compact }\}$$

is the distance of S from the space of all compact linear operators on X. Since the essential spectral radius of S is the infimum of $q_{ess}(S)$ over all norms q on X which are equivalent to the original norm p (see [8]), the condition in the theorem is weaker than Radon's original condition. The essential norm of the operator $(T - \frac{1}{2}I)$ with respect to the usual norm on $L^1(\sigma)$ does not depend on the choice of σ and it is equal to the essential norm of the operator $(N^G\mathcal{U} - \frac{1}{2}I)$ with respect to the usual norm on $\mathcal{C}'(\partial G)$. (We can see it using [19], Proposition 5, the proof of Theorem 9 in [19] and the fact that v_r (see below) is a lower-semicontinuous function (see [14]).) If we denote for a positive r and a point x

$$v_r(x) \equiv \int_{\partial_e G \cap \mathcal{U}(x;r)} \frac{|n^G(y) \cdot (y-x)|}{A|y-x|^m} d\mathcal{H}_{m-1}(y)$$

then

$$v_r(x) = \frac{1}{A} \int_{\partial \mathcal{U}(0;1)} \sum_{y \in \partial_e G \cap \{x+t\theta; 0<t<r\}} 1 \, d\mathcal{H}_{m-1}(\theta)$$

and

$$\|T - \frac{1}{2}I\|_{ess} = \inf_{r>0} \sup_{x \in \partial G} v_r(x) = \inf_{r>0} \|v_r\|_{L^\infty(\sigma)}$$

(see [14]). J. Radon proved that the essential norm (with respect to the usual norm) of the operator $T - \frac{1}{2}I$ is smaller than $\frac{1}{2}$ for domains bounded by finitely many Jordan curves with finite rotation without cusps in the plane ([32], [33]) (particularly for open sets with piecewise-Ljapunov boundary in the plane). I. Netuka proved this property for convex domains and their complements in general Euclidean space ([24]). V. G. Maz'ya and N. V. Grachev proved this condition for several types of sets with "piecewise-smooth" boundary in the general Euclidean space (see [9]-[12]). W. L. Wendland found a domain formed by three three-dimensional cubes for which the essential norm of the operator $T - \frac{1}{2}I$ is greater than $\frac{1}{2}$ (cf.[20]). R. S. Angell, R. E. Kleinman, J. Král and W. L. Wendland ([20], [1]) proved that the essential spectral radius of the operator $T - \frac{1}{2}I$ is smaller than $\frac{1}{2}$ for rectangular domains in three-dimensional Euclidean space. (Rectangular domains are formed from rectangular parallelepipeds.) A. Rathsfeld ([34], [35]) and independently N. V. Grachev and V. G. Maz'ya ([11]) proved this condition for polyhedral open sets in R^3. (Polyhedral set in R^m is an open set whose boundary is locally a hypersurface and is contained in a finite number of hyperplanes.) D. Medková ([22]) proved that this property has a local character, i.e. if for each $x \in \partial G$ there are $r(x) > 0$, an open set D_x with this property and a diffeomorphism $\psi_x : \mathcal{U}(x; r(x)) \to R^m$ of class $C^{1+\alpha}, \alpha > 0$, such that $\psi_x(G \cap \mathcal{U}(x; r(x))) = D_x \cap \psi_x(\mathcal{U}(x; r(x)))$, $D\psi_x(x) = I$, where $D\psi_x(x)$ is the differential of ψ_x at the point x, than G has this property, too.

In spite of knowing several classes of sets for which the essential spectral radius of the operator $T - \frac{1}{2}I$ is smaller than $\frac{1}{2}$ the characterisation of the essential spectral radius of the operator $T - \frac{1}{2}I$ is not known. Since the essential spectral radius of $T - \frac{1}{2}I$ is the infimum of $p_{ess}(T - \frac{1}{2}I)$ over all norms p on $L^1(\sigma)$ which are equivalent to the usual norm we have studied the essential norms of the operator $T - \frac{1}{2}I$. We calculated the essential norm of this operator with respect to one class of norms.

Theorem 2. Let $\sigma(\{y \in \partial G; d_G(y) \neq \frac{1}{2}\}) = 0$, $\sigma(\{y\}) = 0$ for each $y \in \partial G$. Let q be a Baire function on ∂G,

$$0 < \inf_{y \in \partial G} q(y) \leq \sup_{y \in \partial G} q(y) < \infty.$$

Define the norm p on $L(\sigma)$ by

$$p(f) = \int_{\partial G} q|f| \, d\sigma.$$

Then

$$p_{ess}\left(T - \frac{1}{2}I\right) = \inf_{r > 0} \left\| \frac{v_r^q}{q} \right\|_{L^\infty(\sigma)},$$

where

$$v_r^q(x) \equiv \int_{\partial_e G \cap \mathcal{U}(x;r)} \frac{q(y)|n^G(y) \cdot (y-x)|}{A|y-x|^m} d\mathcal{H}_{m-1}(y)$$

$$\left(= \frac{1}{A} \int_{\partial \mathcal{U}(0;1)} \sum_{y \in \partial_e G \cap \{x+t\theta; 0 < t < r\}} q(y) \, d\, \mathcal{H}_{m-1}(\theta) \right).$$

Proof. 1) If p' denotes the norm on $L^\infty(\sigma)$, the dual space of $L^1(\sigma)$, corresponding to p then

$$p'(u) = \left\| \frac{u}{q} \right\|_{L^\infty(\sigma)}.$$

Fix $r > 0$ and choose an infinitely differentiable function γ_r on R^m such that

$$0 \leq \gamma_r \leq 1, \quad \gamma_r(\mathcal{U}(0; r/2)) = \{0\}, \quad \gamma_r(R^m \setminus \mathcal{U}(0; r)) = \{1\}.$$

It has been proved in [16] (cf. Corollaire, pp. 153-154) that

$$[x, y] \mapsto \frac{n^G(x) \cdot (y-x)}{A|y-x|^m} h_G(x)$$

represents a Baire function on $\partial G \times \partial G \setminus \Delta$, where $\Delta = \{[x, x]; x \in \partial G\}$ and that, for each $f \in L^1(\sigma)$, the integral

$$\int \int_{\partial G \times \partial G \setminus \Delta} \frac{|n^G(x) \cdot (y-x)|}{A|y-x|^m} \cdot |f(y)| h_G(x) \, d\sigma(x) \, d\sigma(y)$$

is convergent. Consequently, also the function

$$[x, y] \mapsto \gamma_r(x-y) \frac{n^G(x) \cdot (y-x)}{A|y-x|^m} h_G(x)$$

which we extend by 0 to Δ represents a Baire function on $\partial G \times \partial G$ and, for any $f \in L^1(\sigma)$, the functions

$$T_r f(x) = \int_{\partial G} h_G(x)\gamma_r(x-y)\frac{n^G(x)\cdot(y-x)}{A|y-x|^m}f(y)\,d\sigma(y),$$

$$V_r f(x) = \int_{\partial G} h_G(x)[1-\gamma_r(x-y)]\frac{n^G(x)\cdot(y-x)}{A|y-x|^m}f(y)\,d\sigma(y).$$

are defined for σ-a.e. $x \in \partial G$ and are σ-integrable.

$$(T - \frac{1}{2}I)f(x) = T_r f(x) + V_r f(x)$$

for σ-a.e. $x \in \partial G$. Using the properties of γ_r it is easy to verify the estimates (where $x, y, y_j \in \partial G, j = 1, 2$)

$$\gamma_r(x-y)\frac{|n^G(x)\cdot(y-x)|}{A|y-x|^m} \leq A^{-1}\left(\frac{r}{2}\right)^{1-m},$$

$$|\gamma_r(x-y_1) - \gamma_r(x-y_2)| \leq |y_1 - y_2|\max\{|\nabla\gamma_r(z)|; z \in R^m\},$$

$$\gamma_r(x-y_j)\left|\frac{(y_1-x)}{|y_1-x|^m} - \frac{(y_2-x)}{|y_2-x|^m}\right| \leq (m+1)|y_1-y_2|\left(\frac{r}{4}\right)^{-m} \text{ for } |y_1-y_2| \leq \frac{r}{4}.$$

Denoting by T'_r the adjoint operator of T_r we have for $u \in L^\infty(\sigma)$ and σ-a.e. $y \in \partial G$

$$T'_r u(y) = \int_{\partial G} h_G(x)\gamma_r(x-y)\frac{n^G(x)\cdot(y-x)}{A|y-x|^m}u(x)\,d\sigma(x)$$

$$= \int_{\partial G} \gamma_r(x-y)\frac{n^G(x)\cdot(y-x)}{A|y-x|^m}u(x)\,d\mathcal{H}_{m-1}(x).$$

Hence we conclude that T'_r maps the unit ball in $L^\infty(\sigma)$ into a family of uniformly bounded functions satisfying the Lipschitz condition with the same coefficient on ∂G. By Arzela's theorem, this family is relatively compact in $L^\infty(\sigma)$. We have thus verified that

$$T_r : f \mapsto T_r f$$

is a compact operator on $L^1(\sigma)$. Since T_r is compact, we have

$$p_{ess}(T - \frac{1}{2}I) \leq p(V_r) = p'(V'_r),$$

where V'_r denotes the adjoint operator of V_r sending any $u \in L^\infty(\sigma)$ into a function determined for σ-a.e. $y \in \partial G$ by

$$V'_r u(y) = \int_{\partial G\setminus\{y\}} u(x)[1-\gamma_r(x-y)]\frac{n^G(x)\cdot(y-x)}{A|y-x|^m}h_G(x)\,d\sigma(x).$$

If $u \in L^\infty(\sigma)$ and $p'(u) \leq 1$ then for σ-a.e. $y \in \partial G$

$$|V'_r u(y)| \leq \int_{\partial G \cap \mathcal{U}(y;r)} q(x) \left| \frac{n^G(x) \cdot (y-x)}{A|y-x|^m} \right| d\mathcal{H}_{m-1}(x) = v_r^q(y).$$

Thus

$$p_{ess}\left(T - \frac{1}{2}I\right) \leq \inf_{r>0} \left\| \frac{v_r^q}{q} \right\|_{L^\infty(\sigma)}.$$

2) Fix an arbitrary $\varepsilon > 0$. According to Theorem 10 and Corollary 11 in Chap. VI, §8 in [5] there are mutually disjoint Borel sets $M_1, \ldots, M_n \subset \partial G$ and functions $g_1, \ldots, g_n \in L^1(\sigma)$ such that the finite–dimensional operator

$$K : f \mapsto \sum_{j=1}^n g_j \int_{M_j} f\, d\sigma$$

acting on $L^1(\sigma)$ satisfies

$$p\left(T - \frac{1}{2}I - K\right) < \varepsilon + p_{ess}\left(T - \frac{1}{2}I\right). \tag{1}$$

The operator $(T - \frac{1}{2}I)'$ which is adjoint to $(T - \frac{1}{2}I)$ sends any $u \in L^\infty(\sigma)$ into a function in $L^\infty(\sigma)$ whose values for σ-a.e. $y \in \partial G$ are given by

$$\left(T - \frac{1}{2}I\right)' u(y) = \int_{\partial G} u(x) \frac{n^G(x) \cdot (y-x)}{A|y-x|^m} d\mathcal{H}_{m-1}(x).$$

Denoting by m_j the characteristic function of M_j on ∂G, the operator K' dual to K has the form

$$K' : u \mapsto K'u = \sum_{j=1}^n m_j \int_{\partial G} u g_j d\sigma, \ u \in L^\infty(\sigma).$$

In view of the equality

$$p\left(T - \frac{1}{2}I - K\right) = p'\left(T - \frac{1}{2}I - K\right)' \tag{2}$$

it will suffice to derive a lower estimate for $p'(T - \frac{1}{2}I - K)'$. Choose $c > 0$ small enough to have $c < q$ σ-a.e. on ∂G and fix a $\delta > 0$ such that for any Borel set $M \subset \partial G$,

$$\sigma(M) < \delta \Rightarrow \int_M q|g_j|\, d\sigma < \varepsilon c,\ j = 1, \ldots, n.$$

Since σ does not charge singletons we can fix $r > 0$ small enough to guarantee that

$$y \in \partial G \Rightarrow \sigma(\mathcal{U}(y;r)) < \delta.$$

Observe that any $u \in L^\infty(\sigma)$ with $p'(u) \leq 1$ vanishing outside the ball $\mathcal{U}(y;r)$ centered at an $y \in \partial G$ satisfies

$$|(K'u)(x)| \leq \sum_{j=1}^{n} m_j(x) \int_{\mathcal{U}(y;r)} q|g_j|\, d\sigma < \varepsilon c$$

for σ-a.e. $x \in \partial G$, so that
$$p'(K'u) \leq \varepsilon.$$

Put $H_1 := \{x \in \partial G; \exists r, s, t > 0,\ q(x) < r < s, \sigma(\{y \in \mathcal{U}(x;t), q(y) < s\}) = 0\}$. Then $\sigma(H_1) = 0$. Putting $H_2 := \{x \in \partial G; d_G(x) \neq \frac{1}{2}\}$, $H_0 := H_1 \cup H_2$ we conclude from $\sigma(\{y \in \partial G; d_G(y) \neq \frac{1}{2}\}) = 0$ that $\sigma(H_0) = 0$. Fix now an arbitrary $y \in \partial G \setminus H_0$ and $k > q(y)$. We are looking for some $u \in L^\infty(\sigma)$ with

$$p'(u) \leq 1,\quad u(\partial G \setminus \mathcal{U}(y;r)) = \{0\}$$

such that
$$p'\left(\left(T - \frac{1}{2}I\right)'u\right) \geq \frac{v_r^q(y)}{k} - \varepsilon.$$

We can fix $\rho \in (0, r)$ small enough to have

$$\int_{\partial G \cap [\mathcal{U}(y;r) \setminus \mathcal{U}(y;\rho)]} \frac{q(x)|n^G(x) \cdot (y - x)|}{A|y - x|^m}\, d\mathcal{H}_{m-1}(x) > v_r^q(y) - \varepsilon k.$$

Next define

$$u(x) := \begin{cases} q(x)\,\mathrm{sgn}[n^G(x) \cdot (y - x)] & \text{for } x \in \partial G \cap [\mathcal{U}(y;r) \setminus \mathcal{U}(y;\rho)], \\ 0 & \text{for } x \in \partial G \setminus [\mathcal{U}(y;r) \setminus \mathcal{U}(y;\rho)]. \end{cases}$$

For σ-a.e. $z \in \mathcal{U}(y;\rho) \cap \partial G$ we then have

$$\frac{1}{q(z)}\left(T - \frac{1}{2}I\right)'u(z) =$$

$$\frac{1}{q(z)} \int_{\partial G \cap [\mathcal{U}(y;r) \setminus \mathcal{U}(y;\rho)]} q(x)\mathrm{sgn}[n^G(x) \cdot (y - x)]\frac{n^G(x) \cdot (z - x)}{A|z - x|^m}\, d\mathcal{H}_{m-1}(x).$$

As z approaches y along the set

$$\{z \in \partial G \cap [\mathcal{U}(y;\rho) \setminus H_0];\ q(z) < k\}$$

(which intersects any ball $\mathcal{U}(y;\tau)$ with $\tau \in (0, \rho)$ in a set of positive σ-measure), the corresponding functions

$$x \mapsto \frac{n^G(x) \cdot (z - x)}{A|z - x|^m}$$

converge (even uniformly w.r. to x) in $[\mathcal{U}(y;r) \setminus \mathcal{U}(y;\rho)]$ to
$$x \mapsto \frac{n^G(x) \cdot (y-x)}{A|z-x|^m}$$
whence
$$\int_{\partial G \cap [\mathcal{U}(y;r) \setminus \mathcal{U}(y;\rho)]} q(x) \operatorname{sgn}[n^G(x) \cdot (y-x)] \frac{n^G(x) \cdot (z-x)}{A|z-x|^m} d\mathcal{H}_{m-1}(x) \to$$
$$\to \int_{\partial G \cap [\mathcal{U}(y;r) \setminus \mathcal{U}(y;\rho)]} \frac{q(x)|n^G(x) \cdot (y-x)|}{A|y-x|^m} d\mathcal{H}_{m-1}(x) > v_r^q(y) - \varepsilon k.$$

We see that the function
$$z \mapsto \frac{1}{q(z)} \left(T - \frac{1}{2}I\right)' u(z)$$
remains above the quantity $\frac{v_r^q(y)}{k} - \varepsilon$ on the set
$$\{z \in [\mathcal{U}(y;\tau) \setminus H_0] \cap \partial G;\ q(z) < k\}$$
of positive σ-measure for sufficiently small $\tau \in (0, \rho)$. Consequently,
$$p'\left(\left(T - \frac{1}{2}I\right)'u\right) \geq \frac{v_r^q(y)}{k} - \varepsilon.$$

Since $p'(u) \leq 1$ we have
$$p'\left(\left(T - \frac{1}{2}I\right)' - K'\right) \geq p'\left(\left(T - \frac{1}{2}I - K\right)'u\right) \geq p'\left(\left(T - \frac{1}{2}I\right)'u\right) - p'(K'u)$$
$$\geq \frac{v_r^q(y)}{k} - 2\varepsilon.$$

As k can be chosen arbitrarily close to $q(y)$ we obtain
$$p'\left(\left(T - \frac{1}{2}I - K\right)'\right) \geq \frac{v_r^q(y)}{q(y)} - 2\varepsilon$$
for $y \in \partial G \setminus H_0$, i.e. for σ-a.e. $y \in \partial G$. In view of (1), (2) we arrive at
$$\left\|\frac{v_r^q}{q}\right\|_{L^\infty(\sigma)} \leq p\left(T - \frac{1}{2}I - K\right) + 2\varepsilon \leq p_{ess}\left(T - \frac{1}{2}I\right) + 3\varepsilon,$$
so that
$$\inf_{r>0} \left\|\frac{v_r^q}{q}\right\|_{L^\infty(\sigma)} \leq p_{ess}\left(T - \frac{1}{2}I\right) + 3\varepsilon.$$

This completes the proof of Theorem 2. $\square$

References

[1] R. S. Angell, R. E. Kleinman, J. Král: Layer potentials on boundaries with corners and edges. *Čas. pěst. mat.* **113**, 1988, 387–402.

[2] Yu. D. Burago, V. G. Maz'ya: Some problems of potential theory and function theory for domains with nonregular boundaries (in Russian). *Zapiski Naucnyh Seminarov LOMI* **3**, 1967.

[3] Yu. D. Burago, V. G. Maz'ya: Potential theory and function theory for irregular regions. Seminars in Mathematics V. A. Steklov Mathematical Institute, Leningrad, 1969.

[4] M. Chlebík: Tricomi potentials. Thesis. (in Slovak) Mathematical Institute of the Czechoslovak Academy of Sciences. Praha, 1988.

[5] N. Dunford, J. T. Schwartz, W. G. Bade, R. G. Barth: *Linear Operators*, Part I. Interscience Publishers, New York-London, 1958.

[6] H. Federer: The Gauss-Green theorem. *Trans. Amer. Math. Soc.* **58** (1945), 44–76.

[7] H. Federer: *Geometric Measure Theory*. Springer-Verlag, Berlin, Heidelberg, New York, 1969.

[8] I. Gohberg, A. Marcus: Some remarks on topologically equivalent norms (Russian). *Izvestija Mold. Fil. Akad. Nauk SSSR* **10**(76), 1960, 91–95.

[9] N. V. Grachev, V. G. Maz'ya: On the Fredholm radius for operators of the double layer potential type on piecewise smooth boundaries. *Vest. Leningrad. Univ.* **19**(4), 1986, 60–64.

[10] N. V. Grachev, V. G. Maz'ya: Invertibility of Boundary Integral Operators of Elasticity on Surfaces with Conic Points. Report LiTH-MAT-R-91-07, Linköping Univ., Sweden.

[11] N. V. Grachev, V. G. Maz'ya: Solvability of a boundary integral equation on a polyhedron. Report LiTH-MAT-R-91-50, Linköping Univ., Sweden.

[12] N. V. Grachev, V. G. Maz'ya: Estimates for kernels of the inverse operators of the integral equations of elasticity on surfaces with conic points. Report LiTH-MAT-R-91-06, Linköping Univ., Sweden.

[13] V. Kordula, V. Müller, V. Rakočević: On the semi-Browder spectrum. *Studia Math.* **123**, 1997, 1–13.

[14] J. Král: Integral Operators in Potential Theory. *Lecture Notes in Mathematics* **823**, Springer-Verlag, Berlin, 1980.

[15] J. Král: The Fredholm method in potential theory. *Trans. Amer. Math. Soc.* **125**, 1966, 511–547.

[16] J. Král: Problème de Neumann faible avec condition frontière dans L^1, Séminaire de Théorie du Potentiel (Université Paris VI) No. 9, *Lecture Notes in Mathematics* **1393**, Springer-Verlag, 1989, 145–160.

[17] J. Král: Boundary regularity and normal derivatives of logarithmic potentials. *Proceedings of the Royal Society of Edinburgh* **106 A**, 1987, 241–258.

[18] J. Král, D. Medková: Angular limits of double layer potentials. *Czechoslovak Math. J.* **45**, 1995, 267–292.

[19] J. Král, D. Medková: Essential norms of a potential theoretic boundary integral operator in L^1, Mathematica Bohemica, in print.

[20] J. Král, W. L. Wendland: Some examples concerning applicability of the Fredholm - Radon method in potential theory. *Aplikace matematiky* **31**, 1986, 293–308.

[21] V. G. Maz'ya: Boundary integral equations . *Encyclopedia of Mathematical Sciences.* **27**, Springer-Verlag, 1991.

[22] D. Medková: The third boundary value problem in potential theory for domains with a piecewise smooth boundary. *Czech. Math. J.* **47**, 1997, 651–679.

[23] D. Medková: Solution of the Neumann problem for the Laplace equation. *Czechoslov. Math. J.* **43**, 1998.

[24] I. Netuka: Fredholm radius of a potential theoretic operator for convex sets. *Čas. pěst. mat.* **100**, 1975, 374–383.

[25] I. Netuka: The third boundary value problem in potential theory. *Czech. Math. J.* **2**(97), 1972, 554–580.

[26] I. Netuka: Generalized Robin problem in potential theory. *Czech. Math. J.* **22**(97), 1972, 312–324.

[27] I. Netuka: Smooth surfaces with infinite cyclic variation. (Czech), *Čas. pěst. mat.* **96**, 1971, 86–101.

[28] C. Neumann: *Untersuchungen über das logarithmische und Newtonsche Potential.* Teubner Verlag, Leipzig, 1877.

[29] C. Neumann: Zur Theorie des logarithmischen und des Newtonschen Potentials. *Berichte über die Verhandlungen der Königlich Sächsischen Gesellschaft der Wissenschaften zu Leipzig* **22**, 1870, 49–56, 264–321.

[30] C. Neumann: *Über die Methode des arithmetischen Mittels.* Hirzel, Leipzig, 1887 (erste Abhandlung), 1888 (zweite Abhandlung).

[31] J. Plemelj: *Potentialtheoretische Untersuchungen.* B. G. Teubner, Leipzig, 1911.

[32] J. Radon: Über Randwertaufgaben beim logarithmischen Potential. *Sitzungsber. Akad. Wiss. Wien* **128**, 1919, 1123–1167.

[33] J. Radon: *Über Randwertaufgaben beim logarithmischen Potential.* Collected Works, vol. 1, Birkhäuser, Vienna, 1987.

[34] A. Rathsfeld: The invertibility of the double layer potential in the space of continuous functions defined on a polyhedron. The panel method. *Applicable Analysis* **45**, 1992, 1–4, 135–177.

[35] A. Rathsfeld: The invertibility of the double layer potential operator in the space of continuous functions defined over a polyhedron. The panel method. Erratum. *Applicable Analysis* **56**, 1995, 109–115.

Mathematical Institute of Czech Academy of Sciences,
Žitná 25, 115 67 Praha 1, Czech Republic.
Email:medkova@math.cas.cz

G. MONEGATO and L. SCUDERI

Polynomial Collocation Methods for 1D Integral Equations with Nonsmooth Solutions

1. Introduction

Many papers have been written on the numerical resolution of 1D linear integral equations, particularly when the domain of integration is either an interval or a smooth curve and the data are smooth functions. These equations have kernels that may be weakly singular, strongly singular or even hypersingular. In such cases the solutions are smooth everywhere in the domain of integration, except possibly at its endpoints when the domain is, for example, a bounded interval or an open smooth curve. Classical numerical methods for their resolution, such as product integration, collocation and Galerkin, including discrete versions of the latter two, have been deeply analyzed by several authors (see, for instance, [1], [18]). These methods are either based on global (algebraic or trigonometric) polynomials, or on smooth or nonsmooth piecewise polynomials.

The cases of nonsmooth data or of curves with corners have been properly considered only recently. For instance, several papers have been devoted to the so-called Symm's equation

$$-\frac{1}{\pi}\int_\Gamma \log|\mathbf{x}-\mathbf{y}|u(\mathbf{x})\,d\Gamma_\mathbf{x} = f(\mathbf{y}), \quad \mathbf{y}\in\Gamma, \qquad (1)$$

which arises, for example, from the single-layer potential representation of the solution of the Dirichlet problem for the Laplace equation. The behaviour of the solutions of elliptic boundary value problems defined in domains with edges has been extensively analyzed by Maz'ya (see [12] and the references listed there). Efficient numerical methods to solve equation (1) have been proposed and analyzed. When Γ is a piecewise smooth curve with a finite number of corners, the numerical methods are usually based on the use of adaptive boundary elements, i.e., on piecewise polynomial approximations associated with properly graded meshes (see, for example, [25]).

However, in the case of equation (1) defined on piecewise smooth curves, a prototype problem which has been considered by many authors (see, for instance, [9], [24] and [25] and their references), a new technique has been recently proposed, which allows to use uniform meshes or global polynomial approximations to construct high order methods.

This approach suggests to introduce in (1) a parametrization with a smoothing effect, which transforms the unknown $u(\mathbf{x})$ in a sufficiently smooth (and periodic) new function. In other words, it reduces a nonsmooth problem, defined on a curve Γ with corners, to a smooth problem defined on a reference (bounded) interval. Of course the introduction of the change of variable modifies the form of the kernel of the equation; however this drawback can be overcome both in the theoretical analysis of the method and in its implementation.

The idea of introducing a smoothing change of variable is not new. For instance, it has been already proposed in [11] and in [17]-[19], [23] and more recently in [3]-[8] and [10]; but in our opinion it has never been presented as a general technique for solving several classes of integral equations.

To show the wideness of the applicability of the above-mentioned approach, in [16] and in [20] (see also [21]) it has been used to construct a high order collocation method, based on (global) algebraic polynomial approximations, for solving the well-known generalized airfoil equation for an airfoil with a flap, that is the following Cauchy singular integral equation

$$-\frac{1}{\pi}\oint_{-1}^{1}\sqrt{\frac{1-x}{1+x}}\frac{u(x)}{x-y}\,dx + \frac{1}{\pi}\int_{-1}^{1}\sqrt{\frac{1-x}{1+x}}k(x,y)u(x)\,dx = f(y), \quad -1<y<1, \quad (2)$$

where $k(x,y) = k_1(x,y)\log|x-y| + k_2(x,y)$, k_1 and k_2 being entire functions, and $f(y)$ has a jump at the abscissa where the flap is hinged.

In [13] (see also [21]) the new approach has been applied to construct high order methods of (algebraic and global) polynomial type, such as product integration and collocation, to solve classical second kind weakly singular integral equations defined on bounded intervals, with nonsmooth input functions. Finally, in [15] (see also [21]) a similar approach has been applied to problem (1).

In all three cases mentioned above, our global polynomial collocation method, combined with a smoothing technique, has produced numerical and theoretical results which appear to be very competitive.

In the next section we will define the smoothing function we have proposed to use, as well as a scale of functional spaces where we have examined our equations. In the following two sections we will then consider problems (2) and (1) and describe the main results we have obtained in [20] (see also [16]) and in [15], respectively.

These two equations are both of the first kind, strongly singular (2) and weakly singular (1). As we shall see, after having introduced the smoothing transformation both of them can be reduced to a second kind equation of the form $(I+\widetilde{M})\tilde{z} = \tilde{e}$, $\tilde{z},\tilde{e} \in L^2$, where the operator $\widetilde{M}$ has, locally around its fixed-point singularities, a "two-sided" Mellin convolution behaviour. By using some fundamental results of Elschner and Graham proved in [4], it is possible to show that $I + \widetilde{M}$ is a Fredholm operator of index zero, with a trivial kernel, which satisfies a strongly elliptic estimate. Taking advantage of these properties, as well as further ones, stability and high order convergence estimates have been derived for the proposed polynomial collocation methods. The proofs of these results are quite similar for both equations.

In our survey we do not include the case of second kind integral equations on bounded intervals, since they turn out to be in the form $(I+K)z = e$ with K compact. Therefore the analysis of the corresponding projection methods is quite straightforward and standard (see [13]).

Besides showing how to construct our collocation method, we will describe the main ingredients one needs to derive the theoretical properties of the method, namely stability and convergence estimates.

2. The smoothing function and the functional spaces

Let us consider an integral transform of the type $Ku(y) := \int_{-1}^{1} k(x,y)u(x)\,dx$, where we assume that $u(x)$ is smooth everywhere in $[-1,1]$, except at a finite number of abscissas $-1 \equiv x_0 < x_1 < \cdots < x_r \equiv 1$. If we introduce a nonlinear transformation $x = \gamma(t)$, where $\gamma(t)$ is a sufficiently smooth nondecreasing function mapping $(-1,1)$ onto $(-1,1)$, having $\{x_k\}$ as fixed points, i.e., $x_k = \gamma(x_k)$, and whose leading derivatives vanish at $\{x_k\}$, then the new function $z(t) := u(\gamma(t))\gamma'(t)$ can be made in $[-1,1]$ as smooth as we desire thanks to the factor $\gamma'(t)$. Moreover, as we shall see in the next sections, at least in the case we have considered, we can always write $Ku(\gamma(s)) = Kz(s) + Ez(s)$, where the perturbation operator E has a kernel with fixed-point singularities precisely at the abscissas $\{x_k\}$.

Several smoothing transformations have been proposed in the literature. In [14] we have compared the most known ones; among these the piecewise (generalized) Hermite interpolation polynomial $\gamma_r(t)$ associated with the partition $-1 \equiv x_0 < x_1 < x_2 < \cdots < x_r \equiv 1$ of $[-1,1]$ and defined in each subinterval $[x_k, x_{k+1}]$, $k = 0, \cdots, r-1$, by the conditions

$$\begin{cases} \gamma_r(x_j) = x_j, & j = k, k+1, \\ \gamma_r^{(i)}(x_j) = 0, & j = k, k+1, \ i = 1, \cdots, q_j - 1, \quad q_j \geq 2, \end{cases}$$

seems to be the most efficient one. For its practical construction see [13]. The integers q_j, $j = 0, \cdots, r$, are chosen accordingly to the smoothing effect that $\gamma(t)$ ought to produce at the points $\{x_k\}$. In the sequel we will use $\gamma(t) := \gamma_r(t)$ as smoothing transformation.

The functional spaces where we will analyze our equations, hence derive stability and convergence estimates, are a scale of subspaces of L^2_ω. We denote by L^2_ω the Hilbert space of all square integrable functions on $(-1,1)$ with respect to a Chebyshev weight $\omega(x)$, endowed with the scalar product $(u,v)_\omega := \frac{1}{\pi}\int_{-1}^{1} u(x)\overline{v(x)}\omega(x)\,dx$ and the norm $\|u\|_\omega = \sqrt{(u,u)_\omega}$. By a Chebyshev weight we mean $\omega(t) = (1-t)^\alpha(1+t)^\beta$ with $|\alpha| = |\beta| = \frac{1}{2}$.

If $\{p_n^\omega\}_{n=0}^\infty$ denotes the system of the normalized orthogonal polynomials with respect to the scalar product $(\cdot,\cdot)_\omega$ with positive leading coefficient and $\deg p_n^\omega = n$, then for any real number $r \geq 0$ we define the subspace $L^2_{\omega,r} = \{u \in L^2_\omega : \|u\|_{\omega,r} < \infty\}$, where $\|u\|_{\omega,r} := \sqrt{(u,u)_{\omega,r}}$ and $(u,v)_{\omega,r} := \sum_{m=0}^\infty (1+m)^{2r}(u,p_m^\omega)_\omega\overline{(v,p_m^\omega)_\omega}$. $L^2_{\omega,r}$ is still a Hilbert space and $L^2_{\omega,0} \equiv L^2_\omega$. When r is a positive integer, we have $\|u\|_{\omega,r} \sim \sum_{k=0}^r \|\varphi^k D^k u\|_\omega$, where $\varphi(x) = \sqrt{1-x^2}$ and D^k denotes the operator of differentation of k-th order. Finally, we recall that $L^2_{\omega,r} \subseteq L^2_{\omega,t}$ and $\|u\|_{\omega,t} \leq \|u\|_{\omega,r}$ whenever $t \leq r$. These spaces have been introduced and examined in [2], where several properties have been proved when $\omega(t)$ is one of the four Chebyshev weight functions. Among these properties we recall the following error estimate for the Lagrange interpolation projector P_n^ω associated with Chebyshev nodes

$$\|(I - P_n^\omega)u\|_{\omega,t} \leq c\, n^{t-s}\|u\|_{\omega,s}, \tag{3}$$

which holds whenever $u \in L^2_{\omega,s}$ and $s \geq t \geq 0$, $s > \frac{1}{2}$, with c independent of n.

3. The generalized airfoil equation

We consider equation (2) and rewrite it in operator form

$$(H + K)u = f \tag{4}$$

where, having set $\omega(x) := \sqrt{(1-x)/(1+x)}$, we have defined

$$Hu(y) := -\frac{1}{\pi}\oint_{-1}^{1} \omega(x)\frac{u(x)}{x-y}\,dx, \quad Ku(y) := \frac{1}{\pi}\int_{-1}^{1} \omega(x)k(x,y)u(x)\,dx.$$

If we assume that the airfoil has a smooth profile and a flap hinged at x_0, $-1 < x_0 < 1$, then the right-hand side f in (4) is smooth everywhere except at x_0 where it has a finite jump. In this situation the unknown $u(x)$ is smooth everywhere in $[-1, 1]$, except at x_0 where it admits a $\log|x - x_0|$ singularity (see [16]).

If we consider the operators H and K in the proper spaces defined in Section 2, we have the following mapping properties (see [16]):

$$\|H\|_{L^2_{\omega,r} \to L^2_{\omega^{-1},r}} = 1, \quad \|K\|_{L^2_{\omega,r} \to L^2_{\omega^{-1},r+1}} \leq c, \quad r \geq 0.$$

Moreover, there exists the inverse bounded operator H^{-1}, acting from $L^2_{\omega^{-1},r}$ to $L^2_{\omega,r}$ and given by

$$H^{-1}u(y) := \frac{1}{\pi} \oint_{-1}^{1} \omega^{-1}(x) \frac{u(x)}{x - y} dx, \quad \|H^{-1}\|_{L^2_{\omega^{-1},r} \to L^2_{\omega,r}} = 1, \quad r \geq 0.$$

By introducing in (2) the smoothing transformation $x = \gamma(t)$, $y = \gamma(s)$ defined in Section 2, and on multiplying through by $\gamma'(s)$, we obtain

$$-\frac{1}{\pi} \oint_{-1}^{1} \omega(t) \frac{\gamma'(s)}{\gamma(t) - \gamma(s)} z(t)\, dt + \frac{1}{\pi} \int_{-1}^{1} \omega(t) \gamma'(s) k(\gamma(t), \gamma(s)) z(t)\, dt = g(s),$$

where we have set

$$z(t) := u(\gamma(t)) \frac{\omega(\gamma(t))}{\omega(t)} \gamma'(t), \quad g(s) := f(\gamma(s)) \gamma'(s).$$

The last equation can be rewritten in operator form

$$(H + J)z = g, \quad z \in L^2_\omega,\ g \in L^2_{\omega^{-1}}, \tag{5}$$

where

$$Jz(s) := -\frac{1}{\pi} \int_{-1}^{1} \omega(t) [\frac{\gamma'(s)}{\gamma(t) - \gamma(s)} - \frac{1}{t-s} - \gamma'(s) k(\gamma(t), \gamma(s))] z(t)\, dt,$$

or, equivalently,

$$(I + M)z = e, \quad z, e \in L^2_\omega,$$

where

$$M := H^{-1}J, \quad e := H^{-1}g. \tag{6}$$

To proceed further it is convenient to shift the above operators so that domain and codomain both coincide with L^2. Then, by setting

$$\widetilde{z} := \omega^{\frac{1}{2}} z, \quad \widetilde{g} := \omega^{-\frac{1}{2}} g,$$

$$\widetilde{H} := \omega^{-\frac{1}{2}} H \omega^{-\frac{1}{2}}, \quad \widetilde{J} := \omega^{-\frac{1}{2}} J \omega^{-\frac{1}{2}}$$

our equation (5) becomes
$$(\widetilde{H} + \widetilde{J})\widetilde{z} = \widetilde{g}, \quad \widetilde{z}, \widetilde{g} \in L^2,$$
or, equivalently,
$$(I + \widetilde{M})\widetilde{z} = \widetilde{e}, \tag{7}$$
where
$$\widetilde{M} := \widetilde{H}^{-1}\widetilde{J}, \quad \widetilde{e} := \widetilde{H}^{-1}\widetilde{g}.$$
Further, we rewrite $\widetilde{J} := \widetilde{J}_0 + \widetilde{J}_1$, where, for $\widetilde{z} \in L^2$,
$$\widetilde{J}_0 \widetilde{z}(s) := -\frac{1}{\pi} \int_{-1}^{1} \frac{\omega^{\frac{1}{2}}(t)}{\omega^{\frac{1}{2}}(s)} j_0(t,s) \widetilde{z}(t)\, dt$$
with
$$j_0(t,s) := \frac{\gamma'(s)}{\gamma(t) - \gamma(s)} - \frac{1}{t-s} - \gamma'(s) k_1(\gamma(t), \gamma(s)) \log[\frac{\gamma(s) - \gamma(t)}{s-t}],$$
and
$$\widetilde{J}_1 \widetilde{z}(s) := -\frac{1}{\pi} \int_{-1}^{1} \frac{\omega^{\frac{1}{2}}(t)}{\omega^{\frac{1}{2}}(s)} \gamma'(s) [k_1(\gamma(t), \gamma(s)) \log|s-t| + k_2(\gamma(t), \gamma(s))] \widetilde{z}(t)\, dt.$$

It turns out that the operator $\widetilde{J}_1$ is compact on L^2, while $\widetilde{J}_0$ has a kernel which has a fixed-point singularity at $t = s = x_0$ and local to this point behaves like a two-sided Mellin convolution. This is exploited by carrying out a localization procedure as in [4] to isolate the Mellin behaviour.

Using the Mellin analysis presented by Elschner and Graham in [4] and assuming that the original homogeneous equation $(H + K)v = 0$ has no nontrivial solutions belonging to L_ω^p for any $p > 1$ (this hypothesis will be assumed true from now on), the following main theorem has been proved in [16, Theorem 4.2].

Theorem 1. *For any integer $q \geq 2$, the operator $I + \widetilde{M} : L^2 \longrightarrow L^2$ is a Fredholm operator of index zero with a trivial kernel, which also satisfies the strong ellipticity estimate*
$$Re((I + \widetilde{M} + E_0)\widetilde{v}, \widetilde{v}) \geq c\|\widetilde{v}\|_{L^2}^2, \quad \forall \widetilde{v} \in L^2, \tag{8}$$
for some compact operator E_0 on L^2 and a constant $c > 0$ independent of $\widetilde{v}$.

Therefore equation (7) has in L^2 a unique solution that we shall denote by $\widetilde{z}$. In the sequel we denote by $P_n := P_n^{\omega^{-1}}$ the Lagrange interpolation projector associated with the zeros of $p_{n+1}^{\omega^{-1}}$ (see Section 2). The collocation method we consider (see [16], [20]) is defined by the equation
$$(H + P_n J)z_n = P_n g, \quad z_n \in \Pi_n,\ g \in L^2_{\omega^{-1}}, \tag{9}$$

since $Hz_n \in \Pi_n$, being Π_n the space of all algebraic polynomials of degree n.
Further, we write
$$(I + H^{-1}P_n H H^{-1}J)z_n = H^{-1}P_n H H^{-1}g \tag{10}$$
and define the new operator
$$R_n := H^{-1}P_n H : L^2_\omega \longrightarrow \Pi_n. \tag{11}$$
Notice that R_n is still a projector. From (10), (6) and (11), then we have
$$(I + R_n M)z_n = R_n e.$$

To prove stability we modify our method as it is usually done when dealing with noncompact operators. Thus we consider the following modified collocation method
$$(H + P_n J T^{\frac{i^*}{n}})z_n^\star = P_n g, \tag{12}$$
where T^τ is a truncation operator defined as
$$T^\tau v(t) := \begin{cases} 0, & \text{if } x_0 - \tau \le t \le x_0 + \tau, \\ v(t), & \text{otherwise}, \end{cases} \tag{13}$$
and i^* is a fixed natural number independent of n.

To reformulate (12) in the L^2 setting, we define
$$\widetilde{P}_n := \omega^{-\frac{1}{2}} P_n \omega^{\frac{1}{2}} : L^2 \longrightarrow \omega^{-\frac{1}{2}} \Pi_n, \quad \widetilde{R}_n := \widetilde{H}^{-1} \widetilde{P}_n \widetilde{H}. \tag{14}$$

Notice that $\widetilde{R}_n$ is a well-defined projection operator of L^2 onto $\omega^{\frac{1}{2}} \Pi_n$. It is then straightforward to verify that equation (12) is equivalent to
$$(I + \widetilde{R}_n \widetilde{M} \widetilde{T}^{\frac{i^*}{n}}) \widetilde{z}_n^\star = \widetilde{R}_n \widetilde{e}, \quad \widetilde{z}_n^\star = \omega^{\frac{1}{2}} z_n^\star \in \omega^{\frac{1}{2}} \Pi_n, \tag{15}$$
where
$$\widetilde{T}^{\frac{i^*}{n}} := \omega^{\frac{1}{2}} T^{\frac{i^*}{n}} \omega^{-\frac{1}{2}} \equiv T^{\frac{i^*}{n}}.$$

The following error estimate for the projection operator $\widetilde{R}_n$ is a straightforward consequence of (3) (see [20, Lemma 3.1, Corollary 3.2]).

Lemma 1. Let $\widetilde{v} = \omega^{\frac{1}{2}} v$ and $v \in L^2_{\omega,r}$, $r > 12$. Then, the bounds
$$\|(I - \widetilde{R}_n)\widetilde{v}\|_{L^2} \le \frac{c}{n^r} \|v\|_{\omega,r}, \quad \|(I - \widetilde{R}_n)\widetilde{v}\|_{L^2} \le \frac{c}{n} \|\varphi \widetilde{D} \widetilde{v}\|_{L^2},$$
where $\varphi := \sqrt{1-x^2}$, $\widetilde{D} := \omega^{\frac{1}{2}} D \omega^{-\frac{1}{2}}$, hold with positive constants c independent of n and v.

The starting point of the proof of the stability of our method (15), which is equivalent to that of the original version (12), is the following inequality:
$$\|(I + \widetilde{R}_n \widetilde{M} \widetilde{T}^{\frac{i^*}{n}})\widetilde{v}\|_{L^2} \ge \|(I + \widetilde{M} \widetilde{T}^{\frac{i^*}{n}})\widetilde{v}\|_{L^2} - \|(I - \widetilde{R}_n)\widetilde{M} \widetilde{T}^{\frac{i^*}{n}} \widetilde{v}\|_{L^2}, \quad \widetilde{v} \in L^2. \tag{16}$$

Indeed, for the first term on the right-hand side we obtain

$$||(I + \widetilde{M}T^{\frac{i^\star}{n}})\widetilde{v}||_{L^2} \geq c\, ||\widetilde{v}||_{L^2}, \qquad \widetilde{v} \in L^2,$$

where c is a positive constant. This follows almost immediately from the stability of the so-called finite section (see [4]) of (7), i.e.,

$$||T^\tau(I + \widetilde{M})T^\tau \widetilde{v}||_{L^2} \geq c\, ||T^\tau \widetilde{v}||_{L^2}, \qquad \widetilde{v} \in L^2,$$

as $\tau \to 0$, which in turn follows from the invertibility of $I + \widetilde{M}$ on L^2 and the strong ellipticity estimate (8). Moreover, from Lemma 1, for the last term in (16) we have the bound

$$||(I - \widetilde{R}_n)\widetilde{M}T^{\frac{i^\star}{n}}\widetilde{v}||_{L^2} \leq \varepsilon ||\widetilde{v}||_{L^2}, \qquad \widetilde{v} \in L^2,$$

where $\varepsilon > 0$ can be made arbitrarily small by taking $i^\star$, fixed and sufficiently large, and n arbitrarily large.

The following main result is a trivial generalization of a corresponding one proved in [20, Theorems 3.4, 3.5].

Theorem 2. *Let the smoothing exponent $q \geq 2$ and suppose that $i^\star$ is a sufficiently large integer. Then the estimate*

$$||(I + \widetilde{R}_n\widetilde{M}T^{\frac{i^\star}{n}})\widetilde{v}||_{L^2} \geq c\, ||\widetilde{v}||_{L^2}, \qquad \widetilde{v} \in L^2, \tag{17}$$

holds for all n sufficiently large, where the constant $c > 0$ is independent of n and $\widetilde{v}$. Therefore, method (15) is stable and, moreover, for all n sufficiently large equation (15) has a unique solution $\widetilde{z}_n^\star \in \omega^{\frac{1}{2}}\Pi_n$ satisfying

$$||\widetilde{z} - \widetilde{z}_n^\star||_{L^2} \leq \frac{c}{n^{q-1-\varepsilon}},$$

whenever in (5) $g \in L^2_{\omega^{-1},q-1}$. The constant $c > 0$ depends on $\widetilde{z}$ and $i^\star$, but is independent of n, and $\varepsilon > 0$ is as small as we like.

Remark 1. If we assume, as we did in this section, that the right-hand side f in (4) has only one point singularity at x_0, then when q is odd one could choose the simpler smoothing function $\gamma(t)$ defined in [16], i.e., $\gamma(t) = x_0 + (1-x_0)(\frac{t-t_0}{1-t_0})^q$, where the value of $t_0 := \gamma^{-1}(x_0)$ follows by imposing $\gamma(-1) = -1$. In this case the above-mentioned Mellin behaviour is near to t_0.

Remark 2. The numerical implementation, as well as some numerical tests of method (9), are presented in [16]. As the numerical results show, in our practical computations there was no need to modify the collocation method as suggested in (12). In other words, it seems that our method is stable even if $i^\star = 0$.

4. Symm's equation

Let us consider Symm's equation (1) on a piecewise smooth (open or closed) curve $\Gamma \subset \Re^2$. We recall (see [24, Corollary 1]) that (1) has a unique solution $u \in L^p(\Gamma)$, for any $p > 1$, whenever the so-called transfinite diameter C_Γ of the curve Γ is different from 1; therefore, in the sequel we will assume this last condition. Moreover, we suppose that Γ has r corners, $\mathbf{x}_0, \mathbf{x}_1, \cdots, \mathbf{x}_{r-1}$, and that, for each $j = 0, \cdots, r-1$, the interior angle at $\mathbf{x}_j$ is $(1-\chi_j)\pi$, $0 < |\chi_j| < 1$. We denote by Γ_j the arc joining $\mathbf{x}_j$ to $\mathbf{x}_{j+1}$ and by $|\Gamma_j|$ its length.

To define a parametrization $\nu : [-1,1] \longrightarrow \Gamma$, we choose $r+1$ integers q_j (whose values will be defined below) and introduce $r+1$ points $-1 \equiv S_0 < S_1 < \cdots < S_r \equiv 1$, given by

$$\frac{S_{j+1} - S_j}{2} = \frac{|\Gamma_j|^{\frac{1}{q_j}}}{\sum_{m=0}^{r-1} |\Gamma_m|^{\frac{1}{q_j}}}, \quad j = 0, \cdots, r-1.$$

These will be the preimages of the corner points $\mathbf{x}_j$ under ν. Thus equation (1) can be reformulated as follows

$$-\frac{1}{\pi} \int_{-1}^{1} \log|\nu(\bar{t}) - \nu(\bar{s})| u(\nu(\bar{t})) |\nu'(\bar{t})| d\bar{t} = f(\nu(\bar{s})), \quad -1 \leq \bar{s} \leq 1, \quad (18)$$

where $|\nu'(\bar{t})|$ denotes the determinant of the Jacobian associated with the transformation $\nu(\bar{t})$. At this point we introduce in (18) the change of variable $\bar{t} = \gamma(t)$, $\bar{s} = \gamma(s)$ with the smoothing exponents q_j and defined as in Section 2; thus we have

$$-\frac{1}{\pi} \int_{-1}^{1} \log|\bar{\nu}(t) - \bar{\nu}(s)| u(\bar{\nu}(t)) |\bar{\nu}'(t)| dt = f(\bar{\nu}(s)), \quad -1 \leq s \leq 1, \quad (19)$$

where $\bar{\nu}(t) := \nu(\gamma(t))$, $|\bar{\nu}'(t)| := |\nu'(\gamma(t))|\gamma'(t)$.

In the subsequent analysis it is fundamental to deal with the operator

$$Vz(s) := -\frac{1}{\pi} \int_{-1}^{1} \frac{1}{\sqrt{1-t^2}} \log|t-s| v(t) \, dt \quad (20)$$

since its mapping properties are essential. Therefore, setting $\omega(t) := 1/\sqrt{1-t^2}$, we rewrite equation (19) in the final form

$$-\frac{1}{\pi} \int_{-1}^{1} \omega(t) \log|t-s| z(t) \, dt - \frac{1}{\pi} \int_{-1}^{1} \omega(t) h(t,s) z(t) \, dt = g(s), \quad -1 \leq s \leq 1, \quad (21)$$

where

$$z(t) := \sqrt{1-t^2}\, u(\bar{\nu}(t))\, |\bar{\nu}'(t)|,$$

$$h(t,s) := \log \frac{|\bar{\nu}(t) - \bar{\nu}(s)|}{|t-s|}, \quad g(s) := f(\bar{\nu}(s)).$$

Recalling (20) and introducing the following operator

$$Wz(s) := -\frac{1}{\pi} \int_{-1}^{1} \omega(t) h(t,s) v(t) \, dt,$$

we can rewrite equation (21) in operator form as follows

$$(V+W)z = g. \quad (22)$$

The next two lemmas are about the operator V and its inverse and play a crucial role in the stability and convergence analysis of our method.

Lemma 2. (see, for instance, [22]). *We have*
$$VT_m(s) = \begin{cases} \ln 2 \, T_0(s), & \text{if } m = 0, \\ \frac{1}{m} T_m(s), & \text{if } m \geq 1, \end{cases}$$
where V is defined by (20) and $T_m(\cos\theta) = \cos(m\theta)$, $m \geq 0$ is the classical orthogonal Chebyshev polynomial of the first kind of degree m.

The proof of the next lemma is essentially based on the above-stated lemma (see [15]).

Lemma 3. *The operator*
$$V : L^2_{\omega,r} \longrightarrow L^2_{\omega,r+1}, \quad r \geq 0,$$
defined by (20) is an isomorphism between the two spaces and
$$V^{-1} = HD + I_\omega, \tag{23}$$
where
$$Hv(s) := -\frac{1}{\pi} \oint_{-1}^{1} \omega^{-1}(t) \frac{v(t)}{t-s} dt \quad : L^2_{\omega^{-1},r} \longrightarrow L^2_{\omega,r}, \quad r \geq 0,$$
$$Dv(s) := \frac{dv}{ds} \quad : L^2_{\omega,r+1} \longrightarrow L^2_{\omega^{-1},r}, \quad r \geq 0$$
are bounded operators and
$$I_\omega v(s) := \frac{c_0(v)}{\ln 2} \quad \text{with} \quad c_0(v) := \frac{1}{\pi} \int_{-1}^{1} \omega(t) v(t) \, dt,$$
is bounded and compact from $L^2_{\omega,r+1}$ to $L^2_{\omega,r}$, $r \geq 0$.

To study the solvability of our transformed equation, by proceeding as in Section 3, we rewrite equation (22) as the following second kind equation
$$(I + M)z = e,$$
with
$$M := V^{-1}W, \quad e := V^{-1}g. \tag{24}$$
Recalling (23), the operator M takes the form
$$M = HDW + I_\omega W.$$
In order to use the above-mentioned fundamental results proved in [4], proceeding as in Section 3, we shift the setting spaces from L^2_ω, $L^2_{\omega^{-1}}$ to L^2. Then, by setting
$$\widetilde{z} := \omega^{\frac{1}{2}} z, \quad \widetilde{g} := \omega^{\frac{1}{2}} g,$$

$$\widetilde{V} := \omega^{\frac{1}{2}} V \omega^{-\frac{1}{2}}, \quad \widetilde{W} := \omega^{\frac{1}{2}} W \omega^{-\frac{1}{2}},$$

in L^2 equation (22) becomes

$$(\widetilde{V} + \widetilde{W})\widetilde{z} = \widetilde{g}, \quad \widetilde{z}, \widetilde{g} \in L^2,$$

or, equivalently,

$$(I + \widetilde{M})\widetilde{z} = \widetilde{e}, \quad \widetilde{e} := \omega^{\frac{1}{2}} e, \tag{25}$$

where we have further defined

$$\widetilde{M} := \widetilde{V}^{-1}\widetilde{W} = \widetilde{H}\widetilde{D}\widetilde{W} + \widetilde{I}_\omega \widetilde{W},$$

$$\widetilde{H} := \omega^{\frac{1}{2}} H \omega^{\frac{1}{2}}, \quad \widetilde{D} := \omega^{-\frac{1}{2}} D \omega^{-\frac{1}{2}}, \quad \widetilde{I}_\omega := \omega^{\frac{1}{2}} I_\omega \omega^{-\frac{1}{2}}.$$

To derive some analytical results on equation (25), which are needed in the stability and convergence analysis of our collocation method, we can make use of some similar results, already proved in [4]. The results in [4] are valid for any parametrization satisfying the conditions (A1)-(A4) of [5] (see also [6], [7]). Incidentally we notice that to satisfy condition (A4) in [5], we have to choose the same smoothing exponent q at each point S_j. Therefore, although in the practical computation this choice does not seem to be necessary, in the following we will make this assumption.

After having checked the validity of the above-mentioned assumptions for our transformation $\bar{\nu}(t)$, in [15] we have proved the following main theorem.

Theorem 3. *For any integer $q \geq 2$, the operator $I + \widetilde{M} : L^2 \longrightarrow L^2$ is a Fredholm operator of index zero with a trivial kernel, which also satisfies the strong ellipticity estimate*

$$Re((I + \widetilde{M} + E_0)\widetilde{v}, \widetilde{v}) \geq c\|\widetilde{v}\|_{L^2}^2, \quad \forall \widetilde{v} \in L^2,$$

for some compact operator E_0 on L^2 and a constant $c > 0$ independent of $\widetilde{v}$.

Thus equation (25) has in L^2 a unique solution, that we shall denote by $\widetilde{z}$.

Remark 3. Notice that Theorem 3 is very similar to Theorem 1, although the operators we have denoted by $\widetilde{M}$ are different. In Theorem 1 $\widetilde{M}$ coincides with $\widetilde{H}^{-1}\widetilde{J}$, while in Theorem 3 it is given by $\widetilde{H}\widetilde{D}\widetilde{W}$ plus a compact (in L^2) operator. In the first case we have introduced only a smoothing change of variable; in the second case we had to consider also a parametrization of the curve Γ. Nevertheless, as it is easily seen, the operators $\widetilde{M}$ and their kernels satisfy similar properties.

Denoting by $P_n := P_n^\omega$ the Lagrange interpolation projector associated with the zeros $\{s_i\}$ of the $(n+1)$-th degree Chebyshev polynomial $T_{n+1}(s)$, the collocation method we propose for equation (22) is defined by the projected equation

$$(V + P_n W)z_n = P_n g, \quad z_n \in \Pi_n, \tag{26}$$

being $Vz_n \in \Pi_n$ (see Lemma 2). Further, we write

$$(I + V^{-1}P_n V V^{-1} W)z_n = V^{-1} P_n V V^{-1} g \tag{27}$$

and define the new operator

$$S_n := V^{-1}P_nV : L^2_\omega \longrightarrow \Pi_n. \tag{28}$$

Notice that S_n is still a projector. From (27), recalling (24) and (28), we have

$$(I + S_n M)z_n = S_n e.$$

Finally, setting

$$\widetilde{S}_n := \widetilde{V}^{-1}\widetilde{P}_n\widetilde{V} \tag{29}$$

we obtain for equation (25) the following collocation method

$$(I + \widetilde{S}_n\widetilde{M})\widetilde{z}_n = \widetilde{S}_n\widetilde{e}, \quad \widetilde{z}_n = \omega^{\frac{1}{2}}z_n.$$

Now we state some analytical results on the projector $\widetilde{S}_n$ proved in [15], that are used in the stability and convergence analysis of our collocation method.

Lemma 4. *Let $\widetilde{v} = \omega^{\frac{1}{2}}v$ and $v \in L^2_{\omega,r}$, $r \geq 0$. Then, the bounds*

$$||(I - \widetilde{S}_n)\widetilde{v}||_{L^2} \leq \frac{c}{n^r}||v||_{\omega,r}, \quad ||(I - \widetilde{S}_n)\widetilde{v}||_{L^2} \leq \frac{c}{n}||\widetilde{D}\widetilde{v}||_{L^2},$$

hold with positive constants c independent of n and v.

Also in this case the error estimate for the projection $\widetilde{S}_n$ follows from (3).

Remark 4. Notice that the operators $\widetilde{R}_n$ and $\widetilde{S}_n$ defined by (14) and (29) respectively, give rise to similar error estimates whenever they are applied to $v \in L^2_{\omega,r}$; however, while the estimate for $\widetilde{R}_n$ holds for $r > \frac{1}{2}$, that of $\widetilde{S}_n$ is true also for $r \geq 0$. This last bound, that assures the (uniform with respect to n) boundedness of the operator $\widetilde{S}_n$ in L^2, allows us to give a simpler proof of the convergence estimate of the corresponding collocation method.

As before the stability of the proposed collocation method is proved by modifying it slightly near each corner point. Thus, proceeding as in Section 3, we introduce a corresponding truncation operator T^τ (see (13)) and consider the associated modified collocation method

$$(V + P_n W T^{\frac{i^\star}{n}})z_n^\star = P_n g, \quad z_n^\star \in \Pi_n, \tag{30}$$

for any fixed integer $i^\star$ and for n sufficiently large. Now it is easily seen that (30) is equivalent to

$$(I + \widetilde{S}_n \widetilde{M}\widetilde{T}^{\frac{i^\star}{n}})\widetilde{z}_n^\star = \widetilde{S}_n \widetilde{e}, \tag{31}$$

where

$$\widetilde{T}^\tau := \omega^{\frac{1}{2}}T^\tau \omega^{-\frac{1}{2}} \equiv T^\tau.$$

In the next theorem, proved in [15], stability and convergence of the modified collocation method (31) are given. We remark that the starting point for stability is also in this case inequality (16).

Theorem 4. *Let $q \geq 2$ and suppose that i^* is sufficiently large. Then, the estimate*

$$||(I + \widetilde{S}_n \widetilde{MT}^{\frac{i^*}{n}})\widetilde{v}^*||_{L^2} \geq c||\widetilde{v}||_{L^2}, \quad \widetilde{v} \in L^2$$

holds for all n sufficiently small, where c is independent of n and $\widetilde{v}$. Therefore, method (31) is stable and, moreover, for all n sufficiently large equation (31) has a unique solution $\widetilde{z}_n^* \in \omega^{\frac{1}{2}} \Pi_n$ satisfying

$$||\widetilde{z} - \widetilde{z}_n^*||_\omega \leq \frac{c}{n^l}$$

whenever $f \in H^{l+\frac{5}{2}}(\Gamma)$ and $q > (l + \frac{1}{2})(1 + \chi)$ with $\chi = \max_j |\chi_j|$. The constant $c > 0$ depends on $\widetilde{z}$ and i^* but is independent of n.

Remark 5. Some numerical results obtained by the method (26) have been presented in [15].

References

[1] K. E. Atkinson, *The Numerical Solution of Integral Equations of the Second Kind*, Cambridge University Press, 1997.

[2] D. Berthold, W. Hoppe, B. Silbermann, A fast algorithm for solving the generalized airfoil equation, *J. Comput. Appl. Math.*, **43** (1992), 185–219.

[3] D. Elliott, S. Prössdorf, An algorithm for the approximate solution of integral equations of Mellin type, *Numer. Math.*, **70** (1995), 427–452.

[4] J. Elschner, I. G. Graham, An optimal order collocation method for first kind boundary integral equations on polygons, *Numer. Math.*, **70** (1995), 1–31.

[5] J. Elschner, I. G. Graham, Parametrization methods for first kind integral equations of non-smooth boundaries. In: *Lecture Notes in Pure and Applied Mathematics*, v. **167**, M. Dekker, New York (1995), 81–99.

[6] J. Elschner, I. G. Graham, Quadrature methods for Symm's integral equation on polygons, *IMA J. Numer. Anal.*, **17** (1997), 643–664.

[7] J. Elschner, S. Prössdorf, I. H. Sloan, The qualocation method for Symm's integral equation on a polygon, *Math. Nachr.*, **177** (1996), 81–108.

[8] J. Elschner, E. P. Stephan, A discrete collocation method for Symm's integral equation on curves with corners, *J. Comput. Appl. Math.*, **75** (1996), 131-146.

[9] I. G. Graham, Y. Yan, Piecewise constant collocation for first kind boundary integral equations, *J. Austral. Math. Soc.* Ser. B, **33** (1991), 39–64.

[10] Y. Jeon, I. H. Sloan, E. P. Stephan, J. Elschner, Discrete qualocation methods for logarithmic kernel integral equations on a piecewise smooth boundary, *Adv. Comp. Math.*, **7** (1997), 547–571.

[11] R. Kress, A Nyström method for boundary integral equations in domains with corners, *Numer. Math.*, **58** (1990), 145–161.

[12] V. G. Maz'ya, Boundary Integral Equations. In: *Analysis IV, Encyclopaedia of Mathematical Sciences*, (V. G. Maz'ya and S. M. Nikolskii eds.), Springer, Berlin, v. **27** (1991), 127–222.

[13] G. Monegato, L. Scuderi, High order methods for weakly singular integral equations with non smooth input functions, *Math. Comp.*, **67** (1998), 1493–1515.

[14] G. Monegato, L. Scuderi, Numerical integration of functions with boundary singularities, *J. Comput. Appl. Math.*, to appear.

[15] G. Monegato, L. Scuderi, Global polynomial approximation for Symm's equation on polygons, submitted.

[16] G. Monegato, I. H. Sloan, Numerical solution of the generalized airfoil equation for an airfoil with a flap, *SIAM J. Numer. Anal.*, **34** (1997), 2288–2305.

[17] S. Prössdorf, A. Rathsfeld, Quadrature methods for strongly elliptic Cauchy singular integral equations on an interval, In: *The Gohberg anniversary collection*, vol. **2**, *Topics in Analysis and Operator Theory*, (H. Dym et al., eds.), Birkhäuser Verlag, Basel, (1989), 435–471.

[18] S. Prössdorf, B. Silbermann, *Numerical Analysis for Integral and Related Operator Equations*, Birkhäuser-Verlag, Basel, 1991.

[19] A. Rathsfeld, A quadrature method for a Cauchy singular integral equation with a fixed singularity. In: *Seminar Analysis, Operator Equat. and Numer. Anal.*, 1985/86, Karl-Weierstrasse-Inst. Math., Akad. Wiss. DDR, Berlin (1986), 107–117.

[20] L. Scuderi, A collocation method for the generalized airfoil equation for an airfoil with a flap, *SIAM J. Numer. Anal.*, **35** (1998), 1725–1739.

[21] L. Scuderi, Risoluzione numerica di equazioni integrali lineari con soluzione non regolare, Ph. D. thesis, Università di Napoli "Federico II", 1998.

[22] I. H. Sloan, E. P. Stephan, Collocation with Chebyshev polynomials for Symm's integral equation on an interval, *J. Austral. Math. Soc. Ser. B*, **34** (1992), 199–211.

[23] Y. Yan, Cosine change of variable for Symm's integral equation on open arcs, *IMA J. Numer. Anal.*, **10** (1990), 521–535.

[24] Y. Yan, I. H. Sloan, On integral equations of the first kind with logarithmic kernels, *J. Integral Equations Appl. Anal.*, **1(4)** (1988), 549–579.

[25] Y. Yan, I. H. Sloan, Mesh grading for integral equations of the first kind with logarithmic kernel, *SIAM J. Numer. Anal.*, **26** (1989), 574–587.

Dipartimento di Matematica, Politecnico di Torino, Corso Duca degli Abruzzi 24, 10129 Torino.
Email: monegato@polito.it
        scuderi@polito.it

L. MORINO and G. BERNARDINI

Singularities in Discretized BIE's for Laplace's Equation; Trailing-Edge Conditions in Aerodynamics

1. Introduction

In this paper we address trailing–edge issues connected with the analysis of incompressible potential flows. Specifically, we examine the conditions required to avoid singularities in the boundary integral representation (BIR) for the velocity in a potential incompressible flow around a wing, and use them as the basis for the discussion of the trailing–edge conditions.

The issue of the trailing–edge conditions may be traced back to the beginning of the century, with the pioneering work of Kutta (1902) and Joukowski (1905) addressing the problem of two-dimensional steady incompressible potential flows around isolated airfoils. This corresponds to the exterior Neumann problem for the Laplace equation; for the above problem, the domain is multiply connected, the potential is multivalued, and the solution for the above problem is not unique; for the Neumann boundary condition on the airfoil (which is obtained from the impermeability condition) may be satisfied with different values of the circulation, Γ. Thus, the need for an additional condition, which is appropriately known as the Kutta-Joukowski condition.

To be specific, in the aeronautical community, traditionally "Kutta condition" refers to the condition that the pressure difference vanishes at the trailing–edge, whereas "Joukowski hypothesis" refers to the assumption that the flow is smooth at the trailing–edge. For two-dimensional steady incompressible potential flows, the two conditions are equivalent; hence, the term "Kutta-Joukowski condition". However, for flows more complex than two-dimensional steady flows, the situation is not as simple. Thus, in this paper it is important to distinguish between the two conditions and adopt the traditional terminology: (i) the Joukowski hypothesis of smooth flow at the trailing–edge, and (ii) the Kutta condition of no pressure discontinuity.

Next, note that, already for two-dimensional unsteady flows, the situation is different from that examined by Kutta (1902) and Joukowski (1905), in that Kelvin's theorem applies. Hence, the circulation $\Gamma_\infty(t)$ for a contour of very large radius around the wing remains constant in time and equals its value at time $t = 0$; assuming that the airfoil starts from rest, we have $\Gamma_\infty(0) = 0$, and the potential is no longer multi-valued. However, the non–uniqueness in this case arises from the existence of the wake, a zero-thickness vortex layer, i.e., a line of discontinuity for the potential φ. This yields that $\Gamma(t)$ around the airfoil is determined by the amount of vorticity shed at the trailing–edge; thus the non-uniqueness stems from the need to evaluate the potential discontinuity, $[\varphi]$, as the wake point leaves the trailing–edge.[1]

A similar situation occurs for three–dimensional (steady and unsteady) flows around isolated wings: in this case, the domain is simply–connected, the potential is single-valued, and the wake is a surface of discontinuity for the potential. Again, the non-uniqueness stems from the need to evaluate the potential discontinuity, $[\varphi]$, as the wake point leaves the trailing–edge.

---
[1] As shown in Section 3, $[\varphi]$ remains constant following a wake point.

The trailing–edge issue has been thoroughly studied from a theoretical point of view by Hsiao (1991) and Bassanini et al. (1991, 1996, 1997, 1998), who, in order to obtain appropriate trailing–edge conditions, use the fact that, on an outward dihedral line (i.e., a reentrant corner, one with convexity toward the flow), the flow has a singularity, since $\nabla\varphi = O(r^{-(1-\pi/\alpha)})$, with $\alpha > \pi$ being the outer dihedral angle (Kondrat'ev and Oleinik, 1983).

Here we take a different point of view and following Mangler and Smith (1970), show that, under suitable conditions on the wake geometry at the trailing–edge, the flow does not include reentrant corners, thereby eliminating the possibility of Kondrat'ev and Oleinik (1983) singularity (see Section 3.3). We then proceed to address the condition to remove other types of singularities, known in aerodynamics as vortex–line and edge–jet singularities (Epton, 1992); in doing this, we use the fact that the wake surface, which satisfies no–penetration condition, as a portion of the boundary (Section 3.4).

In contrast to Bassanini et al. (1998) which hold for unsteady, three–dimensional flows around multi–connected domain, this paper is limited to steady incompressible potential flows around simply–connected domains (e.g., an isolated wing).

2. Mathematical preliminaries

In the outer volume $\mathcal{V}$ with boundary $\mathcal{S}$, the BIR (boundary integral representation) for the Laplace equation, with $u = o(1)$ at infinity, is given by (Kress, 1989)

$$u(\mathbf{x}) = \int_{\mathcal{S}} \left(G \frac{\partial u}{\partial n} - u \frac{\partial G}{\partial n} \right) d\mathcal{S}(\mathbf{y}) \qquad \text{for } \mathbf{x} \in \mathcal{V}, \tag{1}$$

where $\partial/\partial n = \mathbf{n}(\mathbf{y}) \cdot \nabla_\mathbf{y}$ ($\mathbf{n}$ being the outward normal to $\mathcal{S}$), whereas $G = -1/4\pi\|\mathbf{x}-\mathbf{y}\|$ is the fundamental solution for the three–dimensional Laplace equation. The integrands in Eq. 1 are weakly singular; thus, the BIR yields that u is bounded.

Next, consider the integral representation for $\mathbf{a}(\mathbf{x}) = \nabla u(\mathbf{x})$ (where ∇ is understood with respect to the $\mathbf{x}$ variable), which is given by

$$\mathbf{a}(\mathbf{x}) = \int_{\mathcal{S}} \left(\nabla G \frac{\partial u}{\partial n} - u \nabla \frac{\partial G}{\partial n} \right) d\mathcal{S}(\mathbf{y}) \qquad \text{for } \mathbf{x} \in \mathcal{V}. \tag{2}$$

The integrand in Eq. 2 is hypersingular. In order to understand the implication of this, we examine what happens if there exists a line $\mathcal{L}$, where $\mathbf{n}$, or u, or $\partial u/\partial n$, or $\partial u/\partial \nu$ (where ν is normal to $\mathcal{L}$ and to $\mathbf{n}$) are discontinuous. In order to do this, note that under appropriate regularity conditions, if $\hat{\mathcal{S}}$ denotes an open surface having boundary $\hat{\mathcal{C}}$, we have (Epton 1992)

$$\int_{\hat{\mathcal{S}}} \left(\nabla G \frac{\partial u}{\partial n} - u \nabla \frac{\partial G}{\partial n} \right) d\mathcal{S}(\mathbf{y}) = \int_{\hat{\mathcal{S}}} \left(G \frac{\partial \mathbf{a}}{\partial n} - \mathbf{a} \frac{\partial G}{\partial n} \right) d\mathcal{S}(\mathbf{y}) \tag{3}$$
$$+ \oint_{\hat{\mathcal{C}}} (G \nabla u - u \nabla G) \times d\mathbf{y} \qquad \text{for } \mathbf{x} \notin \hat{\mathcal{S}}.$$

It is apparent that the second integrand in the line integral (i.e., $u \nabla G$) yields a singularity of the type u_0/r (where 0 denotes evaluation at the point $\mathbf{y}_0$ of $\hat{\mathcal{C}}$ which $\mathbf{x}$

is approaching, and $r = \|\mathbf{x} - \mathbf{y}_0\|$). On the other hand, the first integrand in the line integral (i.e., $G\nabla u$) yields a singular behavior of the type $(\nabla u)_0 \ln r$.[2] As mentioned above, in aerodynamics these are known, respectively, as "vortex–line" and "edge–jet" singularities (Epton, 1992). Of course, the "edge–jet" singularity exists only for the components of the velocity in the direction of $\nabla\varphi$.

To examine the behaviour of line $\mathcal{L}$ where singularities might occur, we divide the surface $\mathcal{S}$ into two subsurfaces, $\hat{\mathcal{S}}_1$ and $\hat{\mathcal{S}}_2$, having $\mathcal{L}$ as a common boundary. Then, adding the contributions of the two surfaces, if φ and $\mathbf{v} = \nabla\varphi$ are continuous across the line $\mathcal{L}$ the line integrals cancel out and one recovers Eq. 2. This is true even if $\mathbf{n}$ is discontinuous (note that in this case $\partial\varphi/\partial n$ might be discontinuous; what matters is that $\nabla\varphi$ be continuous).

The implication of this, in the case of an exterior Neumann problem of interest here, is that if $\mathcal{L}$ is an inward dihedral line (i.e., with concavity toward the flow), then the Neumann boundary conditions on the two sides of $\mathcal{L}$ define $\nabla\varphi \times \boldsymbol{\tau}$ completely along $\mathcal{L}$ ($\boldsymbol{\tau}$ denotes the unit tangent to $\mathcal{L}$). Hence, if φ is also continuous no singularities occur at inward dihedral lines.[3]

3. Theoretical formulation

Consider an impermeable body moving in an incompressible inviscid fluid. Assume that, for $t = 0$, the body and the fluid are at rest, $\boldsymbol{\omega}(\mathbf{x}, 0) = 0$, where $\boldsymbol{\omega}$ denotes the vorticity (e.g., the fluid is initially at rest). In this section, we present a general formulation for these types of flows. The formulation is a refinement of that of Morino (1993) and differs from it only because of the trailing–edge issues. The emphasis is on the steady–state flows, which however are seen as the limiting case of an infinitely long transient.

To study the problem stated above, we may use the principles of conservation of mass and momentum which, in our case, yield respectively, the continuity equation, $\nabla \cdot \mathbf{v} = 0$, and the Euler equation $D\mathbf{v}/Dt = -\nabla(p/\rho)$. The second equation implies that (Kelvin's theorem) $d\Gamma_M/dt = 0$, where

$$\Gamma_M = \oint_{\mathcal{C}_M} \mathbf{v} \cdot d\mathbf{y} = \int_{\mathcal{S}_M} \mathbf{n} \cdot \nabla \times \mathbf{v}\, d\mathcal{S}(\mathbf{y}) \tag{4}$$

is the circulation over a material contour $\mathcal{C}_M$ (boundary of the surface $\mathcal{S}_M$). On the other hand, the assumption that $\boldsymbol{\omega}(\mathbf{x}, 0) = 0$, implies that $\Gamma_M(0) = 0$ for any $\mathcal{C}_M \in \mathcal{V}_F$ (where $\mathcal{V}_F$ denotes the fluid volume). Thus, Kelvin's theorem yields that $\Gamma_M(t) = 0$ for all $\mathcal{C}_M$ that are entirely in the fluid in the time $(0, t)$. This implies that $\nabla \times \mathbf{v} = 0$

---

[2] This is obtained as follows: using $\partial[\ln(y+r)]/\partial y = 1/r$, one obtains $\int_{-b}^{b}(1/r)dy = 2\ln(b+\sqrt{(x^2 + b^2 + z^2)}) - 2\ln(\sqrt{(x^2 + y^2)})$ (as used to obtain the two–dimensional G from the three–dimensional G); for curved edges, line integral minus $\mathbf{a}_0\int_{-b}^{b}(1/r)dy$ (where y is along the tangent to $\mathcal{L}$ in $\mathbf{x}_0$) is not singular.

[3] The behavior at outward dihedral lines, where $\nabla\varphi = O(r^{-(1-\pi/\alpha)})$, α being the outer dihedral angle (see Kondrat'ev and Oleinik, 1983), occur in aerodynamics along outward hinge lines, not considered here. Also of interest is the fact that the traditional approach used in aerodynamic calculations (which consists of including the discontinuity in the boundary conditions, but not in the geometry) yields logarithmic singularities which are wrong for outward hinge lines, and non–physical for inward hinge lines.

(potential flow) for all $\mathbf{x} \in \mathcal{V}_F$ with the exception of those material points that have come in contact with the body.

In the following, we consider a frame of reference in uniform traslation with respect to the undisturbed fluid. For potential flows, the fluid velocity with respect to this frame of reference may be expressed as $\mathbf{v} = \mathbf{v}_\infty + \nabla\varphi$ for $\mathbf{x} \in \mathcal{V}_F$, where $\mathbf{v}_\infty$ is the velocity of the undisturbed fluid with respect to the frame of reference, whereas φ is the velocity perturbation potential. For steady state, the body is rigidly connected to the frame of reference.

In addition, we assume that the body is a wing with a sharp trailing–edge and that the flow leaves the wing at the trailing–edge (attached flow). For the points that emanate from the trailing–edge we need additional information, since as mentioned above Kelvin's theorem does not apply for these points. In the following, we examine two separate cases. In the first we assume that the flow is potential everywhere (potential flows); in the second we assume that the flow is potential everywhere except on the surface (called the wake) composed of the points that come in contact with the trailing–edge: we will refer to these as quasi–potential flows.[4]

3.1. Potential flows

Incompressible potential flows are governed by the Laplace equation

$$\Delta\varphi = 0 \quad \text{for } \mathbf{x} \in \mathcal{V}_F. \tag{5}$$

We assume that the body (wing) is impermeable and denote with $\mathbf{v}_B$ the velocity of $\mathbf{x} \in \mathcal{S}_B$ with respect to the frame of reference indicated above (for steady state $\mathbf{v}_B = 0$). The boundary condition on the body (wing) surface is $\mathbf{v} \cdot \mathbf{n} = \mathbf{v}_B \cdot \mathbf{n}$. Using $\mathbf{v} = \mathbf{v}_\infty + \nabla\varphi$ one obtains,

$$\frac{\partial \varphi}{\partial n} = \chi \quad \text{for } \mathbf{x} \in \mathcal{S}_B, \tag{6}$$

where $\chi = (\mathbf{v}_B - \mathbf{v}_\infty) \cdot \mathbf{n}$. In addition, we have $\varphi = o(1)$ at infinity.

To the contrary of what happens for two–dimensional flows, where the domain is multiply connected and the potential is multivalued, here the solution to the above problem is unique (Kress, 1989) and satisfies Eq. 1. In the limit, as $\mathbf{x}$ approaches $\mathcal{S}$, one obtains an integral equation, the solution of which is also unique (Kress, 1989). If the normal is continuous (e.g., wing with rounded trailing–edge) the solution has no singularities. In the limit, as the trailing–edge radius of curvature tends to infinity, one obtains a solution which exhibits tha type of singularity discussed by Kondrat'ev and Oleinik (1983). Of course, under special conditions (e.g., a wing symmetric with respect to a horizontal plane), the coefficient of the singular terms vanishes and the flow is smooth everywhere.

Once the solution for φ has been obtained, the pressure is obtained from Bernoulli's theorem (in the body frame of reference)

$$\frac{\partial \varphi}{\partial t} + \frac{1}{2}\|\mathbf{v}\|^2 + \frac{p}{\rho} = \frac{1}{2}\|\mathbf{v}_\infty\|^2 + \frac{p_\infty}{\rho}, \tag{7}$$

a first integral of the Euler equation.

---

[4]This is the limiting case of "almost potential flows", a term introduced by Chorin and Marsden (1990) to indicate flows where the vortical region is thin.

3.2. Quasi–potential flows

Unfortunately, the above solution is of no practical interest, because of D'Alambert paradox, which states that the force acting on a body in steady incompressible potential flow equals zero. Thus, here we address the formulation for quasi–potential flows (which incidentally gives results in excellent agreement with experiments in the case of attached flows).

As mentioned above, for these types of flows, we cannot make any assumption regarding the wake; however, the principles of conservation of mass and momentum (Serrin, 1959) must apply across the wake as well. For incompressible flows, these yield: (i) $\mathbf{v} \cdot \mathbf{n} = \mathbf{v}_W \cdot \mathbf{n}$ (where the subscript W denotes wake) and (ii) $p_2 = p_1$ (1 and 2 denote the two sides of the wake surface $\mathcal{S}_W$). For potential flows, the first one implies

$$\frac{\partial \varphi}{\partial n} = (\mathbf{v}_W - \mathbf{v}_\infty) \cdot \mathbf{n} \qquad \text{for } \mathbf{x} \in \mathcal{S}_W; \tag{8}$$

this is true on both sides of the wake. Thus,

$$\left[\frac{\partial \varphi}{\partial n}\right] = 0 \qquad \text{for } \mathbf{x} \in \mathcal{S}_W. \tag{9}$$

>From the second one, $p_2 = p_1$, we obtain, using Bernoulli's theorem,

$$\frac{\partial \varphi_1}{\partial t} + \|\mathbf{v}_1\|^2 = \frac{\partial \varphi_2}{\partial t} + \|\mathbf{v}_2\|^2, \tag{10}$$

or, noting that $\|\mathbf{v}_2\|^2 - \|\mathbf{v}_1\|^2 = (\mathbf{v}_2 + \mathbf{v}_1) \cdot (\mathbf{v}_2 - \mathbf{v}_1)$,

$$\frac{D_W}{Dt}[\varphi] = 0 \qquad \text{for } \mathbf{x} \in \mathcal{S}_W, \tag{11}$$

with $D_W/Dt = \partial/\partial t + \mathbf{v}_W \cdot \nabla$, where $\mathbf{v}_W = \frac{1}{2}(\mathbf{v}_1 + \mathbf{v}_2)$. Equation 11 yields that $[\varphi]$ remains constant following a point $\mathbf{x}_W$ that has velocity $\mathbf{v}_W$.

As mentioned above, the emphasis here is on the steady–state. In this case, Eq. 10 reduces to

$$\|\mathbf{v}_1\|^2 = \|\mathbf{v}_2\|^2 \qquad \text{for } \mathbf{x} \in \mathcal{S}_W. \tag{12}$$

Equation 12 may be rewritten as

$$\mathbf{v}_W \cdot \boldsymbol{\delta} = 0 \qquad \text{for } \mathbf{x} \in \mathcal{S}_W, \tag{13}$$

where $\boldsymbol{\delta} = \mathbf{v}_2 - \mathbf{v}_1$. This, in turn, implies

$$\frac{\partial}{\partial s}[\varphi] = 0 \qquad \text{for } \mathbf{x} \in \mathcal{S}_W, \tag{14}$$

where s is the arclength in direction $\mathbf{v}_W$; this equation yields that $[\varphi]$ remains constant along the trajectory of $\mathbf{x}_W$.

In order to complete the problem we need boundary conditions at the trailing–edge; these are examined in the following subsections.

3.3. Mangler and Smith formulation

In this subsection, we revisit the steady-state formulation of Mangler and Smith (1970) for steady flows around the trailing-edge.[5] Following them, we assume that, near the point $\mathbf{x}_{TE}$ of the trailing-edge, the wake does not leave the trailing-edge outside the wedge $\mathcal{W}_{TE}$ defined by the two tangent planes to $\mathcal{S}_B$ at $\mathbf{x}_{TE}$, because otherwise there would be a Kondrat'ev and Oleinik (1983) singularity, in conflict with Joukowski's hypothesis.

We claim that, if the wake does not leave the trailing-edge outside the wedge $\mathcal{W}_{TE}$, the solution is not singular. In order to see this, we treat the two sides of the wake as portions of the boundary of the problem. Since the wake surface satisfies the no-penetration condition, Eq. 8, and we have assumed that the wake does not leave the trailing-edge outside the wedge $\mathcal{W}_{TE}$ defined above, we conclude that the flow does not exhibit reentrant corners (outward dihedral) and hence does not exhibit Kondrat'ev and Oleinik (1983) singularity.

Having eliminated the possibility of Kondrat'ev and Oleinik (1983) singularities, we still need to address the singularities arising from Eq. 3 (Epton, 1992). This is done in Section 3.3. Here, we continue our review of the work by Mangler and Smith (1970). There remain two possibilities: (A) the trailing-edge behaves like an inward dihedral line on both sides of the wake, or (B) at the trailing-edge there exists a smooth surface on one side and one inward dihedral line on the other side (i.e., the wake at the $\mathbf{x}_{TE}$ is tangent to one of the two sides of the surface $\mathcal{S}_B$). As we will see, case (B) is the norm, case (A) the exception.

Consider first case (B), i.e., assume that the wake is tangent to one of the two sides of $\mathcal{S}_B$. Let $\mathbf{n}$ be the unit normal to $\mathcal{S}_W$, $\boldsymbol{\tau}$ the unit tangent to the trailing-edge, and $\boldsymbol{\nu}$ the unit vector normal to both $\mathbf{n}$ and $\boldsymbol{\tau}$ and pointed towards the wake. Note that, by choice of $\mathbf{n}$, $\mathbf{v}_1 \cdot \mathbf{n} = \mathbf{v}_2 \cdot \mathbf{n} = 0$ (see Eq. 8 with $\mathbf{v}_W = 0$, since we have assumed the flow to be steady in our frame of reference); hence, we may write $\mathbf{v}_W = \frac{1}{2}(\mathbf{v}_1 + \mathbf{v}_2) = v_\tau \boldsymbol{\tau} + v_\nu \boldsymbol{\nu}$ and $\boldsymbol{\delta} = \mathbf{v}_2 - \mathbf{v}_1 = \delta_\tau \boldsymbol{\tau} + \delta_\nu \boldsymbol{\nu}$ and Eq. 12 (which is valid in the limit at the trailing-edge) may be written as

$$\mathbf{v}_W \cdot \boldsymbol{\delta} = v_\tau \delta_\tau + v_\nu \delta_\nu = 0. \tag{15}$$

Next, note the two-dimensional flow in the plane normal to the trailing-edge exhibits a stagnation point on one of the two sides, and hence we have either $v_{1\nu} = 0$ or $v_{2\nu} = 0$; in either case, we have necessarily $v_\nu = (v_{1\nu} + v_{2\nu})/2 > 0$, because, by definition of trailing-edge, the flow leaves $\mathcal{S}_B$.

Thus, we have two subcases: if the wake is tangent to side 1 (case B1), we have $v_{1\nu} > 0$ and $v_{2\nu} = 0$, which implies $\delta_\nu = v_{2\nu} - v_{1\nu} < 0$ and hence (see Eq. 15) $v_\tau \delta_\tau > 0$; similarly, if the flow is tangent to side 2 (case B2), we have $v_{2\nu} > 0$ and $v_{1\nu} = 0$, i.e., $\delta_\nu = v_{2\nu} - v_{1\nu} > 0$, and hence $v_\tau \delta_\tau < 0$. Vice versa, if $v_\tau \delta_\tau > 0$, we are in case B1 and the wake is tangent to side 1, whereas if $v_\tau \delta_\tau < 0$, we are in case B2 and the wake is tangent to side 2.

Next consider case (A), where there exist two stagnation points; when this occurs (e.g., on the vertical plane of symmetry of a wing), we have both $v_\nu = 0$ and $\delta_\nu = 0$. Vice versa, if $v_\tau \delta_\tau = -v_\nu \delta_\nu = 0$, we have either $\delta_\nu = 0$, i.e., $v_{1\nu} = v_{2\nu}$ (which are both zero, because there must be least one stagnation point), or $v_\nu = 0$ (i.e., again

---

[5]For an extension of Mangler and Smith (1970) to unsteady flows see Sipcic and Morino (1985).

$v_{1\nu} = v_{2\nu} = 0$, because neither one can be negative). This corresponds indeed to case (A).

In conclusion, the geometry of the wake at the trailing–edge is determined by the sign of

$$v_\tau \delta_\tau = \left(\mathbf{v}_{\infty_\tau} + \frac{\partial \varphi_A}{\partial \tau}\right)\frac{\partial[\varphi]}{\partial \tau} = \frac{1}{2}\left(\mathbf{v}_{\infty_\tau} + \frac{\partial \varphi_2}{\partial \tau}\right)^2 - \frac{1}{2}\left(\mathbf{v}_{\infty_\tau} + \frac{\partial \varphi_1}{\partial \tau}\right)^2, \qquad (16)$$

where $\mathbf{v}_{\infty_\tau} = \mathbf{v}_\infty \cdot \boldsymbol{\tau}$. If $v_\tau \delta_\tau > 0$, the wake is tangent to side 1; if $v_\tau \delta_\tau < 0$ the wake is tangent to side 2; if $v_\tau \delta_\tau = 0$ (clearly the exception), the wake is in an intermediate position.

Note that the choice of which side is designated as 1 is arbitrary; however, changing the convention changes the sign of $\boldsymbol{\delta}$ thereby yielding equivalent results.

Finally recall that Eq. 15 derives from Eq. 12 which was obtained by imposing $[p] = 0$; hence Eq. 15 is equivalent to the Kutta condition of no pressure discontinuity. Thus, the results presented in this section may be referred to as the "Kutta condition a la Mangler and Smith".

3.4. Integral formulation; further TE conditions

In this subsection, we examine the integral formulation and obtain additional trailing–edge conditions from the Joukowski smooth–flow hypothesis. Next, consider the boundary integral representation for incompressible potential flows. Note that Eq. 5 implies that Eq. 1 applies for quasi–potential flows as well, as long as $\mathcal{S}$ surrounds $\mathcal{S}_B$ and $\mathcal{S}_W$. Thus, using Eq. 9, one obtains, in the limit as $\mathcal{S}$ approaches $\mathcal{S}_B$ and $\mathcal{S}_W$,

$$\varphi(\mathbf{x}) = \int_{\mathcal{S}_B}\left(G\frac{\partial \varphi}{\partial n} - \varphi\frac{\partial G}{\partial n}\right)d\mathcal{S}(\mathbf{y}) - \int_{\mathcal{S}_W}[\varphi]\frac{\partial G}{\partial n}d\mathcal{S}(\mathbf{y}). \qquad (17)$$

In Eq. 17, $\partial \varphi/\partial n$ is known from the Neumann condition, Eq. 6. Also, for the steady case considered here, the wake geometry at the trailing–edge is determined by the Kutta condition a la Mangler and Smith (Section 3.3). In addition, the condition of zero pressure jump across the wake, Eq. 14, determines the geometry of the wake as well as the value of $[\varphi]$ across the wake, provided that the value of $[\varphi]_{TE}$ is known (see Eq. 18 below).

Recall that we have already removed the Kondrat'ev and Oleinik (1983) singularity by appropriate choice of the geometry of the wake at the trailing–edge (Section 3.3). Next, we want to discuss additional conditions required for the flow to be smooth at the trailing–edge (Joukowski hypothesis). On the basis of the results of Section 2, we have to remove $1/r$ (vortex–line) and $\ln r$ (edge–jet) singularities. In order to remove $1/r$, it is sufficient to impose the condition (Morino, 1974, 1993)

$$\lim_{\mathbf{x}_U \to \mathbf{x}_{TE}} \varphi(\mathbf{x}_U) - \lim_{\mathbf{x}_L \to \mathbf{x}_{TE}} \varphi(\mathbf{x}_L) = \lim_{\mathbf{x}_W \to \mathbf{x}_{TE}}[\varphi](\mathbf{x}_W) = [\varphi]_{TE}(\mathbf{x}_{TE}) \qquad (18)$$

(otherwise there would be a $1/r$ singularity, as if a concentrated vortex existed at the trailing–edge). This determines the required value of $[\varphi]_{TE}$ at the trailing–edge.

In order to examine the regularity condition necessary to avoid logarithmic ("edge–jet") singularities, let us rewrite Eq. 17 as

$$\varphi(\mathbf{x}) = \int_{\mathcal{S}}\left(G\frac{\partial \varphi}{\partial n} - \varphi\frac{\partial G}{\partial n}\right)d\mathcal{S}(\mathbf{y}) \qquad (\mathcal{S} = \mathcal{S}_B \cup \mathcal{S}_1 \cup \mathcal{S}_2), \qquad (19)$$

where S_k denotes the side k of the wake (Eq. 19 is equivalent to Eq. 17 because $\mathbf{n}_W = \mathbf{n}_2 = -\mathbf{n}_1$ and because of Eq. 8).

Thus, the issues of smoothness in the boundary integral representation for $\mathbf{v} = \nabla\varphi$ may be discussed by matching the k-th side of the body with S_k. In the following (in contrast with Section 3.2, where side 1 is an arbitrary choice), let side 1 denote the side to which the wake is tangent. Then, on side 2 of the trailing–edge there is an inward dihedral line and the two Neumann boundary conditions (i.e., $\partial\varphi_2/\partial n = -\mathbf{v}_\infty \cdot \mathbf{n}$, with $\mathbf{n}$ corresponding to different vectors on the two sides of the dihedral line) yield that, along the dihedral line, $(\mathbf{v}_\infty + \nabla\varphi) \times \boldsymbol{\tau} = 0$.

In addition, on side 1 the union of the body and wake surfaces exhibits no normal singularity; hence, the lack of edge–jet singularity is guaranteed by the continuity of $\partial\varphi/\partial\nu$, as required for any line where the surface is smooth (see comments at the end of Section 2). However, $\partial\varphi/\partial\nu$ on the wake is not known. On the other hand, $\partial\varphi_1/\partial s = \partial\varphi_2/\partial s$ (where s is the arclength along $\mathbf{v}_W$), because of Eq. 14, which is valid, in the limit, at the trailing–edge. Hence, noting that $\partial\varphi_{W_2}/\partial s = -\cos\theta\,\mathbf{v}_\infty\cdot\boldsymbol{\nu}+\sin\theta\,\partial\varphi_2/\partial\tau$ (with $\cos\theta = \boldsymbol{\nu}\cdot\mathbf{e}_s$ and $\sin\theta = \boldsymbol{\tau}\cdot\mathbf{e}_s$), we have

$$\frac{\partial\varphi_1}{\partial s} = -\cos\theta\,\mathbf{v}_\infty\cdot\boldsymbol{\nu} + \sin\theta\,\frac{\partial\varphi_2}{\partial\tau}. \tag{20}$$

In summary, we have obtained that all types of singularities have been removed, in line with the Joukowski hypothesis. Specifically: (i) the Kondrat'ev and Oleinik (1983) singularity has been removed by having the wake leave $\mathbf{x}_{TE}$ within the wedge defined by two tangent planes to S_B in $\mathbf{x}_{TE}$; (ii) the $1/r$ (vortex–line) singularity has been removed by imposing Eq. 18; (iii) the $\ln r$ (edge–jet) singularity has been removed by imposing Eq. 20. In addition, on side 2 of the body we have $\mathbf{v}_2 \cdot \boldsymbol{\nu} = 0$, or

$$\partial\varphi_2/\partial\nu = -\mathbf{v}_\infty \cdot \boldsymbol{\nu}. \tag{21}$$

The above conditions are necessary to ensure that the flow is smooth at the trailing–edge.

Note that the Kutta condition, $[p]_{TE} = 0$, has been used in obtaining the geometry of the wake at the trailing–edge, a la Mangler and Smith (1970), as well as in obtaining Eq. 20 (it may be used that the Kutta condition, $[p]_{TE} = 0$, is a consequence of: (i) $[p]_W = 0$ and (ii) uniform continuity of the pressure in the field).

Next, we try to give a mathematical interpretation of our results, and distinguish between those conditions that are necessary to make the solution unique and those which are automatically satisfied by the solution. Of course, Eq. 18 is necessary otherwise we cannot link $[\varphi]$ to the solution on the body (and non–uniqueness would arise). But are the others necessary for the uniqueness of the solution, or does the solution automatically satisfy the wake-geometry condition by Mangler and Smith (1970)? What about Eq. 20 and Eq. 21?

A theoretical answer to the above questions is beyond the scope of this paper and we leave these as open questions. However, the issues have been explored numerically (Section 4).

4. Numerical results

The issue of the correct trailing–edge condition is important not only from a theoretical but also from a numerical point of view and has been raised in connection with the numerical results obtained in the past by the authors, where Eq. 18 appears sufficient to obtain a unique solution that satisfies the Kutta condition, $[p]_{TE} = 0$. This fact has been sometimes attributed to the low-order (zeroth order, *i.e.*, piecewise constant) numerical implementation, typically used in the past by the authors and their collaborators (in this case, the unknown is evaluated at the centers of the elements and $[\varphi]$ at the trailing–edge is obtained from the values of φ at the centers of the adjacent elements). It should be pointed out however, that numerical results with pressure discontinuities at the trailing–edge have been observed and remedies have been proposed (*e.g.*, include the values of $[\varphi]_{TE}$ among the unknowns and impose the Kutta condition $[p]_{TE} = 0$, D'Alascio *et al.*, 1997).

In order to get some insight we have explored the issues from a numerical point of view using a formulation that allows to impose all the above conditions (Section 4). This formulation is an extension (pertaining only trailing–edge issues) of the high-order formulation (bi-cubic over each element, for geometry as well as φ and $\partial\varphi/\partial n$) introduced by Gennaretti *et al.* (1998), which yields a velocity distribution that is smooth at the grid nodes. Specifically, the surface is divided into topologically quadrilateral elements. The function φ is described (through a two–dimensional Hermite interpolation), in terms of: (*i*) the corner values of the function, (*ii*) its two first–order derivatives, and (*iii*) the mixed second–order derivative (the same interpolation scheme is used for $\partial\varphi/\partial n$ and for the geometry). The derivatives of φ in turn are expressed in terms of the nodal values of φ, through suitable finite–difference approximations.

Next, we note that the above formulation yields another issue connected to the trailing–edge: the existence of the trailing–edge potential discontinuity implies that at any trailing–edge node there exist two unknowns, but only one collocation point. In Gennaretti *et al.* (1998) this problem is circumvented by having the collocation points coincide with the nodes, except for the trailing–edge, where two collocation points suitably located slightly ahead of the trailing–edge are employed. Of course, placing the collocation points at a very small distance from the trailing–edge yields an ill–conditioned matrix. In order to avoid this, here we use the limit (as $\mathbf{x}$ from the field approaches $\mathbf{x}_{TE}$), of Eq. 17 and of its derivative in direction of the wake normal at $\mathbf{x}_{TE}$.[6] This yields, respectively,

$$\frac{1}{2}\left[\varphi_u + \left(1 - \frac{\gamma}{\pi}\right)\varphi_l\right] = \int_S \left(G\frac{\partial\varphi}{\partial n} - \varphi\frac{\partial G}{\partial n}\right) dS(\mathbf{y}) \quad (22)$$
$$- \int_{S_W} [\varphi]\frac{\partial G}{\partial n} dS(\mathbf{y})$$

(where γ denotes the dihedral angle) and

$$\left(1 - \frac{\gamma}{2\pi}\right)\frac{\partial\varphi}{\partial n_x} = \frac{\partial}{\partial n_x}\left[\int_S \left(G\frac{\partial\varphi}{\partial n} - \varphi\frac{\partial G}{\partial n}\right) dS(\mathbf{y}) - \int_{S_W} [\varphi]\frac{\partial G}{\partial n} dS(\mathbf{y})\right] \quad (23)$$

with $\partial/\partial n_x = \mathbf{n}\cdot\nabla$, where $\mathbf{n}$ is normal to the wake at the trailing–edge.

---

[6]This is equivalent to taking the limit of the semi–sum and of the difference (divided by the distance) of the two equations used by Gennaretti *et al.* (1998), as the collocation points approach in a suitable manner the trailing–edge.

Next, note that as mentioned above the two first–order derivatives and the second–order mixed derivative, introduced by the Hermite interpolation, are approximated in terms of the nodal values of φ. However, a special approach is required for the evaluation of $\partial\varphi/\partial\xi^1$ at the trailing–edge (where the ξ^1-line is the grid line away from the trailing–edge). In Gennaretti et al. (1998), $\partial\varphi/\partial\xi^1$ at the trailing–edge is approximated by backward differences, on either side of the wing. Here, this requirement provides a very natural way to impose the following two additional conditions: (a) the stagnation–point condition in the plane normal to the trailing–edge, Eq. 21, which determines $\partial\varphi/\partial\xi^1$ on the inward dihedral line, and (b) the Joukowski smooth–flow condition, Eq. 20, which determines $\partial\varphi/\partial\xi^1$ on the other side.

Preliminary numerical results are presented in Fig. 1, which depicts $\varphi_0(x) = \varphi(x, 0)$ for a rectangular wing with aspect ratio one, angle of attack $\alpha = 3°$, and NACA0012 airfoil. Four different numerical schemes are used. Curve 0 was obtained using the zeroth–order (piecewise–constant) formulation (see, e.g., Morino, 1993). Curve 1 was obtained with the scheme of Gennaretti et al. (1998), i.e., with two collocation points slightly ahead of the trailing–edge and $\partial\varphi/\partial\xi^1$ at the trailing–edge evaluated by backward finite differences. Curve 2 was obtained by using Eqs. 22 and 23, but with $\partial\varphi/\partial\xi^1$ on the body at the trailing–edge still evaluated by backward finite differences. Finally, curve 3 is obtained by using Eqs. 22 and 23 and by evaluating $\partial\varphi/\partial\xi^1$ at the trailing–edge by using Eqs. 20 and 21 (i.e., conditions (a) and (b) discussed in the preceding paragraph). For all the results, the upper–right portion of the wing is divided into N_1 elements streamwise and N_2 elements spanwise, with $N_1 = N_2 = 16$. For all four cases, the wake geometry is prescribed and is tangent to the lower side, since the numerical results indicate that $v_\tau \delta_\tau > 0$ (here 2 denotes the upper side of the wing, consistently with the convention used in Subsection 3.4). The regularity conditions obtained from the Joukowski hypothesis are used in full (i.e., including conditions (a) and (b) above) only in the most sophisticated scheme (curve 3). The tangency of the wake to the lower side is the only place where the Mangler and Smith (1970) results are used.

The curves are relatively close each other to draw a specific conclusion on which approach is better (indeed further analysis appears desirable, including a convergence analysis, a free–wake analysis, and applications to more complex configurations).

Nonetheless, it is worth noting that in none of the schemes the Kutta condition, $[p]_{TE} = 0$, is imposed explicitly (although it is used to determine the side to which the wake is tangent, as well as Eq. 20). Moreover, the conditions (a) and (b) (stemming from the Joukowski hypothesis) are not used for Curves 1 and 2; therefore, they appear not to be conditions necessary to have a unique solution. Finally, for all the schemes, the pressure difference appears to go to zero (without having to impose this condition explicitly). This is not surprising because, as stated above, $[p] = 0$ is imposed on the wake and therefore it is expected to hold, in the limit, at the trailing–edge as well.

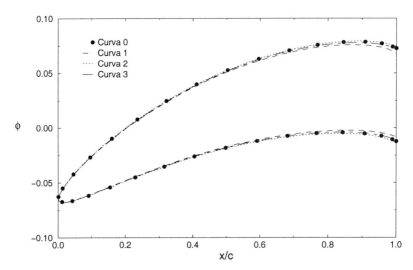

Figure 1: Root section: Distribution of the potential.

References

[1] Bassanini, P., Casciola, C.M., Lancia, M.R., Piva, R.: A boundary integral formulation for the kinetic field in aerodynamics. I. Mathematical analysis, *Eur. J. Mech., B/Fluids*, **10** (1991), 605-627.

[2] Bassanini, P., Casciola, C.M., Lancia, M.R., Piva, R.: On the trailing edge singularity and Kutta condition for 3D airfoils, *Eur. J. Mech., B/Fluids*, **15** (1996), 809-830.

[3] Bassanini, P., Casciola, C.M., Lancia, M.R., Piva, R.: On the removal of the trailing edge singularity in 3D flows . In: Morino, L., and Wendland, W.L., (Eds.): *Boundary Integral Methods for Nonlinear Problems*, Kluwer Academic Publishers, Dordrecht, The Netherlands, 161–166, 1997.

[4] Bassanini, P., Casciola, C.M., Lancia, M.R., Piva, R.: Uniqueness of the bounded flow solution in aerodynamics, *Computational Mechanics*, **22** (1998), 12-18.

[5] Bassanini, P., Casciola, C.M., Lancia, M.R., Piva, R.: *Edge singularities and Kutta condition for 3D unsteady flows in aerodynamics*, (extended–abstract). Proceedings of the IABEM Symposium, Paris, 1998.

[6] Chorin, A.J., Marsden, J.E.: *A Mathematical Introduction to Fluid Mechanics*. Springer–Verlag, New York, 1990.

[7] D'Alascio, A., Visingardi, A., Renzoni, P.: *Explicit Kutta condition correction for rotary wing flows*, Proceedings of the 19th World Conference on the Boundary Element Method, Rome, Italy, Sept. 1997.

[8] Epton, M.A.: Integration by parts formulas for boundary–element methods, *AIAA J.*, **30** No. 2 (1992), 496-504.

[9] Gennaretti, M., Calcagno, G., Zamboni, A., Morino, L.: A high order boundary element formulation for potential incompressible aerodynamics, *The Aeronautical Journal*, **102** No. 1014 (1998), 211-219.

[10] Hsiao, G.C.: Solution of boundary value problem by integral equations of the first kind – an update. In: L. Morino and R. Piva (Eds.): *Boundary Integral Methods – Theory and Applications*, Springer–Verlag, Heidelberg, 1991.

[11] Kondrat'ev, V.A., Oleinik, O.A.: Boundary value problems for partial differential equations in nonsmooth domains, *Russian Math. Surveys*, **38** (1983), 1-86.

[12] Kress, R.: *Linear Integral Equations*. Springer–Verlag, Berlin, 1989.

[13] Kutta, M.W.: Auftriebskräfte in strömenden Flüssigkeiten,' *Illustrierte Aeronautische Mitteilungen*, **6** (1902), 133-135 (in German).

[14] Joukowski, N.: On the adjunct vortices, *Obshchestvo liubitelei estestvoznaniia, antropologii i etnografii*, Transactions of the Physical Section, **XII**, (1905).

[15] Mangler, K.W., Smith J.H.B.: Behaviour of the vortex sheet at the trailing edge of a lifting wing, *The Aeronautical Journal*, **74** (1970), 906-908.

[16] Morino, L.: *A General Theory of Unsteady Compressible Potential Aerodynamics.* NASA CR-2464, 1974.

[17] Morino, L.: Boundary integral equation in aerodynamics, *Appl. Mech. Rev.*, **46**, (1993).

[18] Serrin, J.: Mathematical principles of classical fluid mechanics. In: S. Flügge (Ed.), *Encyclopedia of Physics*, **VIII/1**, Springer–Verlag, Berlin (1959) 125-263.

[19] Sipcic, S.R., Morino, L.: Wake dynamics for incompressible and compressible flows. In: L. Morino (Ed.): *Computational Methods in Potential Aerodynamics*, Springer–Verlag, Heidelberg, 1985.

Università "Roma Tre", Dipartimento di Ingegneria Meccanica e Industriale,
Via della Vasca Navale 79, I–00146 Roma, Italy
Email: luigi@thames.dimi.uniroma3.it
         berno@thames.dimi.uniroma3.it

D. NATROSHVILI, A-.M.SÄNDIG and W.L.WENDLAND

Fluid-Structure Interaction Problems

1. Introduction

This paper is concerned with three-dimensional mathematical problems of fluid-structure interaction and their generalizations in the case of general piecewise smooth *Lipschitz* domains satisfying *the cone condition*.

The physical fluid-solid acoustic interaction problem leading to the generalized mathematical model considered in the present paper can be described as follows. We consider an acoustic wave propagation in an anisotropic inviscid fluid domain of infinite extent where a bounded inhomogeneous anisotropic elastic structure is immersed. The elastic body is coupled to the fluid via the interface surface where the so-called *kinematic and dynamic boundary conditions* are given: the normal velocity of the fluid on one side of the boundary must match the normal velocity of the solid on the other side and the resultant force acting on all parts of the interface has to vanish. We suppose that a time-harmonic acoustic wave is incident upon the elastic target and the problem is to determine the scattered acoustic pressure in the fluid domain and the displacement field in the elastic structure (for the detailed physical background we refer to [3] and references therein).

We apply the so-called non-local approach which, in fact, is the coupling of the boundary integral equations methods and the functional-variational methods.

The basic idea of the non-local approach is the reduction of the original interface problem to an equivalent problem for a bounded domain (occupied by an elastic body). To this end, firstly, with the help of a special integral representation formula we derive *the generalized Steklov-Poincaré relations* between the Dirichlet and the Neumann type data for the exterior scalar (acoustic pressure) field. Secondly, applying the Green identities for the elastic field in a bounded domain and taking into account the transmission conditions and the Steklov-Poincaré formulae on the interface, we derive the equivalent weak functional-variational formulation in a bounded domain for the original generalized fluid-structure interaction problem. Thirdly, we show that for the sesquilinear form, involved in the weak formulation, the Gårding inequality holds. Then we apply the functional-variational approach developed in [5]. On the basis of the Lax-Milgram theorem and the results obtained in [8], [13], [1], [9], [11], [10] we establish the necessary and sufficient conditions of solvability for the general interface problem.

In particular, we have proved that the physical fluid-solid acoustic interaction problem is always solvable for arbitrary values of the frequency parameter.

2. Elastic field

Let $\Omega_1 = \Omega^+ \subset \mathbb{R}^3$ be a bounded domain (diam $\Omega_1 < +\infty$) with a connected, piecewise smooth, nonselfintersecting Lipschitz boundary $S = \partial \Omega_1$; $\Omega_2 = \Omega^- = \mathbb{R}^3 \setminus \overline{\Omega}_1$, $\overline{\Omega}_1 = \Omega_1 \cup S$ (see [11], [13], [9]). We assume that the domains Ω_j ($j = 1, 2$) satisfy the cone condition (see, e.g., [2]).

The domain $\overline{\Omega}_1$ is supposed to be filled up by an inhomogeneous anisotropic material with the elastic coefficients $c_{kjpq}(x) = c_{pqkj}(x) = c_{jkpq}(x)$, $k, j, p, q = 1, 2, 3$, and the density $\rho_1(x) > 0$. For simplicity, throughout the paper we assume that

$c_{kjpq}, \rho_1 \in C^1(\overline{\Omega}_1)$.

The stress tensor $\{\tau_{kj}\}$ and the strain tensor $\{\varepsilon_{kj}\}$ are related by Hook's law $\tau_{kj} = c_{kjpq}(x)\varepsilon_{pq}$, $\varepsilon_{kj} = 2^{-1}(D_k u_j + D_j u_k)$, where $u = (u_1, u_2, u_3)^\top$ is the displacement vector and $D_p = D_{x_p} = \partial/\partial x_p$, $p = 1, 2, 3$; here and in what follows the summation over repeated indices is meant from 1 to 3; the superscript $\top$ denotes transposition.

As usual, the quadratic form corresponding to the potential energy is supposed to be positive definite in the symmetric variables $\varepsilon_{kj} = \varepsilon_{jk}$ (see, e.g., [2])

$$E(u, \overline{u}) = \tau_{kj}\overline{\varepsilon_{kj}} = c_{kjpq}(x)\varepsilon_{kj}\overline{\varepsilon_{pq}} \geq \delta_1 \varepsilon_{kj}\overline{\varepsilon_{kj}}, \quad \delta_1 = \text{const} > 0, \quad x \in \overline{\Omega}_1, \quad (1)$$

where an overbar denotes complex conjugation, and where $\varepsilon_{kj} = \varepsilon'_{kj} + i\varepsilon''_{kj}$ are complex variables (complex strain tensor) corresponding to the complex vector $u = u' + iu''$.

The elastic field in $\Omega^+ = \Omega_1$ is described by the following steady state oscillation equation

$$A_\omega(x, D)u(x) = F(x), \quad x \in \Omega_1, \quad (2)$$

where $F = (F_1, F_2, F_3)^\top$ is a given body force, $D = \nabla = (D_1, D_2, D_3)$, u is the complex-valued amplitude of displacements, and

$$A_\omega(x, D) := [\,(A_\omega(x, D))_{kp}\,]_{3\times 3}, \quad (A_\omega(x, D))_{kp} = (A(x, D))_{kp} + \delta_{kp}\rho_1(x)\omega^2,$$
$$(A(x, D))_{kp} = D_j(c_{kjpq}(x)D_q);$$

here $\omega > 0$ is the frequency parameter and δ_{kp} is the Kronecker's delta.

The inequality (1) implies that $A_\omega(x, D)$ is a strongly elliptic, formally self-adjoint matrix differential operator in $\overline{\Omega}_1$ (see [2], Theorem 5.11).

By $T(D_x, n)u(x)$ we denote the stress vector acting on a surface element with the unit normal vector $n = (n_1, n_2, n_3)$:

$$[T(D_x, n(x))u(x)]_k = \tau_{kj}n_j(x) = c_{kjpq}(x)n_j(x)D_q u_p(x), \quad k = 1, 2, 3. \quad (3)$$

3. Scalar field

We assume that the exterior domain Ω_2 is filled up by a homogeneous anisotropic (fluid) medium with the constant density ρ_2. Moreover, let $a_{pq} = a_{qp}$ be real constants defining a positive definite matrix, i.e., $\widetilde{a} := [a_{pq}]_{3\times 3}$ and $\widetilde{a}\zeta \cdot \zeta = a_{pq}\zeta_q\overline{\zeta}_p \geq \delta_2 |\zeta|^2$, $\delta_2 = \text{const} > 0$, for arbitrary $\zeta \in \mathbb{C}^3$.

Further, let some physical process (say the propagation of acoustic waves) in Ω_2 be described by a complex-valued scalar function (scalar field) $w(x)$ being a solution of the homogeneous "wave equation" (generalized Helmholtz equation)

$$\mathcal{A}_\omega(D)w(x) := a(D)w(x) + \rho_2 \omega^2 w(x) = 0, \quad x \in \Omega^- = \Omega_2, \quad (4)$$

where $a(D) = a_{pq}D_p D_q$, and where ω is again the frequency parameter.

To derive the Sommerfeld type radiation condition at infinity for the scalar field defined by the equation (4), let us introduce the characteristic function corresponding to the operator $\mathcal{A}_\omega(D)$ (cf. [12]): $\Phi_A(\xi, \omega) = a_{pq}\xi_p\xi_q - \rho_2\omega^2$, $\xi \in \mathbb{R}^3$. Denote by S_ω the characteristic surface given by the equation $\Phi_A(\xi, \omega) = \widetilde{a}\xi \cdot \xi - \rho_2\omega^2 = 0$, $\xi \in \mathbb{R}^3$.

It is evident that for an arbitrary vector $\eta \in \mathbb{R}^3$ with $|\eta| = 1$ there exists only one point $\xi(\eta) \in S_\omega$ such that the outward unit normal vector $n(\xi(\eta))$ to S_ω at the point $\xi(\eta)$ has the same direction as η, i.e., $n(\xi(\eta)) = \eta$.

It can be easily verified that

$$\xi(\eta) = \omega\sqrt{\rho_2}\,(\tilde{a}^{-1}\eta \cdot \eta)^{-1/2}\,\tilde{a}^{-1}\eta, \tag{5}$$

where $\tilde{a}^{-1}$ is the matrix inverse to $\tilde{a}$.

Now we are in the position to define the class $\mathrm{Som}_r(\Omega^-)$ of complex-valued functions satisfying the generalized Sommerfeld type radiation conditions.

A function w belongs to $\mathrm{Som}_r(\Omega^-)$, where either $r = 1$ or $r = 2$, if $w \in C^1(\Omega^-)$ and for sufficiently large $|x|$

$$w(x) = O(|x|^{-1}), \quad \frac{\partial w(x)}{\partial x_k} + \mathrm{i}(-1)^r \xi_k(\eta) w(x) = O(|x|^{-2}), \quad k = 1,2,3, \tag{6}$$

where $\xi(\eta) \in S_\omega$ corresponds to the vector $\eta = x/|x|$. Obviously, the conditions (6) are equivalent to the classical Sommerfeld radiation conditions for the Helmholtz equation if $a(D)$ is the Laplace operator.

4. Formulation of the basic interface problem

Now we can formulate the basic transmission problem corresponding to the interaction of the fields introduced above. We shall work with the Sobolev scale $\{H^s\}_{-1 \leq s \leq 1}$.

Problem $(P^{(\omega,r)})$. Find a vector $u \in H^1(\Omega_1)$ and a radiating function $w \in H^1_{\mathrm{loc}}(\Omega_2) \cap \mathrm{Som}_r(\Omega_2)$ which are solutions in the distributional sense of equations (2) and (4), respectively, and satisfy the following coupling conditions on $S = \partial\Omega^\pm$:

$$[u(x) \cdot n(x)]^+ = d_1\,[\partial_n w(x)]^- + f_0(x); \tag{7}$$

$$[T(D,n)u(x)]^+ = d_2\,[w(x)]^- n(x) + f(x), \quad f = (f_1, f_2, f_3)^\top, \tag{8}$$

where d_1 and d_2 are given complex constants satisfying the conditions $d_1 d_2 \neq 0$ and $\mathrm{Im}[\overline{d_1} d_2] = 0$, $T(D,n)u$ is the stress vector given by formula (3) and $\partial_n w = a_{pq} n_p D_q w$ is the so-called co-normal derivative, the symbols $[\cdot]^\pm$ denote generalized traces on S from $\Omega^\pm$ (see, e.g., [1]). Here and throughout this paper $n(x)$ denotes the unit outward normal vector to S at the point $x \in S$. Note that $f_0, f_j \in H^{-1/2}(S)$, and $F_j \in L^2(\Omega_1)$ ($j = 1,2,3$) are given functions (functionals).

We remark that the real physical fluid-solid acoustic interaction problem is the particular case of the problem $(P^{(\omega,r)})$. In fact, in that case $w(x) = p^{sc}(x)$ is the scattered acoustic pressure and

$$d_1 = [\rho_2 \omega^2]^{-1}, \quad d_2 = -1, \quad f_0(x) = f_0^{inc}(x) = [\rho_2 \omega^2]^{-1} \partial_n p^{inc}(x),$$
$$f(x) = f^{inc}(x) = -p^{inc}(x)\,n(x), \quad F = 0, \tag{9}$$

where p^{inc} is the given incident field, e.g.,

$$p^{inc}(x) = \exp\{\mathrm{i}\,d \cdot x\} \quad \text{with} \quad d = \omega\sqrt{\rho_2}\,(\tilde{a}^{-1}\eta' \cdot \eta')^{-1/2}\,\tilde{a}^{-1}\eta', \tag{10}$$

where η' is an arbitrary unit vector. Clearly, the p^{inc} defined by formula (10) solves the equation (4) but it does not represent a radiating function.

5. Integral representation formula and potentials

The fundamental functions (solutions) of the class $\operatorname{Som}_r(\mathbb{R}^3\backslash\{0\})$ for the operator $\mathcal{A}_\omega(D)$ reads

$$\gamma_A(x,\omega,r) = -\frac{\exp\{\mathrm{i}(-1)^{r+1}\omega\sqrt{\rho_2}(\tilde{a}^{-1}x\cdot x)^{1/2}\}}{4\pi|\tilde{a}|^{1/2}(\tilde{a}^{-1}x\cdot x)^{1/2}}, \quad r=1,2, \tag{11}$$

where $|\tilde{a}| = \det\tilde{a}$. For sufficiently large $|x|$ we have

$$\gamma_A(x-y,\omega,r) = -\frac{|\tilde{a}\xi|}{4\pi\omega\,(\rho_2|\tilde{a}|)^{1/2}}\frac{\exp\{\mathrm{i}(-1)^{r+1}(\xi\cdot(x-y))\}}{|x|} + O(|x|^{-2}), \tag{12}$$

where y varies in a bounded subset of $\mathbb{R}^3$ and $\xi = \xi(\eta) \in S_\omega$ corresponds to the direction $\eta = x/|x|$ (see (5)); the asymptotic formula (12) can be differentiated any times with respect to x and y.

Note that if w is a solution of the homogeneous equation (4), then w is an analytic function of the real variable x in the domain Ω_2. Moreover, if, in addition, $w \in H^1(\Omega_2) \cap \operatorname{Som}_r(\Omega_2)$, then the following representation formula holds in Ω_2 (cf. [12])

$$w(x) = \int_S \gamma_A(x-y,\omega,r)[\partial_n w(y)]^- dS_y - \int_S [\partial_{n(y)}\gamma_A(y-x,\omega,r)][w(y)]^- dS_y. \tag{13}$$

Lemma 1. (**Analogue of Rellich's lemma**). *Let $w \in \operatorname{Som}_r(\Omega^-)$ be a solution of (4) in Ω^- and let*

$$\lim_{R\to+\infty} \operatorname{Im} \int_{\Sigma_R} \overline{w(x)}\,\partial_n w(x)\,d\Sigma_R = 0,$$

where Σ_R is the sphere centered at the origin and radius R, ∂_n denotes the co-normal differentiation. Then $w(x) = 0$ in Ω^-.

Let us introduce the single- and double-layer potentials corresponding to the differential operator $\mathcal{A}_\omega(D)$:

$$\begin{aligned}V(g)(x) &= \int_S \gamma_A(x-y,\omega,r)\,g(y)\,dS_y, \quad x\in\mathbb{R}^3\backslash S,\\ W(g)(x) &= \int_S [\partial_{n(y)}\gamma_A(y-x,\omega,r)]\,g(y)\,dS_y, \quad x\in\mathbb{R}^3\backslash S,\end{aligned} \tag{14}$$

where g is a scalar density function.

In what follows we essentialy will use the following properties of these potentials.

Lemma 2. *Let $-1/2 \leq \sigma \leq 1/2$. Then*
i) the operators (14) possess the following mapping properties

$$\begin{aligned}V &: H^{-1/2}(S) \to H^1(\Omega^+), \quad [H^{-1/2}(S) \to H^1_{loc}(\Omega^-) \cap \operatorname{Som}_r(\Omega^-)],\\ W &: H^{1/2}(S) \to H^1(\Omega^+), \quad [H^{1/2}(S) \to H^1_{loc}(\Omega^-) \cap \operatorname{Som}_r(\Omega^-)];\end{aligned}$$

ii) *for arbitrary $g \in H^\sigma(S)$ the following jump relations hold on S:*

$$[V(g)]^\pm = \int_S \gamma_A(z-y,\omega,r)\, g(y)\, dS_y =: \mathcal{H}\, g,$$

$$[\partial_{n(z)} V(g)]^\pm = \mp 2^{-1} g + \int_S [\partial_{n(z)} \gamma_A(z-y,\omega,r)]\, g(y)\, dS_y =: [\mp 2^{-1} I + \mathcal{K}^{(1)}]\, g,$$

$$[W(g)]^\pm = \pm 2^{-1} g + \int_S [\partial_{n(y)} \gamma_A(y-z,\omega,r)]\, g(y)\, dS_y =: [\pm 2^{-1} I + \mathcal{K}^{(2)}]\, g,$$

$$[\partial_{n(z)} W(g)]^+ = [\partial_{n(z)} W(g)]^- =: \mathcal{L}\, g(z),$$

where I stands for the identical operator;

iii) *the operators*

$$\mathcal{H} : H^{-1/2+\sigma}(S) \to H^{1/2+\sigma}(S), \quad \mathcal{K}^{(1)} : H^{-1/2+\sigma}(S) \to H^{-1/2+\sigma}(S),$$
$$\mathcal{K}^{(2)} : H^{1/2+\sigma}(S) \to H^{1/2+\sigma}(S), \quad \mathcal{L} : H^{1/2+\sigma}(S) \to H^{-1/2+\sigma}(S),$$

are continuous;

iv) *the following operator equations*

$$\mathcal{H}\mathcal{K}^{(1)} = \mathcal{K}^{(2)}\mathcal{H}, \quad \mathcal{L}\mathcal{K}^{(2)} = \mathcal{K}^{(1)}\mathcal{L},$$
$$\mathcal{H}\mathcal{L} = -4^{-1} I + [\mathcal{K}^{(2)}]^2, \quad \mathcal{L}\mathcal{H} = -4^{-1} I + [\mathcal{K}^{(1)}]^2,$$

hold in appropriate functional spaces;

v) *the operators $\mathcal{H}$ and $\mathcal{L}$ are strongly elliptic, i.e. there exist compact operators*

$$\mathcal{C}_\mathcal{H} : H^{-1/2}(S) \to H^{1/2}(S), \quad \mathcal{C}_\mathcal{L} : H^{1/2}(S) \to H^{-1/2}(S),$$

and positive constants $\lambda_\mathcal{H}$ and $\lambda_\mathcal{L}$ such that

$$\mathrm{Re}\,\langle (\mathcal{H}+\mathcal{C}_\mathcal{H})\psi, \overline{\psi} \rangle_S \geq \lambda_\mathcal{H} ||\psi, H^{-1/2}(S)||^2 \quad \text{for all} \quad \psi \in H^{-1/2}(S),$$
$$\mathrm{Re}\,\langle (\mathcal{L}+\mathcal{C}_\mathcal{L})\varphi, \overline{\varphi} \rangle_S \geq \lambda_\mathcal{L} ||\varphi, H^{1/2}(S)||^2 \quad \text{for all} \quad \varphi \in H^{1/2}(S),$$

where the brackets $\langle \cdot, \cdot \rangle_S$ denote the natural duality betwen a Sobolev space $H^s(S)$ and its dual $H^{-s}(S)$;

vi) *the inclusions*

$$\psi \in H^{-1/2}(S), \quad \mathcal{H}\psi \in H^{1/2+\sigma}(S), \quad \varphi \in H^{1/2}(S), \quad \mathcal{L}\varphi \in H^{-1/2+\sigma}(S),$$

imply $\psi \in H^{-1/2+\sigma}(S)$, $\varphi \in H^{1/2+\sigma}(S)$, and there hold the a priori estimates

$$||\psi, H^{-1/2+\sigma}(S)|| \leq C_1 (||\mathcal{H}\psi, H^{1/2+\sigma}(S)|| + ||\psi, H^{-1/2}(S)||),$$
$$||\varphi, H^{1/2+\sigma}(S)|| \leq C_2 (||\mathcal{L}\varphi, H^{-1/2+\sigma}(S)|| + ||\varphi, H^{1/2}(S)||).$$

with some positive constants C_1 and C_2.

Lemma 3. *Let $g \in H^{1/2}(S)$ and $w(x) = W(g)(x) + p_0 V(g)(x)$, $x \in \Omega^- = \Omega_2$, where $p_0 = p_1 + ip_2$ is a complex number with $p_2 \neq 0$. If w vanishes in Ω^-, then $g = 0$ on S.*

The proofs of these lemmata are verbatim of the proofs of the similar propositions in [13], [1], [9], [10], [4]).

6. Steklov-Poincaré type relations

. Here we formulate the exterior Dirichlet and Neumann BVPs for the equation (4).
Find a complex-valued scalar function $w \in H^1_{loc}(\Omega^-) \cap \mathrm{Som}_r(\Omega^-)$ satisfying the equation (4) in $\Omega^- = \Omega_2$ and the following boundary conditions on S:

$$\text{Dirichlet Problem } (D^{(\omega,r)})^- : \quad [w]^- = \varphi, \tag{15}$$

$$\text{Neumann Problem } (N^{(\omega,r)})^- : \quad [\partial_n w]^- = \psi, \tag{16}$$

where $\varphi \in H^{1/2}(S)$ and $\psi \in H^{-1/2}(S)$.

Lemma 4. *The homogeneous problems $(D^{(\omega,r)})^-$ and $(N^{(\omega,r)})^-$ [$\varphi = 0$, $\psi = 0$] admit only the trivial solution.*

The proof follows from the analogue of Rellich's lemma (cf. [4]) and the equation

$$\langle [\partial_n w]^-, [v]^- \rangle_S := \int_{\Omega^-} [\rho_2 \omega^2 \, w \, v - a_{pq} D_p w \, D_q v] \, dx.$$

Next, we introduce the operators

$$\mathcal{N} g := \left[(-2^{-1} I + \mathcal{K}^{(2)}) + p_0 \, \mathcal{H} \right] g \tag{17}$$

and

$$\mathcal{M} g := \left[\mathcal{L} + p_0 \, (2^{-1} I + \mathcal{K}^{(1)}) \right] g, \tag{18}$$

where I, $\mathcal{K}^{(1)}$, $\mathcal{K}^{(2)}$, $\mathcal{H}$, $\mathcal{L}$, and p_0 are as in Lemmata 2 and 3.

Lemma 5. *The operators*

$$\mathcal{N} : H^{1/2}(S) \to H^{1/2}(S), \quad \mathcal{M} : H^{1/2}(S) \to H^{-1/2}(S) \tag{19}$$

are isomorphisms.

Proof follows from Lemmata 2 and 3.

Lemma 6. *Let $\varphi \in H^{1/2}(S)$. Then the Dirichlet problem $(D^{(\omega,r)})^-$ is uniquely solvable and the solution is representable in the form of linear combination of double- and single-layer potentials*

$$w(x) = W(g)(x) + p_0 V(g)(x), \quad x \in \Omega_2, \quad p_0 = p_1 + i p_2, \quad p_2 \neq 0, \tag{20}$$

where the density $g \in H^{1/2}(S)$ is the unique solution of the integral equation on S

$$\mathcal{N} g(x) = \varphi(x). \tag{21}$$

Lemma 7. *Let $\psi \in H^{-1/2}(S)$. Then the Neumann problem $(N^{(\omega,r)})^-$ is uniquely solvable and the solution is representable in the form of (20) where the density $g \in H^{1/2}(S)$ is the unique solution of the equation*

$$\mathcal{M}g(x) = \psi(x).$$

Proof follows from Lemmata 1–4. >From Lemmata 5,6, and 7 it follows that the radiating solution $w \in H^1_{loc}(\Omega_2) \cap \text{Som}_r(\Omega_2)$ of the problems $(D^{(\omega,r)})^-$ and $(N^{(\omega,r)})^-$ can be represented as

$$w(x) = [W + p_0 V] \, (\mathcal{N}^{-1}\varphi)(x), \quad x \in \Omega_2, \tag{22}$$

and

$$w(x) = [W + p_0 V] \, (\mathcal{M}^{-1}\psi)(x), \quad x \in \Omega_2, \tag{23}$$

respectively, where $[w]^- =: \varphi \in H^{1/2}(S)$ and $[\partial_n w]^- =: \psi \in H^{-1/2}(S)$.

>From the above results it follows that the Dirichlet and the Neumann data for an arbitrary radiating solution $w \in H^1_{loc}(\Omega_2) \cap \text{Som}_r(\Omega_2)$ of the equation (4) are related on S by the following *Steklov-Poincaré type equation*

$$\mathcal{M}^{-1}[\partial_n w]^- = \mathcal{N}^{-1}[w]^-,$$

where $[w]^- \in H^{1/2}(S)$ and $[\partial_n w]^- \in H^{-1/2}(S)$.

7. Weak formulation of the interaction problem

As we have shown the sought-for scalar function $w \in H^1_{loc}(\Omega_2) \cap \text{Som}_r(\Omega_2)$ is uniquely representable in the form (20) with some density $g \in H^{1/2}(S)$. Moreover, $\mathcal{N}g = [w]^- \in H^{1/2}(S)$ and $\mathcal{M}g = [\partial_n w]^- \in H^{-1/2}(S)$. Therefore, we can recast the interface conditions (7) and (8) on S as follows

$$[u \cdot n]^+ = d_1 \mathcal{M} g + f_0, \quad [Tu]^+ = d_2 \, n(x) \, \mathcal{N} g + f. \tag{24}$$

Further, applying Green formula for the operator $A_\omega(x, D)$ and taking into account equations (24) we get

$$\int_{\Omega_1} \{E(u, \overline{v}) - \rho_1 \omega^2 \, u \cdot v\} \, dx - d_2 \langle \mathcal{N}g, [\overline{v}]^+ \cdot n \rangle_S = \langle f, [\overline{v}]^+ \rangle_S - \int_{\Omega_1} F \cdot v \, dx, \tag{25}$$

$$\langle [u \cdot n]^+, \overline{h} \rangle_S - d_1 \langle \mathcal{M}g, \overline{h} \rangle_S = \langle f_0, \overline{h} \rangle_S. \tag{26}$$

Multiplying (26) by $(-\overline{d}_1)$ and adding termwise to (25) we arrive at the equation

$$\mathcal{B}(u, g; v, h) = \mathcal{F}(v, h), \tag{27}$$

where

$$\mathcal{B}(u, g; v, h) = \int_{\Omega_1} E(u, \overline{v}) \, dx - \omega^2 \int_{\Omega_1} \rho_1 \, u \cdot v \, dx - d_2 \int_S \mathcal{N}g \, ([\overline{v}]^+ \cdot n) \, dS - \overline{d}_1 \int_S [u \cdot n]^+ \overline{h} \, dS + |d_1|^2 \langle \mathcal{M}g, \overline{h} \rangle_S, \tag{28}$$

$$\mathcal{F}(v, h) = \langle f, [\overline{v}]^+ \rangle_S - \overline{d}_1 \langle f_0, \overline{h} \rangle_S - \int_{\Omega_1} F \cdot v \, dx. \tag{29}$$

Now we can formulate the so-called non-local setting of the basic interaction problem.

Problem ($\widetilde{\mathcal{P}}^{(\omega,r)}$): Find a pair $(u, g) \in [H^1(\Omega_1)]^3 \times H^{1/2}(S)$ satisfying the equation (27) for arbitrary $(v, h) \in [H^1(\Omega_1)]^3 \times H^{1/2}(S)$, where the sesquilinear form $\mathcal{B}$ and the functional $\mathcal{F}$ are defined by (28) and (29), respectively.

If (u, g) is a solution of the problem ($\widetilde{\mathcal{P}}^{(\omega,r)}$), then u describes the elastic field of displacements in Ω_1, while the scalar field w is constructed with the help of the g by formula (20).

8. Uniqueness theorem. Jones modes and Jones frequencies

Let us consider the homogeneous version of the problem ($\widetilde{\mathcal{P}}^{(\omega,r)}$) (i.e., $F = 0$, $f = 0$, $f_0 = 0$) and let $(u, g) \in [H^1(\Omega_1)]^3 \times H^{1/2}(S)$ be its arbitrary solution. Then

$$\mathcal{B}(u, g; v, h) = 0 \quad \text{for} \quad \forall (v, h) \in [H^1(\Omega_1)]^3 \times H^{1/2}(S), \tag{30}$$

whence (see (28))

$$\int_{\Omega_1} \{E(u, \bar{v}) - \rho_1 \omega^2 u \cdot v\} \, dx = d_2 \int_S (\mathcal{N}g) n \cdot [v]^+ dS \quad \text{for} \quad \forall v \in [H^1(\Omega^+)]^3, \tag{31}$$

and

$$\langle [u \cdot n]^+, \bar{h} \rangle_S = d_1 \langle \mathcal{M}g, \bar{h} \rangle_S \quad \text{for} \quad \forall h \in H^{1/2}(S). \tag{32}$$

These equations imply $[Tu]^+ = d_2 n(\mathcal{N}g)$ and $[u \cdot n]^+ = d_1 \mathcal{M}g$ on S. The both equations are understood in the sense of the functional space $H^{-1/2}(S)$. Consequently,

$$[Tu]^+ = d_2 [w]^- n, \quad [u \cdot n]^+ = d_1 [\partial_n w]^-. \tag{33}$$

With the help of Lemma 1 we can derive that $w = 0$ in Ω_2. Therefore, for the elastic field u we obtain the following eigenvalue problem

$$A_\omega(x, D)u = 0 \text{ in } \Omega_1, \quad [u \cdot n]^+ = 0 \text{ and } [Tu]^+ = 0 \text{ on } S. \tag{34}$$

We denote by $J(\Omega_1)$ the set of values of the frequency parameter $\omega > 0$ for which the problem (34) admits a nontrivial solution. Such solutions are called *Jones modes*, while the corresponding values of ω are called *Jones eigenfrequencies* (cf. [6], [4]). Note that $J(\Omega_1)$ is at most enumerable, and for each $\omega \in J(\Omega_1)$ the space of associated Jones modes is of finite dimension.

Lemma 8. *Let*

$$\omega \notin J(\Omega_1), \quad \text{Im}(\bar{d}_1 d_2) = 0, \quad d_1 d_2 \neq 0. \tag{35}$$

Then the homogeneous problem ($\widetilde{\mathcal{P}}^{(\omega,r)}$) *($f = 0$, $f_0 = 0$, $F = 0$) has only the trivial solution.*

9. Existence theorem

To apply the well-known functional-variational methods to the equation (27) we have to study some properties of the sesquilinear form $\mathcal{B}$ and the anti-linear functional $\mathcal{F}$. In particular, we will establish the boundedness of $\mathcal{B}$ and $\mathcal{F}$, and the Gårding type inequality for $\mathcal{B}$, in appropriate function spaces (see, e.g., [5], Chapter 2, Section 9, [2], [11].) To this end, we recall the well-known Ehrling inequalities (see, e.g., [5], Chapter 1, §16, Theorem 16.3; [7], Chapter 14, §3). Let $u \in H^1(\Omega^+)$, $S = \partial \Omega^+$, and $[u]^+$ be the trace of u on S. Then for arbitrary $\varepsilon > 0$ there exist positive constants $c'_1(\varepsilon)$ and $c'_2(\varepsilon)$ such that

$$||u \,;\, H^{s_2}(\Omega_1)|| \leq \varepsilon ||u \,;\, H^{s_1}(\Omega_1)|| + c'_1(\varepsilon)||u \,;\, H^{s_3}(\Omega_1)||,$$
$$||[u]^+ \,;\, H^{t_2}(S)|| \leq \varepsilon ||u \,;\, H^{t_1}(\Omega_1)|| + c'_2(\varepsilon)||u \,;\, H^{t_3}(\Omega_1)||,$$
$$1 \geq s_1 > s_2 > s_3 \geq 0, \quad 0 \leq t_3 < t_2 + \frac{1}{2} < t_1 \leq 1, \quad t_2 \geq 0. \tag{36}$$

Lemma 9. *Let F, f, and f_0 be as above. Then the anti-linear functional $\mathcal{F}(\cdot,\cdot)$ and the sesquilinear form $\mathcal{B}(\cdot,\cdot\,;\,\cdot,\cdot)$ defined by (29) and (28), respectively, are bounded on $[H^1(\Omega_1)]^3 \times H^{1/2}(S)$. Moreover, there exist constants $c_1 > 0$ and $c_2 > 0$ such that*

$$\mathrm{Re}\,[\mathcal{B}(u,g;u,g)] \geq c_1 \left\{||u \,;\, H^1(\Omega_1)||^2 + ||g \,;\, H^{1/2}(S)||^2\right\} -$$
$$c_2 \left\{||u \,;\, L_2(\Omega_1)||^2 + ||g \,;\, L_2(S)||^2\right\} \tag{37}$$

for arbitrary $u \in [H^1(\Omega_1)]^3$ and $g \in H^{1/2}(S)$.

Proof is based on the above Ehrling's inequalities, the well-known Korn's inequality (see [2]), and the ellipticity of the operator $\mathcal{L}$ (see Lemma 2).

Theorem 1. *Let the conditions (35) be fulfilled. Then the non-local problem $(\widetilde{\mathcal{P}}^{(\omega,r)})$ is uniquely solvable.*

Now let ω be a Jones eigenfrequency for the domain Ω_1 and let us consider the fluid-solid acoustic interface problem:

$$A_\omega(x,D)u(x) = 0 \text{ in } \Omega_1, \quad u \in H^1(\Omega_1),$$
$$\mathcal{A}_\omega(D)w(x) = 0 \text{ in } \Omega_2, \quad w \in H^1_{loc}(\Omega_2) \cap \mathrm{Som}_r(\Omega_2),$$
$$[Tu]^+ = d_2[w]^- n + f^{inc}, \quad [u \cdot n]^+ = d_1 [\partial_n w]^- + f_0^{inc} \text{ on } S, \tag{38}$$

where d_1, d_2, f^{inc} and f_0^{inc} are given by (9) and (10). Clearly, the non-local formulation of the problem (38) reads as the equation (27) where $\mathcal{B}$ and $\mathcal{F}$ are represented by (28) and (29) with $F = 0$, $f = f^{inc}$, and $f_0 = f_0^{inc}$.

Denote by $X_\omega(\Omega_1)$ the null-space of the corresponding homogeneous problem (34), i.e., the space of Jones modes. Obviously, if the interface problem is solvable, a solution of the non-homogeneous problem is then defined within the summand $(u_0, w_0)^\top$ where $u_0 \in X_\omega(\Omega_1)$ and $w_0 = 0$. Let us show that the nonhomogeneous problem (38) is always solvable.

To establish the necessary and sufficient conditions for the problem (38) to be solvable, we have to formulate the so-called "adjoint" interface problem.

To this end, we note that the problem (38) can be equivalently rewritten as:

$$A_\omega(x, D)u(x) = 0 \quad \text{in} \quad \Omega_1, \tag{39}$$

$$[Tu]^+ - d_1^{-1}d_2\{\mathcal{N}\mathcal{M}^{-1}[u \cdot n]^+\}n = f^{inc} - d_1^{-1}d_2\{\mathcal{N}\mathcal{M}^{-1}f_0^{inc}\}n, \quad \text{on} \quad S \tag{40}$$

$$w(x) = W(g)(x) + p_0 V(g)(x), \quad x \in \Omega_2, \quad p_0 = p_1 + ip_2, \quad p_2 \neq 0, \tag{41}$$

$$g = d_1^{-1}\mathcal{M}^{-1}([u \cdot n]^+ - f_0^{inc}), \tag{42}$$

where $\mathcal{N}$ and $\mathcal{M}$ are given by (17) and (18).

Thus, the original interface problem is completely split into the two problems: the first one is the boundary value problem (39)-(40) for the elastic field u in Ω_1, and the second one is the construction of the scalar field w in Ω_2 by formula (41), where g is given by the equation (42). Clearly, the w is uniquely defined by the boundary function $[u \cdot n]^+$. So we need only to solve the problem (39)-(40). It is evident that the problems (39)-(40) and (38) are simultaneously solvable or unsolvable.

In just the same way as in [4] we can show that the sought-for homogeneous adjoint boundary value problem can be formulated as follows

$$A_\omega(x, D)v(x) = 0 \quad \text{in} \quad \Omega_1, \quad [Tv]^+ - d_1^{-1}d_2[\mathcal{N}\mathcal{M}^{-1}[v \cdot n]^+]n = 0 \quad \text{on} \quad S. \tag{43}$$

Obviously, the problem (39)-(40) is solvable if and only if (see, e.g., [5], Chapter 2)

$$\int_S \langle f^{inc} - d_1^{-1}d_2[\mathcal{N}\mathcal{M}^{-1}f_0^{inc}]n, \ [v]^+\rangle \, dS = 0, \tag{44}$$

where v is an arbitrary solution of the adjoint problem (43) and $\langle a, b\rangle = \sum_{k=1}^3 a_k b_k$ for $a, b \in \mathbb{C}^3$.

It can also be shown that the problem (43) coincides with the homogeneous problem of Jones modes (34):

$$A_\omega(x, D)v(x) = 0 \quad \text{in} \quad \Omega_1, \quad [Tv]^+ = 0 \quad \text{and} \quad [v \cdot n]^+ = 0 \quad \text{on} \quad S.$$

The condition (44) is then converted into the equation

$$\int_S \langle f^{inc}, [v]^+\rangle \, dS = 0 \quad \text{for all} \quad v \in X_\omega(\Omega_1), \tag{45}$$

which represents the necessary and sufficient condition for the problem (39)-(40) (and, consequently, for the problem (38)) to be solvable.

Since in the fluid-solid acoustic interaction problem $f^{inc} = -p^{inc}(x)n(x)$ due to (9), where $p^{inc}(x)$ is a scalar function, we have

$$\langle f^{inc}, [v]^+\rangle = -p^{inc}(x) \langle n, [v]^+\rangle = -p^{inc}(x) [v \cdot n]^+ = 0$$

for an arbitrary Jones mode v. From this fact it follows that (45) is satisfied and the problem (38) is always solvable (cf. [4]).

Thus, the fluid-solid acoustic interaction problem is solvable for arbitrary value of the frequency parameter ω. If $\omega \notin J(\Omega_1)$ then the problem is uniquely solvable, and, if $\omega \in J(\Omega_1)$, then solutions are defined modulo $(u_0, 0)$, where $u_0 \in X_\omega(\Omega_1)$ is an arbitrary Jones mode.

References

[1] Costabel, M.: *Boundary integral operators on Lipschitz domains: elementary results.* SIAM J. Math. Anal., **19**, **3**(1988), 613-626.

[2] Fichera, G.: *Existence Theorems in Elasticity.* Handb. der Physik, Bd. 6/2, Springer-Verlag, Heidelberg, 1973.

[3] Hsiao, G.C., Kleinman, R.E. and Roach, G.F.: *Weak Solution of Fluid-solid Interaction Problems.* Technische Hochschule Darmstadt, Fachbereich Mathematik, Preprint-Nr.1917, May, 1997.

[4] Jentsch, L. and Natroshvili, D.: *Non-local approach in mathematical problems of fluid-structure interaction.* To appear in: *Mathematical Methods in the Applied Sciences* **21**, 1998.

[5] Lions, J.-L. and Magenes, E.: *Problèmes aux limites non homogènes et applications*, Vol. 1. Dunod-Paris, 1968.

[6] Luke, C.J. and Martin, P.A.: Fluid-solid interaction: acoustic scattering by a smooth elastic obstacle. SIAM J. Appl. Math., **55**, 4(1995), 904-922.

[7] Maurin, K.: *Methods of Hilbert Spaces.* Polish Scientific Publishers, Warsaw, 1959.

[8] Maz'ya, V.G.: Boundary integral equations. In: R.V.Gamkrelidze (Ed.), *Encyclopedia of Mathematical Sciences*, **27**, Springer-Verlag: Berlin, Heidelberg, (1991), 127-222.

[9] Mitrea, D., Mitrea, M. and Pipher, J.: Vector potential theory on nonsmooth domains in $\mathbb{R}^3$ and applications to electromagnetic scattering. *The Journal of Fourier Analysis and Applications*, **3**, **2**(1997), 131-192.

[10] Mitrea, D.: The method of layer potentials for non-smooth domains with arbitrary topology. *Integral Equations and Operator Theory*, **29** , (1997), 320-338.

[11] Nečas, J.: *Les Méthodes Directes en Théorie des Équations Élliptique.* Academia, Prague, 1967.

[12] Vainberg, B.R.: The radiation, limiting absorption and limiting amplitude principles in the general theory of partial differential equations. *Usp. Math. Nauk*, **21**, **3(129)** (1966), 115-194 (Russian).

[13] Verchota, G.: Layer potentials and regularity for the Dirichlet problem for Laplace's equation in Lipschitz domains. *Journal of Functional Analysis*, **59**(1984), 572–611.

D. Natroshvili: Department of Mathematics, Georgian Technical University
Kostava str. 77, Tbilisi 380075, Republic of Georgia
Email: natrosh@hotmail.com

A-.M.Sändig and W.L.Wendland: Mathematical Institute A, University of Stuttgart
Pfaffenwaldring 57, 70569 Stuttgart, Germany
Email: anna@mathematik.uni-stuttgart.de;
         wendland@mathematik.uni-stuttgart.de

H. SCHULZ, Ch. SCHWAB and W.L. WENDLAND

Extraction, Higher Order Boundary Element Methods and Adaptivity

1. Introduction

The computation of potentials, displacements, stresses and their derivatives via potential representations on and near to the boundary is one of the difficult tasks in boundary element methods. Here we present a procedure which allows us to compute these quatities by repeatedly solving the underlying boundary integral equation with modified right–hand sides. Additional recovery improves the original order of convergence on one hand and provides us with pointwise a–posteriori error estimates for driving adaptive mesh refinement, on the other hand.

We consider in a bounded two- or three-dimensional domain Ω with boundary $\Gamma := \partial \Omega$ the formally positive, selfadjoint elliptic system of second order

$$\mathcal{P}\mathcal{U} = -\sum_{k,\ell=1}^{n} \frac{\partial}{\partial x_\ell}\left(a_{k\ell}\frac{\partial}{\partial x_k}\mathcal{U}\right) + c\mathcal{U} = 0 \quad \text{in } \Omega \subset \mathbb{R}^n, \quad n=2,3 \tag{1}$$

with the Cauchy data

$$\mathcal{U}|_\Gamma =: \phi, \quad \sum_{k,\ell=1}^{n} n_\ell a_{k\ell}\frac{\partial \mathcal{U}}{\partial x_k} = T\mathcal{U}|_\Gamma =: \psi \quad \text{on the boundary } \Gamma. \tag{2}$$

In traditional mixed boundary value problems, ϕ is given on one part Γ_D of the boundary and ψ on the complementary part $\Gamma_N = \Gamma \backslash \Gamma_D$. However, for simplicity we consider here the Dirichlet problem with given ϕ on $\Gamma = \Gamma_D$. By using a fundamental solution $\mathcal{G}(x,y)$ of $\mathcal{P}$, the solution $\mathcal{U}$ can be represented by

$$\mathcal{U}(x) = (V\psi)(x) - (W\phi)(x) \quad \text{for } x \in \Omega \tag{3}$$

with a single layer potential V and a double layer potential W. For $x \in \Gamma$ one obtains with the so-called jump relations the integral equation of the first kind for ψ,

$$(V\psi)(x) = (\frac{1}{2}I + K)\phi(x), \quad x \in \Gamma. \tag{4}$$

With given ϕ, the right-hand side $F(x) := (\frac{1}{2}I + K)\phi(x)$ is given and $K\phi$ is the direct Cauchy principal value double layer integral operator if $\partial\Omega$ is smooth which needs to be modified at edge and corner points.

It is well known that on either side of $\partial\Omega$ the potentials are smooth (even analytic if the data and Γ are analytic) up to the boundary. Hence, there should be a "nonsingular" method for their accurate evaluation near to the boundary. If the field point $x \in \Omega$ is sufficiently close to $\partial\Omega$, then the potential $\mathcal{U}(x)$ may be expanded into a Taylor series about some boundary point x_0 near to x, [12]. Due to the smoothness of the potential, this Taylor expansion has good accuracy and the precision increases as $|x - x_0| \to 0$. This is paid for, however, with the requirement to know higher order normal derivatives of $\mathcal{U}$ at x_0. Using the differential equation, these normal derivatives can be calculated recursively in terms of the tangential derivatives of the solution at $x_0 \in \Gamma$. The tangential derivatives can be computed in a stable and efficient fashion in two steps:

1. The recursive application of the tangential operator to the original boundary integral equation (4) provides a triangular system for the tangential derivatives of the solution with a right–hand side which contains derivatives of the original right–hand side $F(x)$ and commutators of the boundary integral operator with the tangential derivative. To solve this system one only needs an efficient method for the approximation of the original boundary integral equation.

2. The recovery technique, based on the Riesz transform and a corresponding pointwise representation formula on the boundary in terms of the tangential derivatives of the solution, gives a superconvergent approximation, if superconvergence in lower norms is provided.

Such a method was proposed first in [12], numerically tested for the Laplacian in [10] and for the system of elasticity in [11]. In addition, here we present some new error estimates for the recovered solutions on the boundary. Based on the recovery technique, an a-posteriori error estimator can be developed, which can be evaluated globally as well as locally.

Up to now we restricted our considerations to "sufficiently smooth" boundaries only. In Section 7, however, we present some ideas to overcome the difficulties occurring at corner points.

2. Computation of derivatives on the boundary

For our method we need the representation of the differential operator $\mathcal{P}$ in a tubular neighbourhood of the boundary in tangential and normal coordinates,

$$-\mathcal{P}\mathcal{U} = \sum_{k,\ell=1}^{n} \frac{\partial}{\partial x_\ell}\left(a_{k\ell}\frac{\partial}{\partial x_k}\mathcal{U}\right) - c\mathcal{U} = \left\{\mathcal{P}_2 + \mathcal{P}_1\frac{\partial}{\partial n} + \mathcal{P}_0\left(\frac{\partial}{\partial n}\right)^2\right\}\mathcal{U} \quad (5)$$

with tangential differential operators $\mathcal{P}_m$ of orders m. These operators can be expressed in terms of the curvature, the components of the Riemann-tensor and corresponding derivatives of the coefficients a_{jk}, see [3, Chap. III] or the appendix of [12]. For the case of linear elasticity see the appendix of [11]. Due to ellipticity, the matrix $\mathcal{P}_0$ is invertible.

Theorem 1. [3, 11, 12] *Suppose that $\mathcal{P}\mathcal{U} = 0$, Dirichlet boundary conditions ϕ and a decomposition (5) are given. Then all normal derivatives $(\frac{\partial}{\partial n})^\ell \mathcal{U}$ and tangential derivatives $\partial_u^\alpha \mathcal{U}$ are computable on the boundary Γ.*

Proof. Solving the Dirichlet boundary value problem gives the second Cauchy datum $\psi = T\mathcal{U}$ on Γ which fulfills the boundary integral equation (4). Using (4) one obtains with $A := V$ and $F := (\frac{1}{2}I + K)\phi$ for arbitrary tangential derivatives $\partial_1^k \partial_2^\ell \psi$ up to the order $k + \ell$, the triangular system

$$\begin{aligned}
A\psi &= F \\
A(\partial_1\psi) &= \partial_1 F - A_{(1,0)}\psi \\
A(\partial_2\psi) &= \partial_2 F - A_{(0,1)}\psi \\
&\vdots \\
A(\partial_1^k \partial_2^\ell \psi) &= \partial_1^k \partial_2^\ell F - \sum_{\substack{i,j=0 \\ 0<i+j}}^{k,\ell} \binom{k}{i}\binom{\ell}{j} A_{(i,j)}\left(\partial_1^{k-i}\partial_2^{\ell-j}\psi\right)
\end{aligned} \qquad (6)$$

where $A_{(0,0)} := A$ and $A_{(i+1,j)} := [\partial_1, A_{(i,j)}]$, $A_{(i,j+1)} := [\partial_2, A_{(i,j)}]$.

The system (6) can be solved recursively if A is invertible. The brackets $[\partial, A]$ denote the commutator $(\partial A - A\partial)$. Then, using

$$T\mathcal{U} = \psi = \mathcal{P}_0 \frac{\partial \mathcal{U}}{\partial n} + \alpha_1 \frac{\partial \phi}{\partial u_1} + \alpha_2 \frac{\partial \phi}{\partial u_2} \quad \text{with } \alpha_\ell(x) = \sum_{j,k=1}^{3} a_{jk} n_j \frac{\partial \chi_k}{\partial u_\ell}, \qquad (7)$$

the first normal derivative of $\mathcal{U}$ can be computed. The vector function $\chi = \chi(s_1, s_2)$ denotes a (local) representation of the boundary surface. Using the differential equation $\mathcal{P}\mathcal{U} = 0$, the application of Cauchy's algorithm gives

$$\frac{\partial^2 \mathcal{U}}{\partial n^2} = -\mathcal{P}_0^{-1}\left\{\mathcal{P}_1 \frac{\partial \mathcal{U}}{\partial n} + \mathcal{P}_2 \mathcal{U}\right\},$$

$$\frac{\partial^3 \mathcal{U}}{\partial n^3} = -\mathcal{P}_0^{-1}\left\{\mathcal{P}_1 \frac{\partial^2 \mathcal{U}}{\partial n^2} + \mathcal{P}_2 \frac{\partial \mathcal{U}}{\partial n}\right\} - \left(\frac{\partial}{\partial n}\mathcal{P}_0^{-1}\mathcal{P}_1\right)\frac{\partial \mathcal{U}}{\partial n} - \left(\frac{\partial}{\partial n}\mathcal{P}_0^{-1}\mathcal{P}_2\right)\mathcal{U}$$

$$\vdots$$

for the normal derivatives. Since $\mathcal{P}_m$ with $m = 0, 1, 2$, are tangential differential operators, this completes the proof. □

The numerical algorithm works in exactly the same manner. We emphasize that for the computation of the tangential derivative always the same stiffness matrix corresponding to the single layer or the double layer potential can be used which in a usual method is already implemented and computed. The calculation of the commutators can be done in a straightforward and simple way by using the chain rule and the local kernel representation of the boundary integral operators, see [3, 12].

Theorem 2. *Using a Galerkin-Bubnov method for solving system (6), one obtains for the tangential derivatives the error estimate*

$$\|\partial_u^\varrho \psi - \psi_h^{(\varrho)}\|_{H^p(\Gamma)} \leq c(\varrho) h^{d-p} \|\psi\|_{H^{d+|\varrho|}(\Gamma)}, \qquad (8)$$

provided $-d + 2\alpha \leq p \leq \alpha \leq d$, *where* 2α *is the order of the operator A and d is the approximation order of the trial- and test-space.*

The proof is based on strong ellipticity of A; for more details see [1, 3, 8, 10, 12].

Remark 1. In Equation (4), the single layer potential $A = V$ is of order $2\alpha = -1$ and ψ is the desired boundary traction in (2).

3. Evaluation of the solution near to the boundary

For x near to the boundary, the convergence of the traditional evaluation (3) deteriorates significantly. But with the derivatives on the boundary Γ we can evaluate $\mathcal{U}$ at $x \in \Omega$ via the Taylor formula:

$$\mathcal{U}(x) = \phi(x_0) - v\partial_n\mathcal{U}(x_0) + \frac{1}{2}v^2\partial_n^2\mathcal{U}(x_0) - \frac{1}{3!}v^3\partial_n^3\mathcal{U}(x_0) + \ldots \tag{9}$$

about the point $x_0 \in \Gamma$ with minimal distance to x. Then, $v := |x - x_0|$. Obviously the accuracy of (9) strongly depends on v. For the corresponding error estimates and for a comparison of the numerical errors of (3) and (9), see [10].

4. Recovery techniques

The approximate solutions of the triangular system (6) are possibly discontinuous whereas in (9) we need point values, which we compute via the techniques in [10, 12].

Let $y = \chi(v_1, v_2)$ be a local parametric boundary representation and let $x_0 = \chi(u_1, u_2)$. For the two-dimensional case, u is only one real parameter and can be chosen as the arc length of Γ. Then, for any distribution ψ which is smooth at $x_0 \in \Gamma$ one has [12]

$$(\partial_u^\gamma \psi)(u) = \sum_{0 \le |\beta| \le M} \int_\Gamma d_\beta(v-u)(\partial^{\gamma+\beta}\psi)(v) H_M(v-u)\, dv \tag{10}$$

where H_M satisfies

$$H_M: \quad (-\Delta)^{\frac{M}{2}} H_M(v-u) = \delta(v-u) \quad \text{for } M \text{ even} \tag{11}$$

and $H_M(v-u) := \nabla_v H_{M+1}(v-u)$ for M odd. In particular, for M even one finds

$$H_M(v-u) = \begin{cases} \frac{(-1)^{\frac{M}{2}}}{2(M-1)!}|v-u|^{M-1} & \text{for } \Gamma \text{ a curve,} \\ \frac{(-1)^{\frac{M}{2}}}{2^{M-1}\pi\left(\left(\frac{M}{2}-1\right)!\right)^2}|v-u|^{M-2}\ln|v-u| & \text{for } \Gamma \text{ a surface,} \end{cases}$$

see [13, p. 288] for the three-dimensional case. The functions $d_\beta(v-u)$ are given in terms of derivatives of a sufficiently smooth cut-off function $\rho(v-u)$ with $\rho(0) = 1$ and are defined via the Leibniz formula. The integration in (10) is done with respect to the local coordinate system. Using on the right–hand side the Galerkin solution $\psi_h^{(\varrho)}$ of system (6) instead of the exact derivatives $\partial^{\gamma+\beta}\psi$, one obtains the *superconvergent recovery approximation*

$$\partial_u^\gamma \psi \approx \tilde{\psi}_h^{(\gamma)}(u) := \sum_{0 \le |\beta| \le M} \int_\Gamma d_\beta(v-u)(\psi_h^{(\gamma+\beta)})(v) H_M(v-u)\, dv. \tag{12}$$

Theorem 3. *For the error of the recovered solution there holds the pointwise estimate:*

$$\left|(\partial_u^\gamma \psi)(u) - \tilde{\psi}_h^{(\gamma)}(u)\right| \le c\, h^{\ell+m} \|\psi\|_{H^{\ell+|\gamma|+M+\alpha}(\Gamma)}, \tag{13}$$

provided $M > 1 - \alpha$ and $0 \leq \ell \leq \min\{d - \alpha, M - 1 + \alpha - \varepsilon\}$, $\varepsilon > 0$; $0 \leq m \leq d - \alpha$, $\ell + m \leq 2d - 2\alpha$. Also, there holds the Sobolev space estimate

$$||\partial_u^\gamma \psi - \widetilde{\psi}_h^{(\gamma)}||_{H^s(\Gamma)} \leq \tilde{c}\, h^{\ell-s+M}||\psi||_{H^{\ell+M+|\gamma|}}, \quad \text{if } 2\alpha - d \leq s - M \leq \alpha \leq \ell \leq d. \quad (14)$$

Proof. For the proof of (13), see [12]. To prove (14), we define operators

$$(V_{M,\beta}\psi)(u) := \int_\Gamma H_M(v - u) d_\beta(v - u) \psi(v) \, dv. \quad (15)$$

With the properties of $H_M(v - u)$ it can be shown that $V_{M,\beta} : H^s(\Gamma) \longrightarrow H^{s+M}(\Gamma)$ are continuous mappings. From (15), the estimate (8) and the representations

$$(\partial_u^\gamma \psi)(u) = \sum_{0\leq|\beta|\leq M} (V_{M,\beta}\partial_u^{\gamma+\beta}\psi)(u), \quad \widetilde{\psi}_h^{(\gamma)}(u) = \sum_{0\leq|\beta|\leq M} (V_{M,\beta}\psi_h^{(\gamma+\beta)})(u)$$

one obtains

$$||\partial_u^\gamma\psi - \widetilde{\psi}_h^{(\gamma)}||_{H^s(\Gamma)} \leq c_1 \sum_{0\leq|\beta|\leq M} ||\partial_u^{\gamma+\beta}\psi - \widetilde{\psi}_h^{(\gamma+\beta)}||_{H^{s-M}(\Gamma)} \leq c_2\, h^{\ell-s+M}\,||\psi||_{H^{\ell+|\gamma|+M}(\Gamma)},$$

provided $2\alpha - d \leq s - M \leq \ell \leq \alpha \leq d$; i.e. (14). $\square$

Remark 2. Near the boundary estimate (13) implies a high convergence order if

$$\widehat{\mathcal{U}}_h(x) := \phi(x_0) - v\mathcal{P}_0^{-1}(\widetilde{\psi}_h^{(0)} - \alpha\, \partial_u\phi)$$

is used as an approximation of $\mathcal{U}(x)$, see [10]. Estimate (14) shows that to compute $\mathcal{U}$ via the representation formula (3) far away from the boundary one can use the recovered solution instead of the Galerkin solution. In both cases one obtains the optimal convergence order of $2d + 1$.

5. Computation of the full stress tensor in linear elasticity

As a special case of (1), we consider the Dirichlet problem for the Lamè system for homogeneous isotropic materials,

$$\mathcal{PU} := -\mu\Delta\mathcal{U} - (\lambda + \mu)\,\text{grad div}\,\mathcal{U} = f \quad \text{in } \Omega \quad \text{and} \quad \mathcal{U} = \phi \text{ on } \Gamma, \quad (16)$$

where μ and $\lambda > -\mu$ are the Lamè constants and [3, 15]

$$a_{jk\ell m} = \mu \delta_{jk}\delta_{\ell m} + \frac{1}{2}(\lambda + \mu)\,(\delta_{j\ell}\delta_{km} + \delta_{jm}\delta_{k\ell}). \quad (17)$$

After computing the boundary traction ψ, we use the representation for the stress tensor

$$\sigma_{j\ell} = \sum_{k,m=1}^n a_{jk\ell m} \frac{\partial \mathcal{U}_m}{\partial x_k}. \quad (18)$$

For simplicity let us consider the two-dimensional case and let us suppose that Γ is given by the parametric representation $x = \chi(s)$ with $s \in [0, L]$ the arc length.

Theorem 4. *Suppose that $\Gamma \subset \mathbb{R}^2$ is a closed curve given by $x = \chi(s) \in C^1$ and that the Cauchy data ϕ and ψ are known. Then, the stress tensor $\sigma_{j\ell}$ is given on Γ by*

$$\sigma_{k\ell} = n_k \psi_\ell + \dot{\chi}_k \tau_\ell; \quad k, \ell = 1, 2, \tag{19}$$

where τ_ℓ is obtained from ψ and ϕ as

$$\tau_\ell = \left(\mu - \frac{(\lambda + \mu)^2}{4(\lambda + 2\mu)} \right) \dot{\phi}_\ell + (\lambda + \mu) \dot{\chi}_\ell \sum_{j=1}^{2} \dot{\chi}_j \dot{\phi}_j + \frac{\lambda + \mu}{2(\lambda + 2\mu)} \dot{\chi}_\ell \sum_{j=1}^{2} n_j \psi_j$$

$$+ \frac{\lambda + \mu}{2\mu} n_\ell \sum_{j=1}^{2} \dot{\chi}_j \psi_j - \frac{(\lambda + \mu)^3}{4\mu(\lambda + 2\mu)} n_\ell \sum_{j=1}^{2} n_j \dot{\phi}_j .$$

Here $\dot{}$ denotes the derivative $\frac{\partial}{\partial s}$.

Proof. We begin with (18):

$$\sum_{j=1}^{2} \dot{\chi}_j \sigma_{j\ell} = \mu \sum_{j,k,m=1}^{2} \delta_{jk} \delta_{\ell m} \frac{\partial \mathcal{U}_m}{\partial x_k} \dot{\chi}_j + \frac{1}{2}(\lambda + \mu) \sum_{j,k,m=1}^{2} \left(\delta_{j\ell} \delta_{km} \frac{\partial \mathcal{U}_m}{\partial x_k} \dot{\chi}_j + \delta_{jm} \delta_{k\ell} \frac{\partial \mathcal{U}_m}{\partial x_k} \dot{\chi}_j \right)$$

$$= \mu \dot{\phi}_\ell + \frac{1}{2}(\lambda + \mu) \sum_{m=1}^{2} \left(\frac{\partial \mathcal{U}_m}{\partial x_m} \dot{\chi}_\ell + \frac{\partial \mathcal{U}_m}{\partial x_\ell} \dot{\chi}_m \right). \tag{20}$$

By using the relations $\frac{\partial \mathcal{U}_\ell}{\partial x_k} = \dot{\phi}_\ell \dot{\chi}_k + \frac{\partial \mathcal{U}_\ell}{\partial n} n_k$ and writing the boundary conditions in the form[3],

$$R_1 \frac{\partial \mathcal{U}}{\partial n} + R_0 \gamma_0 \mathcal{U} = \psi, \qquad R_0 \gamma_0 \mathcal{U} = R_{00} \dot{\phi} \quad \text{with} \tag{21}$$

$$((R_{00}))_{\ell m} := \frac{1}{2}(\lambda + \mu)(n_\ell \dot{\chi}_m + n_m \dot{\chi}_\ell) \quad \text{and} \quad ((R_1))_{\ell m} := \mu \delta_{\ell m} + (\lambda + \mu)(n_\ell n_m),$$

we obtain from (20) the equation

$$\sum_{j=1}^{2} \dot{\chi}_j \sigma_{j\ell} = \mu \dot{\phi}_\ell + (\lambda + \mu) \dot{\chi}_\ell \sum_{m=1}^{2} \dot{\chi}_m \dot{\phi}_m + \left(R_{00} \frac{\partial \mathcal{U}}{\partial n} \right)_\ell . \tag{22}$$

To complete the proof we have to evaluate $R_{00} \frac{\partial \mathcal{U}}{\partial n}$. From (21) we obtain

$$\left(R_{00} \frac{\partial \mathcal{U}}{\partial n} \right)_\ell = \left(R_{00} R_1^{-1}(\psi - R_{00} \dot{\phi}) \right)_\ell .$$

By using
$$((R_1^{-1}))_{\ell m} = \frac{1}{\mu(2\mu + \lambda)} (\mu \delta_{\ell m} + (\lambda + \mu) \dot\chi_\ell \dot\chi_m)$$
there follows
$$(R_1^{-1}\psi)_j = \frac{1}{\lambda + 2\mu}\psi_j + \frac{\lambda + \mu}{\mu(\lambda + 2\mu)}\dot\chi_j \sum_{m=1}^{2} \dot\chi_m \psi_m$$
and, using in particular the orthogonality $n \cdot \dot\chi = 0$ and
$$\begin{pmatrix} n_1 & n_2 \\ \dot\chi_1 & \dot\chi_2 \end{pmatrix}^{-1} = \begin{pmatrix} n_1 & \dot\chi_1 \\ n_2 & \dot\chi_2 \end{pmatrix}, \tag{23}$$
we get with elementary manipulations
$$(R_{00} R_1^{-1} \psi)_\ell = \frac{\lambda + \mu}{2(\lambda + 2\mu)} \dot\chi_\ell \sum_{j=1}^{2} n_j \psi_j + \frac{\lambda + \mu}{2\mu} n_\ell \sum_{j=1}^{2} \dot\chi_j \psi_j. \tag{24}$$

Similarly, we obtain for $R_1^{-1} R_{00} \dot\phi$ the equation
$$(R_1^{-1} R_{00} \dot\phi)_k = \frac{\lambda + \mu}{2(\lambda + 2\mu)} n_k \sum_{j=1}^{2} \dot\chi_j \dot\phi_j + \frac{\lambda + \mu}{2\mu} \dot\chi_k \sum_{j=1}^{2} n_j \dot\phi_j. \tag{25}$$

Using (25) and employing the orthogonality $n \cdot \dot\chi = 0$ again, there follows
$$(R_{00} R_1^{-1} R_{00} \dot\phi)_\ell = \frac{(\lambda + \mu)^2}{4(\lambda + 2\mu)} \sum_{j=1}^{2} \dot\chi_j \dot\chi_\ell \dot\phi_j + \frac{(\lambda + \mu)^2}{4\mu} \sum_{j=1}^{2} n_j n_\ell \dot\phi_j. \tag{26}$$

Applying $\dot\chi_j \dot\chi_\ell = \delta_{j\ell} - n_j n_\ell$ corresponding to (23), from (26) one obtains
$$(R_{00} R_1^{-1} R_{00} \dot\phi)_\ell = \frac{(\lambda + \mu)^2}{4(\lambda + 2\mu)} \dot\phi_\ell + \frac{(\lambda + \mu)^3}{4\mu(\lambda + 2\mu)} n_\ell \sum_{j=1}^{2} n_j \dot\phi_j. \tag{27}$$

By combining (22) with (24) and (27) we find the system
$$\sum_{j=1}^{2} \dot\chi_j \, \sigma_{j\ell} = \tau_\ell, \quad \sum_{j=1}^{2} n_j \, \sigma_{j\ell} = \psi_\ell,$$
whereas the second equation holds by definition. With (23), this implies (19). □

6. Adaptivity

In finite element analysis, one of the main applications of superconvergence properties is the construction of error estimators and local error indicators. A similar approach is also possible with the recovered solution described in Section 4. Applying (12), we get an approximation for the error of the triangular system (6) by using

$$\eta^{(\gamma)}(u) := \psi_h^{(\gamma)}(u) - \widetilde{\psi}_h^{(\gamma)}(u), \quad \gamma = (\ell, k) \quad |\gamma| \leq M. \tag{28}$$

With Theorem 3, one obtains asymptotic equivalence between error and our estimator:

Theorem 5. *For the Galerkin-Bubnov error of (6) and the error estimator defined by (28) there holds the pointwise estimate*

$$|\eta(u)| - \varepsilon_1(\psi) \leq |\partial^\gamma \psi - \psi_h^{(\gamma)}|(u) \leq |\eta(u)| + \varepsilon_2(\psi), \tag{29}$$

provided $M > 1 - \alpha$ *and* $0 \leq \ell \leq \min\{d - \alpha, M - 1 + \alpha - \varepsilon\}$, $\varepsilon > 0$; $0 \leq m \leq d - \alpha$, $\ell + m \leq 2d - 2\alpha$. *In addition,*

$$||\eta||_{H^s(\Gamma)} - \widetilde{\varepsilon}_1(\psi) \leq ||\partial^\gamma \psi - \psi_h^{(\gamma)}||_{H^s(\Gamma)} \leq ||\eta||_{H^s(\Gamma)} + \widetilde{\varepsilon}_2(\psi), \tag{30}$$

provided $2\alpha - d \leq s - M \leq \alpha \leq \ell \leq d$. *The perturbation terms are of higher order than the error itself and given by*

$$\varepsilon_{1,2}(\psi) = c_{1,2}\, h^{\ell+m} ||\psi||_{H^{\ell+|\gamma|+M+\alpha}(\Gamma)}, \qquad \widetilde{\varepsilon}_{1,2}(\psi) = \widetilde{c}_{1,2}\, h^{\ell-s+M} ||\psi||_{H^{\ell+M+|\gamma|}(\Gamma)}.$$

Proof. The proof is an immediate consequence of Theorem 3. □

7. Piecewise smooth Lipschitz boundaries

For non-smooth domains, the solution of (1) might not be smooth at edges, corners or interfaces [4], whereas the generalization of our method is only applicable to the smooth parts. Hence, special techniques for treating the singularities first with augmented methods or scaled meshes are necessary (see e.g. [5, 9, 14]). For simplicity, let us consider a domain $\Omega \subset \mathbb{R}^2$ with only one corner corresponding to the arc length s_0. Then, the solution of (1) can be represented by a singular and a regular part,

$$\mathcal{U}(x) = \mathcal{U}_{\text{sing}}(r, \theta) + \mathcal{U}_{\text{reg}}(x), \tag{31}$$

where (r, θ) are polar coordinates with center in $\chi(s_0)$. The singular part $\mathcal{U}_{\text{sing}}$ can be represented by powers of r up to some order and logarithmical terms multiplied by the stress intensity factors and trigonometric functions. For its computation see e.g. [2, 5, 6, 7]. Then our technique is applicable to the boundary integral equation for $\mathcal{U}_{\text{reg}}$ coupled with the stress intensity factors via the Maz'ya-Plamenevskii functionals [6, 7] which need to be computed with high accuracy [2]. As a preliminary step in this procedure, we neglect now $\mathcal{U}_{\text{sing}}$ of the solution (31) and assume that this part is equal to zero. But, by definition, even if $\mathcal{U}$ is globally smooth, the co-normal derivative $\psi = T\mathcal{U}$ has a jump at $\chi(s_0)$. If one ignores the jumps, pollution effects occur and the method fails, see Figure 3. Therefore, $T\mathcal{U}$ must be evaluated in the sense of distributions. Since the densities are piecewise smooth, we split ψ in a *jump part* ψ_1 and in a *continuous part* ψ_2,

$$\psi = \psi_1 + \psi_2 \quad \text{with } \psi_1 := [\psi]_{|s_0} \cdot \zeta(s) \quad \text{and } \psi_2 := \psi - \psi_1, \tag{32}$$

where $\zeta(s)$ is a fixed local truncation function with the jump

$$\lim_{\varepsilon \to +0} \zeta(s-\varepsilon) = -\frac{1}{2}, \quad \lim_{\varepsilon \to +0} \zeta(s+\varepsilon) = \frac{1}{2}.$$

Using (32) we have for the tangential derivative $\partial_s \psi$

$$\partial_s \psi = \partial_s \psi_1 + \partial_s \psi_2 = [\psi]_{|s_0} \partial_s \zeta + \partial_s \psi_2. \tag{33}$$

To obtain $\partial_s \psi$, we have to compute the jump $[\psi]_{|s_0}$ and the differentiated continuous part $\partial_s \psi_2$. For their derivation, one has to solve the following coupled system, which is a natural extension of (6),

$$V\psi = V([\psi]_{|s_0}\zeta + \psi_2) = F \tag{34}$$

$$V(\partial_s \psi_2) = \partial_s F - [\partial_s, V]\psi_2 - [\psi]_{|s_0}\partial_s V\zeta \tag{35}$$

$$-[\psi]_{|s_0}\left(\varphi(s_0) + \int_{\Gamma\setminus\{s_0\}} (\partial_s \zeta)(s)\,\varphi(s)\,ds\right) = \int_{\Gamma} ((\partial_s \psi_2)(s)\,\varphi(s) + \psi(s)\,(\partial_s \varphi)(s))ds \tag{36}$$

with some fixed smooth test function φ satisfying $\varphi(s_0) + \int_{\Gamma\setminus\{s_0\}} (\partial_s \zeta)(s) \cdot \varphi(s)\,ds \neq 0$.

The notation $(\partial_s \zeta)(s)$ means the piecewise tangential derivative of $\zeta(s)$ for $s \neq s_0$.

Equation (35) follows immediately from (34) by applying the tangential derivative and using the commutator $[\partial_s, V]$. Equation (36) is obtained by piecewise integration by parts and (33).

Remark 3. Because of the smoothness of ψ_2, for the computation of the commutator term $[\partial_s, V]\psi_2$, the representation of the commutator kernel from [12, Section 3.3] can be used without any modification. The term $\partial_s V\zeta$ is easily computable, e.g., by integration by parts $(\partial_s V\zeta, v)_{L^2(\Gamma)} = -(V\zeta, \partial_s v)_{L^2(\Gamma)}$ and then standard techniques.

Remark 4. For the numerical example given in Section 8, we have used (34, 35) only. For the jumps $[\psi]$ we have substituted the jumps of the Galerkin solution $[\psi_h]$ which, however, causes a loss of convergence order.

Remark 5. For the numerical computation of the singular part $\mathcal{U}_{\text{sing}}$ as presented e.g., in [2] one has to incorporate the corresponding additional equations for the stress intensity factors into the system (34, 35, 36).

Remark 6. For recovery of the smooth part $\mathcal{U}_{\text{reg}}$ on piecewise smooth domains, one has to add jump terms to the recovery functional (10). For the simplest case $M = 1$ on a piecewise smooth curve with one corner at $\chi(s_0)$ we obtain for $s' \neq s_0$

$$\psi(s') = \int_{\Gamma\setminus\{s_0\}} (\partial_s \psi)(s)w(s-s')\,ds + \int_{\Gamma\setminus\{s'\}} \psi(s)\partial_s w(s-s')\,ds + [\psi]_{s_0} \cdot w(s_0 - s') \tag{37}$$

with $w(s - s') := \rho(s - s') \cdot H_1(s - s')$ using the notations of Section 4.

8. Numerical examples

8.1. A smooth domain: the ellipse

We consider the Lamè system with Dirichlet boundary conditions,

$$-\mu\Delta\mathcal{U} - (\lambda+\mu)\operatorname{grad}\operatorname{div}\mathcal{U} = 0 \quad \text{in } \Omega \subset \mathbb{R}^2,$$
$$\mathcal{U}_j(x)|_\Gamma = \phi_j(x) := \mathcal{G}_{1j}(x,y), \quad y = (-0.5, -0.5)^\top,$$

on the elliptical domain $\Omega := \{x \in \mathbb{R}^2 : x_1^2/a^2 + x_2^2/b^2 < 1\}$ $a = 0.6$, $b = 0.3$. The Lamè constants are chosen as

$$\lambda = \frac{E\nu}{(1+\nu)(1-2\nu)}, \quad \mu = \frac{E}{2(1+\nu)}, \quad E = 200\,000 \text{ MPa}, \quad \nu = 0.3,$$

that means we consider a material like iron. Table 1 shows errors and computing times for the computation of ψ_h, $(\partial_s\psi)_h$ and $\widetilde{\psi}_h$. Figure 1 shows the approximations and the errors for the components of the boundary tractions using 32 boundary elements.

N	$\|\psi-\psi_h\|_{L^2}$	$\|\psi-\psi_h\|_{L^\infty}$	cpu (a)	$\|\psi'-\psi'_h\|_{L^2}$	$\|\psi'-\psi'_h\|_{L^\infty}$	$\|\psi-\widetilde{\psi}_h\|_{L^2}$	$\|\psi-\widetilde{\psi}_h\|_{L^\infty}$	cpu (b)
16	0.2246	0.2983	0.16	0.5674	0.8606	0.1258	0.1974	0.33
32	0.1145	0.1727	0.51	0.2713	0.5414	0.0252	0.0373	1.12
64	0.0577	0.0970	1.81	0.1388	0.3065	0.0053	0.0103	4.01
128	0.0289	0.0485	7.25	0.0695	0.1561	1.24e-3	0.0024	15.94
256	0.0144	0.0242	28.94	0.0347	0.0781	2.98e-4	5.92e-4	63.45
512	0.0072	0.0121	115.80	0.0173	0.0390	7.32e-5	1.44e-4	252.58
1024	0.0036	0.0060	466.36	0.0086	0.0195	1.81e-5	3.57e-5	1016.27
τ	≈ 1	$0.8 - 1$		≈ 1	$0.7 - 1$	≈ 2	≈ 2	

cpu (a): CPU time in seconds for the computation of ψ_h.
cpu (b): CPU time in seconds for the whole algorithm, including the evaluation of the recovered solution in $10 \cdot N$ points for the computation of the errors.

Table 1: Errors, computing time and convergence rates

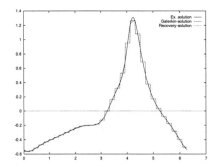
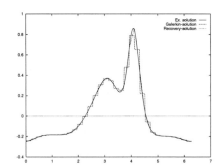

Figure 1: First and second component of the boundary tractions: Exact solution, Galerkin solution and recovered solution ($N = 32$)

8.2. A non-smooth domain: the square

Here, we consider the Laplacian with Dirichlet boundary conditions,

$$-\Delta \mathcal{U} = 0 \quad \text{in } \Omega, \quad \mathcal{U}(x)|_\Gamma = \phi(x) := \log|x - y|, \quad y = (-1, 1)^\top,$$

on the square domain $\Omega := \{x \in \mathbb{R}^2 : 0 \leq x_1 \leq 1/2, 0 \leq x_2 \leq 1/2\}$. Figure 2 shows the results for the recovery algorithm when using 32 degrees of freedom. Figure 3 shows the computed tangential derivative with and without splitting into the jump part and the continuous part.

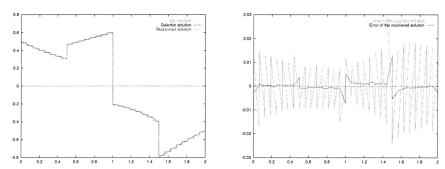

Figure 2: Exact solution, Galerkin solution and recovered solution:
Galerkin error and recovery error ($N = 32$)

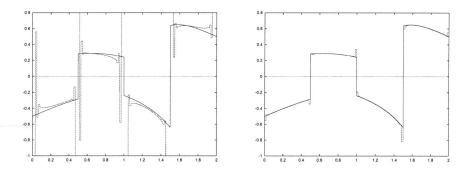

Figure 3: Computed tangential derivative without and with splitting
into the jump part and the continuous part

References

[1] D. N. Arnold, W. L. Wendland: On the asymptotic convergence of collocation methods. *Math. Comp.* **41** (1983) 349-381.

[2] H. Blum, M. Dobrowolski: On the finite element method for elliptic equations on domains with corners. *Computing* **28** (1982) 53-63.

[3] G.C. Hsiao, W.L. Wendland: *Variational Methods for Boundary Integral Equations*. In preparation, Springer-Verlag Berlin.

[4] L. Jentsch: On the elastic potentials at corners. *Asymptotic Analysis* **14** (1997) 73–95.

[5] V.A. Kozlov, V.G. Maz'ya, J. Rossmann: *Elliptic Boundary Value Problems in Domains with Point Singularities*. American Mathematical Society, Providence, RI, 1997.

[6] V.G. Maz'ya, B.A. Plamenevskii: Coefficients in the asymptotics of the solution of elliptic boundary value problems. *Sov. Math.* **9** (1978) 750–764.

[7] V.G. Maz'ya, S.A. Nazarov, B.A. Plamenevskij: *Asymptotische Theorie elliptischer Randwertaufgaben in singulär gestörten Gebieten, Teil I und II*. Akademie Verlag, Berlin, 1991.

[8] S. Prössdorf and B. Silbermann: *Numerical Analysis for Integral and Related Operator Equations*. Birkhäuser–Verlag, Basel, 1991.

[9] A.-M. Sändig, W.L. Wendland: Asymptotic expansions of elastic fields in domains with boundary and structural singularities. In: *Boundary Element Topics* (W.L. Wendland ed.), Springer-Verlag Berlin 1997, 419-444.

[10] H. Schulz, Ch. Schwab, W.L. Wendland: The computation of potentials near and on the boundary by an extraction technique for boundary element methods. *Comput. Methods Appl. Mech. Engrg.* **157** (1998) 225-238.

[11] H. Schulz, Ch. Schwab, W.L. Wendland: On the computation of derivatives up to the boundary and recovery techniques in BEM. *Proceedings of the IUTAM Conference 1997 "Discretization Methods in Structural Mechanics II"*, Kluwer Academic Publishers, to appear.

[12] Ch. Schwab, W.L. Wendland: On the extraction technique in boundary integral equations. *Math. Comp.* **68** (1999) 91-122.

[13] L. Schwartz: *Théorie des Distributions*. Hermann, Paris 1966 (3rd. ed.).

[14] E.P. Stephan, W.L. Wendland: An augmented Galerkin procedure for the boundary integral method applied to two-dimensional screen and crack problems. *Appl. Anal.* **18** (1985), 183-219.

[15] W.L. Wendland: On boundary integral equations and applications. In: *Tricomi's Ideas and Contemporary Applied Mathematics*. Accad. Naz. dei Lincei, Rom 1998.

H. Schulz, Institute of Mathematics A, University of Stuttgart,
D-70569 Stuttgart, Germany, Email:schulz@mathematik.uni-stuttgart.de.

Ch. Schwab, Seminar für Angewandte Mathematik, ETH Zürich,
CH-8092 Zürich, Switzerland, Email:schwab@sam.math.ethz.ch.

W.L. Wendland, Institute of Mathematics A, University of Stuttgart,
D-70569 Stuttgart, Germany, Email:wendland@mathematik.uni-stuttgart.de.

A. SELLIER

Asymptotic Solution of Boundary Integral Equations

1. Introduction

As it is well-known a wide class of linear problems arising in different fields such as Elasticity, Potential or Stokes flows, Electrostatics may be solved by inverting a well-posed boundary integral equation (BIE) on the boundary $\partial \mathcal{A}'$ of an open, simply connected and bounded subset $\mathcal{A}'$ of $\mathbb{R}^3$. More precisely, the BIE is often a Fredholm integral equation of the first (set $\lambda := 0$) or the second (set $\lambda := cste \neq 0$) kind and reads

$$\lambda q(M) + pv \int_{\partial \mathcal{A}'} q(P) K(M,P) dS'_P = d(M), \text{ for } M \in \partial \mathcal{A}'. \tag{1}$$

Here, $\partial \mathcal{A}'$ is a sufficiently smooth boundary whilst smooth functions or vectors q and d respectively denote the unknown density and a quantity given in a neighborhood of $\mathcal{A}'$. More precisely, for $\lambda := 0$ the kernel $K(M,P)$ is regular and the symbol pv is omitted. For $\lambda \neq 0$, the kernel is weakly singular and writes (see [11]) $K(M,P) = k(M, \Theta)/PM^2$ with $\Theta := \mathbf{MP}/PM$ and $\int_0^{2\pi} k(M, \Theta) d\Theta = 0$. In such circumstances, the symbol pv indicates the Cauchy principal value, i.e.

$$pv \int_{\partial \mathcal{A}'} q(P) K(M,P) dS'_P := \lim_{\mu \to 0, \mu > 0} \int_{\partial \mathcal{A}' \setminus D_\mu(M)} q(P) K(M,P) dS'_P \tag{2}$$

where the removed neighborhood $D_\mu(M)$ is defined as $D_\mu(M) = \{P \in \partial \mathcal{A}'; MP < \mu\}$.

Possibly except for specific body shapes $\partial \mathcal{A}'$ and data g, only a numerical treatment of (1) is thinkable. However, as soon as the body is thin or slender it remains of prime interest to gain an asymptotic solution. This is mainly motivated by the need of tractable asymptotic models and the numerical solution of (1) unfortunately fails in predicting these models. This work focuses on the case of a slender body $\partial \mathcal{A}'$ (see the Figure in Section 2): the slenderness ratio ϵ comparing the typical length $O'E'$ to the main radius is very small. Under this assumption, we asymptotically expand and invert the governing BIE (1) with respect to ϵ. This use of (1) has been disregarded by all the previous asymptotic studies in the field. Actually, two other approaches were employed: the well-known method of matched asymptotic expansions ([3, 4]) which unfortunately imposes tremendous matching rules at high orders of approximation and, only in case of a slender body of revolution, the use of an alternative integral equation bearing on singularities to be put on the body axis. This latter method, which circumvents the tedious matching procedure, has been actually pioneered by [12] and extensively applied by other authors (see successively [14, 8, 9, 19, 5, 6, 10, 1]). However, as clearly highlighted in [2], such an alternative integral equation may be sometimes ill-posed for the electrostatic field around a slender body and this bad feature also casts serious doubts on the legitimacy of this approach in other fields.

The author gratefully thanks Prof. W. L. Wendland for fruitful discussions during the IABEM 98 Conference.

The aim of this paper is to detail our asymptotic treatment of the well-posed equation (1). Such a task actually requires to expand, with respect to the small parameter ϵ and up to high orders of approximation if necessary, a large class of one-dimensional integrals. This key step makes use of the fruitful concept of Hadamard's finite part integrals and rests on a systematic formula which has been established in a previous work. However, this really technical point is the only mathematical price to pay when starting from the well-posed formulation (1). One is thereafter lead to a pyramidal set of two-dimensional integral equations and obtains a solution free from the drawbacks encountered by the other approaches.

The paper is organized as follows. In Section 2 we clarify our assumptions and outline the general steps to achieve in asymptotically solving (1). Section 3 is entirely devoted to the concept of integration in the finite-part sense of Hadamard. It also reports the general formula of interest for this paper. The last Section details an application to the case of a slender dielectric body embedded in an arbitrary electrostatic potential. Finally, several concluding remarks close the paper.

2. A general procedure

This Section presents our notations and assumptions. It also details and discusses the different steps to achieve when dealing with (1) for a slender body $\partial \mathcal{A}'$.

2.1. Body shape assumptions

For convenience the set (O', x', y', z') of cartesian co-ordinates and the set of cylindrical coordinates (r', θ, z') are introduced. The body $\mathcal{A}'$ (see the Figure below), whose length is L, admits rounded ends O' and E' with $\underline{e}_z = \overline{O'E'}/L$.

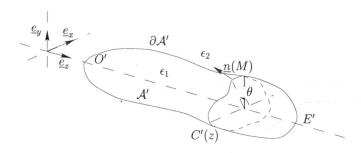

Figure 1: A slender body with an indication of our notations.

If $e := \text{Max}(r')$ for $M \in \partial \mathcal{A}'$ then $\epsilon = e/L \ll 1$ and $\mathcal{A}'$ is "straight" in the sense it collapses to the segment $O'E'$ as ϵ goes to zero. If the new non-dimensional coordinates (x, y, r, z) obey $x' = ex, y' = ey, r' = er$ and $z' = Lz$, the boundary $\partial \mathcal{A}$ is described by a smooth and positive body shape function f such that $M(x', y', z') \in \partial \mathcal{A}'$ also writes $M(ef(\theta, z), \theta, Lz)$. The function f satisfies $f(\theta, 0) = f(\theta, 1) = 0$ and $f(\theta, z) = f(\theta + 2\pi, z)$. Moreover, we impose the function $f^2(\theta, .)$ to be analytic in $[0, 1]$ with the following behaviors

$$f^2(\theta, z) = \sum_{n\geq 1} c_n(\theta) z^n; \quad 2[ff_z^1](\theta, z) = \sum_{n\geq 1} nc_n(\theta) z^{n-1}, \text{ as } z \to 0^+, \quad (3)$$

$$f^2(\theta, z) = \sum_{n\geq 1} b_n(\theta)(1-z)^n; \quad 2[ff_z^1](\theta, z) = -\sum_{n\geq 1} nb_n(\theta)(1-z)^{n-1}, \text{ as } z \to 1^-, \quad (4)$$

where $\partial_v^i f := \partial^i f/\partial v^i$ for $i \in \mathbb{N}, v \in \{\theta, z\}$ and $0 < c_1(\theta) = O(1), 0 < b_1(\theta) = O(1)$.

2.2. The proposed approach

Taking into account the previous assumptions the following steps are achieved:

(i) **Change of variables**: This task consists in rewriting (1) in terms of the non-dimensional variables in order to exhibit the small parameter ϵ. If the given quantity $d(M)$ easily takes the form $d[\epsilon f(\theta, z), \theta, z]$ for $M(x', y', z') \in \partial \mathcal{A}'$, the integral arising on the left-hand side of (1) requires a special attention. Actually, in most cases there exists a real number R such that the kernel reads

$$K(M, P) = \sum_{l,m,n} a_{lmn}(M)[x'_P - x']^l [y'_P - y]^m [z'_P - z']^n / PM^{(R+l+m+n)} \quad (5)$$

with $(l, m, n) \in \mathbb{N}^3, R = 2$ for a weakly singular kernel and $R < 2$ for a regular one (see the Section 4). Using the next relation and definitions

$$s_\epsilon := \left\{1 + (f^{-1} f_\theta^1)^2 + (\epsilon f_z^1)^2\right\}^{1/2}; \quad dS'_P = eL[fs_\epsilon](\theta_P, z_P) d\theta_P dz_P, \quad (6)$$

$$H(\theta_P, z_P, \theta, z) := \left\{f^2(\theta_P, z_P) + f^2(\theta, z) - 2\cos(\theta_P - \theta) f(\theta, z) f(\theta_P, z_P)\right\}^{1/2}, (7)$$

one thereafter obtains, via the change of variables $(x', y', z') \to (r, \theta, z)$,

$$\epsilon^{R-1} pv \int_{\partial \mathcal{A}'} q(P) K(M, P) dS'_P = \sum_{l,m,n} a_{lmn}(M) \epsilon^{R+l+m-1}$$

$$\times pv \int_0^{2\pi} \left[\int_0^1 \frac{[x_P - x]^l [y_P - y]^m [z_P - z]^n [efs_\epsilon q](\theta_P, z_P) dz_P}{[(z_P - z)^2 + \epsilon^2 H^2(\theta_P, z_P, \theta, z)]^{(R+l+m+n)/2}}\right] d\theta_P \quad (8)$$

where $x_P(\theta_P, z_P) = f(\theta_P, z_P) \cos \theta_P; y_P(\theta_P, z_P) = f(\theta_P, z_P) \sin \theta_P$ and the symbol pv on the right-hand side of (8) only concerns the one-dimensional integral bearing on the variable θ_P and is defined right after definition 2. Since $H > 0$ as soon as $\theta_P \neq \theta$, observe that each integral on the domain $[0, 1]$ is indeed regular on $[0, 2\pi] \setminus \{\theta\}$. For a regular kernel (8) is justified and the symbols pv vanish whilst the proof of (8) is not so trivial for a weakly singular kernel. This proof actually employs the special properties of weakly singular and multi-dimensional Hadamard's finite part integrals (the reader is referred to the material developed in [17]).

(ii) **Asymptotic approximation of** (1): It is easy to expand the term $d(M)$. The difficult point comes from the asymptotic expansion of the integrals arising on the right-hand side of (8). By virtue of the shift $z_P = z + u$, each integration on the variable z_P looks like the following integral

$$I_h^\epsilon[w] = \int_{-z}^{1-z} \frac{w(u)du}{[u^2 + \epsilon^2 h^2(u)]^\gamma} \tag{9}$$

with $\epsilon > 0, \gamma \in \mathbb{R}$ and smooth enough functions w and h such that $h(0) > 0$. Under such assumptions the integral $I_h^\epsilon[w]$ is indeed regular for $\epsilon > 0$. If $\epsilon = 0$, note that $I_h^0[w]$ may become hypersingular (choose for instance $w(0) \neq 0$ and $\gamma > 1/2$). This feature suggests troubles in establishing the asymptotic behavior of $I_h^\epsilon[w]$ as ϵ goes to zero. The usual procedures indeed fail and the method of matched asymptotic expansions yields tremendous matching rules at high orders of approximation. The theorem 1 stated in Section 3 brings the answer at a reasonable mathematical cost and for any order of approximation.

(iii) **Asymptotic solution**: We denote by $C(z)$ the non-dimensional cross-section defined as $C(z) := \{N(x, y, z) \text{ such that } M(ex, ey, Lz) \in \partial \mathcal{A}'\}$. The associated dimensional cross-section is $C'(z)$ (see the Figure). As the theorem 1 shows, the asymptotic behavior of $I_h^\epsilon[w]$ actually involves three kind of terms: integrals of functions w and h over the whole set $]-z, 1-z[$, local terms depending on the derivatives of w at 0 and local terms only depending on w at 0. By integrating over θ_P in (8) one thereafter respectively gets fully three-dimensional (the unknown density q is involved over the whole boundary $\partial \mathcal{A}'$), weakly three-dimensional (the unknown density q is involved over cross-sections close to $C'(z)$) and "two-dimensional" (the density q is taken on $C'(z)$ only) terms. Note that the leading term arising in equality (8) is not necessarily a "two-dimensional" one.

In the cross-section $C(z)$ the line element dl_P reads $dl_P = [fs_0(\theta_P, z_P)]d\theta_P$ and it is worth introducing the lineic density t by $efs_\epsilon q = fs_0 t$. One seeks the function t in the form $t = \sum_{i \geq 0} \mu_i(\epsilon) t_i$ with $\mu_{i+1}(\epsilon) = o(\mu_i(\epsilon))$. Even if the leading term in (8) is weakly or strongly three-dimensional it is always possible to deduce for the unknown functions t_i a pyramidal set of two-dimensional boundary integral equations in the cross-section $C(z)$. Contrary to the BIE (1), these integral equations are not always the boundary counterparts of really two-dimensional boundary value problems (this is due to exotic behaviors far from the cross-section $C(z)$) but are much more pleasant to solve.

3. The key mathematical tool: Hadamard's finite part integrals

This section draws the reader's attention to the powerful Hadamard's finite-part concept and its basic applications to strongly singular integrals and asymptotic approximation of such integrals depending upon a small parameter.

3.1. Hadamard's finite part concept and integrals

Definition 1. *Consider a complex function f such that there exist $a > 0, N \in \mathbb{N}$, a family of positive integers $(M(n))$, two complex families $(\gamma_n), (f_{nm})$ and a complex function F such that:* $\operatorname{Re}(\gamma_N) < ... < \operatorname{Re}(\gamma_0) := 0$, $f_{00} := 0$ *if* $\gamma_0 = 0$, $\lim_{\mu \to 0, \mu > 0} F(\mu)$ *exists and* $f(\mu) = \sum_{n=0}^{N} \sum_{m=0}^{M(n)} f_{nm} \mu^{\gamma_n} \log^m \mu + F(\mu)$ *for* $\mu \in]0, a[$. *Then* $fp[f(\mu)] := \lim_{\mu \to 0, \mu > 0} F(\mu)$ *is called the Hadamard's finite part of the quantity* $f(\mu)$.

This definition authorizes us to define Hadamard's partie finie integrals. As an example of one-dimensional and hypersingular integrals, one gets (see also [7, 13, 15]):

Definition 2. *If* $]x_0 - \eta, x_0 + \eta[\subset \mathcal{D} \subset \mathbb{R}$ *with* $\eta > 0$ *and* $g \in L^1_{loc}(\mathcal{D} \setminus \{x_0\}, \mathbb{C})$ *satisfies the decompositions* $g[x_0 + (-1)^i u] = \sum_n^N \sum_m^{M(n)} g^i_{nm} u^{\alpha^i_n} \log^m u + G_i(u)$ *for* $i \in \{1, 2\}$ *and* $0 < u < \eta$ *with* $\operatorname{Re}(\alpha^i_N) < ... < \operatorname{Re}(\alpha^i_0) := -1$ *and* $G_i \in L^1_{loc}(]0, \eta[, \mathbb{C})$, *then*

$$fp \int_{\mathcal{D}} g(x)dx := fp \left[\int_{\mathcal{D} \setminus]x_0 - \mu, x_0 + \mu[} g(x)dx \right]. \tag{10}$$

Thus the function g can present different and singular behaviors near x_0 on the left ($i = 1$) or on the right ($i = 2$). If g is singular at x_0 but $\lim_{\mu \to 0, \mu > 0} \int_{x_0 - \mu}^{x_0 + \mu} g(x)dx$ exists then one notes, as for (8), $pv \int_{\mathcal{D}} g(x)dx := fp \int_{\mathcal{D}} g(x)dx$. A widely encountered case is associated to $\alpha^i_n = -\delta_{n0}$ and $g^i_{nm} = (-1)^i \delta_{n0} \delta_{m0}$ with δ_{ij} denoting throughout the paper the usual Kronecker delta. Note that a change of variables may generate extra terms for the above integration (see [16] for a change of scale).

3.2. A systematic formula

The previous concept of integration in the finite-part sense of Hadamard makes it possible to derive the asymptotic expansion of hypersingular integrals with respect to a large or a small parameter. For details the reader is referred to [16]. Here we only state a theorem for integrals involving a Q pseudo-homogeneous kernel $L(u, v)$ such that $L(tx, ty) = S(t)t^Q L(x, y)$ for $t \neq 0$ with Q a positive or negative integer and two possible circumstances: $S(t) = \operatorname{sgn}(t)$ for $t \in \mathbb{R}^*$ or $S(t) = 1$ for $t \in \mathbb{R}^*$.

Theorem 1. *For* $-\infty < a < 0 < b < \infty$, *smooth enough* Q *pseudo-homogeneous kernel* L *and functions* g, h *with* $h(0) > 0$ *then the following asymptotic estimate holds, for* $N \in \mathbb{N}$ *and strictly positive* ϵ *going to zero,*

$$fp \int_a^b g(u) L[u, \epsilon h(u)] du = \sum_{n=0}^N \frac{\partial_2^n L(1,0)}{n!} \left[fp \int_a^b \frac{S(u) g(u) [h(u)]^n du}{u^{n-Q}} \right] \epsilon^n$$

$$+ \sum_{m=0}^{N-Q-1} \sum_{l=0}^m \sum_{i=0}^{m-l} \frac{g^{(l)}(0) c^i_{m-l-i}}{l! i!} \left[fp \int_{-\infty}^{\infty} \partial_2^i L[t, h(0)] t^m dt \right] \epsilon^{Q+m+1} \tag{11}$$

$$-[1 - S(-1)] \sum_{n=0}^N \sum_{l=0}^n \sum_{i=0}^{n-l} \sum_{j=0}^{l-Q-1} \frac{[h(0)]^l g^{(j)}(0) c^i_{l-Q-j-1}}{l! i! j!} \partial_2^n L(1,0) \epsilon^n \log \epsilon + O(\epsilon^{N+1} \log \epsilon),$$

where $\partial_2^n L(x, y) := [\partial^n L / \partial y^n](x, y)$; $g^{(n)}$ denotes the derivative of order n and, for $(i, p) \in \mathbb{N} \times \mathbb{N}$, the coefficients c^i_p only depend on the behavior of function h near zero in the following fashion: $c^0_p := \delta_{p0}$ and if $i \geq 1$

$$\left[\frac{h(u) - h(0)}{u} \right]^i = \sum_p c^i_p u^p, \quad \text{as } u \to 0. \tag{12}$$

This key theorem is proved in [16]. As previously announced, three sums arise on the right-hand side of formula (11). In the terminology of matched asymptotic expansions (see [20]) the first one comes from the outer domain ($|u| \gg \epsilon$) and involves the whole set of integration whereas the second one comes from the inner domain ($|u| = O(\epsilon)$) and only requires the functions g and h near the critical point $u = 0$. Both domains contribute to the last sum which vanishes as soon as $S(t) \neq \text{sgn}(t)$. However, no matching procedure is needed in using (11) which applies to hypersingular integrals and provides, in terms of the sequence ($\epsilon^m \log \epsilon$) and without too much effort, an asymptotic expansion up to any order of approximation. The only task consists in carefully calculating the occurring integrals on the right-hand side of (11). Those integrals are to be understood in the finite-part sense of Hadamard even if the integral to expand is regular.

4. The example of a slender dielectric body

Here we briefly specify the proposed procedure to the case of a slender dielectric body embedded in an arbitrary electrostatic potential ϕ_0. For additional details the reader is directed to [18].

4.1. The governing BIE

Both media $\mathcal{A}'$ and $\mathbb{R}^3 \setminus (\mathcal{A}' \cup \partial \mathcal{A}')$ are linear, homogeneous and isotropic dielectrics with ϵ_1 and ϵ_2 respectively denoting the associated dielectric constants (see the Figure). As an external electrostatic potential ϕ_0 is imposed a polarization surface-charge density q takes place on $\partial \mathcal{A}'$ and the total potential ϕ_t writes, near $\mathcal{A}'$,

$$\phi_t(M) = \phi_0(M) + \int_{\partial \mathcal{A}'} \frac{q(P) dS'_P}{4\pi \epsilon_0 PM} \tag{13}$$

if ϵ_0 designates the free space permittivity. The unknown density is determined by imposing the continuity of the displacement field across $\partial \mathcal{A}'$, namely

$$\epsilon_1 \underline{E}_1(M).\underline{n}(M) = \epsilon_2 \underline{E}_2(M).\underline{n}(M); \quad \text{for } M \in \partial \mathcal{A}', \tag{14}$$

where $\underline{E}_1(M)$ and $\underline{E}_2(M)$ respectively designate the limit value of the electrostatic field $\underline{E}(P) = -[\underline{\text{grad}}\phi_t](P)$ as P approaches the point M of $\partial \mathcal{A}'$ from the internal or the external side of the body surface and $\underline{n}(M)$ denotes (see the Figure once more) the outer unit normal. If $\delta := \epsilon_2/\epsilon_1$, this results in the following BIE:

$$\frac{\delta + 1}{\delta - 1} \frac{q(M)}{2\epsilon_0} + \int_{\partial \mathcal{A}'} \frac{\underline{PM}.\underline{n}(M)q(P)dS'_P}{4\pi \epsilon_0 PM^3} = [\underline{n}.\underline{\text{grad}}\phi_0](M); \quad M \in \partial \mathcal{A}'. \tag{15}$$

We consider the case $\delta > 0$. The case of a potential flow around the body $\mathcal{A}'$ is recovered by choosing $\delta \to \infty$. Note that (15) is a Fredholm integral equation of the second kind whose regular kernel fulfills (5) with $R = 2$.

4.2. Asymptotic expansions

We first expand the right-hand side of (15). According to [6], the imposed potential ϕ_0 actually admits near $\mathcal{A}'$ the general representation

$$\phi_0(M) = A_0(\epsilon^2 r^2, z) + \sum_{n=1}^{\infty} \epsilon^n r^n \{A_n(\epsilon^2 r^2, z) \cos n\theta + B_n(\epsilon^2 r^2, z) \sin n\theta\} \quad (16)$$

with functions $A_n(u, z)$ or $B_n(u, z)$ regular near $u = 0$ for $0 \leq z \leq 1$. By superposition this authorizes us to restrict the study to the basic case

$$\phi_0(M) = \phi_0^n(r, \theta, z) := \epsilon^n r^n \psi(\epsilon^2 r^2, z) \cos n\theta, \quad n \geq 0 \quad (17)$$

where the smooth function $\psi(u, z)$ admits partial derivatives $\partial_u^i \partial_z^j \psi(u, z) = O(1)$ for u near zero and $z \in [0, 1]$. If $M \in \partial \mathcal{A}'$ we also introduce $\underline{n}^{2d}(M)$ as the outer unit vector which is normal to the closed path $C(z)$ in the plane $z_P = z$. Observe that

$$\underline{n}(M) = \frac{\underline{e}_r - f^{-1} f_\theta^1 \underline{e}_\theta - \epsilon f_z^1 \underline{e}_z}{s_\epsilon(\theta, z)}; \quad \underline{n}^{2d}(M) = \frac{\underline{e}_r - f^{-1} f_\theta^1 \underline{e}_\theta}{s_0(\theta, z)}. \quad (18)$$

Thus and contrary to the case of $\underline{n}^{2d}(M)$, the vector $\underline{n}(M)$ depends on ϵ. However, the product $fs_\epsilon \underline{n}(M)$ admits a uniform asymptotic expansion with respect to ϵ (remind assumptions (3)-(4)). Consequently, we adopt the relation (17) and expand $[ef s_\epsilon \underline{n}.\mathrm{grad}\phi_0^n](M)$ for $M(\theta, z) \in \partial \mathcal{A}'$. Invoking a Taylor expansion near $u = 0$, one immediately gets

$$[ef s_\epsilon \underline{n}.\mathrm{grad}\phi_0^n](M) = \epsilon^n \{a_0^n(M) + a_1^n(M)\epsilon^2 + [a_2^n(M) + \delta_{n0} O(\epsilon^2)]\epsilon^4\} \quad (19)$$

with $a_i^n(M) = O(1)$ for $i \in \{0, 1, 2\}$ and the useful relations:

$$a_0^n(M) = n\psi(0, z)[f^n \cos n\theta + f^{n-1} f_\theta^1 \sin n\theta] = \psi(0, z)\partial_\theta^1[f^n \sin n\theta], \quad (20)$$
$$(n+2)a_1^n(M) = n\partial_u^1 \psi(0, z)\partial_\theta^1[f^{n+2} \sin n\theta] - \partial_z^1[\partial_z^1 \psi(0, z) f^{n+2}] \cos n\theta, \quad (21)$$
$$4a_2^0(M) = -\partial_z^1[\partial_u^1 \partial_z^1 \psi(0, z) f^4]. \quad (22)$$

By virtue of (8) one immediately obtains

$$[ef s_\epsilon](M) \int_{\partial \mathcal{A}'} \frac{\underline{PM}.\underline{n}(M)q(P)dS'_P}{PM^3} = \sum_{i=1}^{2} \epsilon^2 \int_0^{2\pi} A_{\theta_P, \epsilon}^{\theta, z}[h_i(\theta_P, z_P, \theta, z) ef s_\epsilon q]d\theta_P \quad (23)$$

with the following definitions

$$\Delta(P, M) = \Delta(\theta_P, z_P, \theta, z) = \{[x - x(\theta_P, z_P)]\underline{e}_x + [y - y(\theta_P, z_P)]\underline{e}_y\}.\underline{n}^{2d}(M), \quad (24)$$
$$h_1(\theta_P, z_P, \theta, z) = \Delta(\theta_P, z_P, \theta, z); \quad h_2(\theta_P, z_P, \theta, z) = (z_P - z)[f f_z^1](\theta, z) \quad (25)$$

and

$$A_{\theta_P, \epsilon}^{\theta, z}[v] = \int_0^1 \frac{v(\theta_P, z_P)dz_P}{[(z_P - z)^2 + \epsilon^2 H^2(\theta_P, z_P, \theta, z)]^{3/2}} = \int_{-z}^{1-z} \frac{w(u)du}{[u^2 + \epsilon^2 h^2(u)]^{3/2}} \quad (26)$$

Thus, one employs theorem 1 with $L(x, y) = [x^2 + y^2]^{-3/2}, Q = -3$ and $S(t) = \text{sgn}(t)$ to expand the left-hand side of (15). Keeping the link $efs_\epsilon q = f s_0 t$, one obtains

$$[efs_\epsilon \underline{n}.\text{grad}\phi_0^n](M) = [fs_0](M)\left\{\frac{\delta+1}{\delta-1}[\frac{t}{2\epsilon_0}](M) + I_0^{\theta,z}[\Delta f s_0 t]4\pi\epsilon_0\right\}$$
$$+ \frac{\epsilon^2 \log \epsilon}{4\pi\epsilon_0}\left\{[fs_0](M)I_1^z[\Delta f s_0 t] + [ff_z^1](M)I_1^z[(z_P-z)fs_0 t]\right\} \quad (27)$$
$$+ \frac{\epsilon^2}{4\pi\epsilon_0}\left\{[fs_0](M)I_2^{\theta,z}[\Delta f s_0 t] + [ff_z^1](M)I_2^{\theta,z}[(z_P-z)fs_0 t]\right\} + O(fs_0 t \epsilon^4 \log \epsilon)$$

where the operators $I_0^{\theta,z}, I_1^z$ and $I_2^{\theta,z}$ obey

$$I_0^{\theta,z}[v] = \int_0^{2\pi} \frac{2v(\theta_P, z)d\theta_P}{H^2(\theta_P, z, \theta, z)}; I^z[v] = \int_0^{2\pi} v(\theta_P, z)d\theta_P; I_1^z[v] = -\frac{d^2}{dt^2}\left[I^t[v]\right]_{t=z}, \quad (28)$$

$$I_2^{\theta,z}[v] = fp \int_0^1 \frac{I^t[v]dt}{|t-z|^3} + (\log 2 - 1)\frac{d^2}{dt^2}\left[I^t[v]\right]_{t=z}$$
$$- \int_0^{2\pi} \frac{d^2}{dt^2}\left[\log\frac{H(\theta_P, z_P, \theta, z)}{2} v(\theta_P, z)\right]_{t=z} d\theta_P. \quad (29)$$

The combination of results (19) and (27) provides the required asymptotic expansion of BIE (15).

4.3. Discussion and asymptotic solution

In view of (27)-(28) it is worth considering, for $M(\theta, z) \in \partial \mathcal{A}' \setminus \{O', E'\}$, the boundary integral equation

$$\mathcal{L}_{0,\delta}^{\theta,z}[u] = \frac{\delta+1}{\delta-1}\frac{u(M)}{2\epsilon_0} + \frac{1}{2\pi\epsilon_0}\oint_{C(z)}\frac{PM.\underline{n}^{2d}(M)u(P)dl_P}{PM^2} = a(\theta, z), \quad (30)$$

for a given function a and an unknown solution u defined on the closed path $C(z)$ in the non-dimensional $x-y$ plane. Any possible solution u of (30) is readily found to satisfy

$$\delta K^z[u] := \delta \oint_{C(z)} u(P)dl_P = [\delta - 1]\oint_{C(z)} a(P)dl_P. \quad (31)$$

Thus, for $\delta > 0$ and except if $\delta = 1$ there is no reason for the total "charge" spread on $C(z)$, i. e. $K^z[u]$, to be zero as it would be the case for the polarization charge pertaining to a real two-dimensional "dielectric" boundary value problem. If $\delta = 0$ the integral equation (30) may also admit a solution only if the compatibility condition $K^z[a] = 0$ holds and the associated homogeneous problem $\mathcal{L}_{0,0}^{\theta,z}[u] = 0$ is known (see [21]) to have as solutions any constant function u. Moreover, the link

$$\mathcal{L}_{0,\delta}^{\theta,z}[u] = \mathcal{L}_{0,0}^{\theta,z}[u] + \frac{\delta}{\delta-1}\frac{u(M)}{\epsilon_0} \quad (32)$$

also shows that, as δ goes to zero, the leading term on the right-hand side of (27) becomes $[fs_0](M)\mathcal{L}_{0,0}^{\theta,z}[t]$. These features prevent us from building, for small values of δ, an asymptotic approximation of t by resorting to the expansion (27). Here we restrict our attention to the cases $\delta - 1 \geq O(1)$ or $\delta = O(1)$ and $0 < 1 - \delta = O(1)$. In such circumstances, the asymptotic solution of (27) is sought in the following form

$$fs_0 t = \epsilon^{n+2\delta n_0}\{fs_0 t_0 + \epsilon^2 \log \epsilon [fs_0 t_1] + \epsilon^2 [fs_0 t_2] + O(t_1 \epsilon^4 \log^2 \epsilon)\}. \qquad (33)$$

Accordingly, under the links $4\pi\epsilon_0 \mathcal{L}_1^z[u] = I_1^z[fs_0 u]$ and $4\pi\epsilon_0 \mathcal{L}_2^{\theta,z}[u] = I_2^{\theta,z}[fs_0 u]$, one deduces the pyramidal set of boundary integral equations

$$[fs_0](M)\mathcal{L}_{0,\delta}^{\theta,z}[t_0] = a_{0+\delta n_0}^n(M), \qquad (34)$$

$$[fs_0](M)\mathcal{L}_{0,\delta}^{\theta,z}[t_1] = -[fs_0](M)\mathcal{L}_1^z[\Delta t_0] - [ff_z^1](M)\mathcal{L}_1^z[(z_P - z)t_0], \qquad (35)$$

$$[fs_0](M)\mathcal{L}_{0,\delta}^{\theta,z}[t_2] = a_{1+\delta n_0}^n(M)$$
$$- [fs_0](M)\mathcal{L}_2^{\theta,z}[\Delta t_0] - [ff_z^1](M)\mathcal{L}_2^{\theta,z}[(z_P - z)t_0] \qquad (36)$$

to be solved from the top to the bottom. The validity of these equations has been checked in [18] by comparing the proposed asymptotic solution (33) to the asymptotic behavior of the available and exact solution for a slender ellipsoid embedded in a linear potential ϕ_∞. Other values of δ are also addressed in this latter paper.

5. Concluding remarks

The method discussed in this paper rests on the Hadamard's finite-part concept. This powerful concept actually admits a wider range of applications and extensions. Our general procedure applies to several well-posed BIE and the derived asymptotic approximation of the density q authorizes us to deduce its moments and also, via integral representations outside or inside $\mathcal{A}'$, related functions or fields. Such a point of view is also likely to deal with the "opposite" case of a thin body but this enterprise would require to derive the asymptotic expansion of another class of integrals depending upon a small parameter. This question however does not fall within the scope of this paper.

References

[1] R. Barshinger, J. Geer: The electrostatic field about a slender dielectric body. *SIAM J. Appl. Math.* **4**, 605-623, 1987.

[2] R. Cade: On integral equations of axisymmetric potential theory. *IMA J. Appl. Math.* **53**, 1-25, 1994.

[3] R. G. Cox: The motion of long slender bodies in a viscous fluid. Part 1. General theory. *J. Fluid. Mech.* **44**, 791-810, 1970.

[4] R. G. Cox: The motion of long slender bodies in a viscous fluid. Part 2. Shear flow. *J. Fluid. Mech.* **45**, 625-657, 1971.

[5] J. Geer: Uniform asymptotic solutions for potential flow about a slender body of revolution. *J. Fluid. Mech.* **67**, 817-827, 1975.

[6] J. Geer: Stokes flow past a slender body of revolution. *J. Fluid. Mech.* **78**, 577-600, 1976.

[7] J. Hadamard: *Lecture on Cauchy's problem in linear differential equations.* New York: Dover. 1932.

[8] A. Handelsman, J. B. Keller: Axially symmetric potential flow around a slender body. *J. Fluid. Mech.* **28**, 131-147, 1967.

[9] A. Handelsman, J. B. Keller: The electrostatic field around a slender conducting body of revolution. *SIAM J. Appl. Math.* **15**, 824-842, 1967.

[10] R. E. Johnson: An improved slender-body theory for Stokes flow. *J. Fluid. Mech.* **99**, 411-431, 1980.

[11] V. D. Kupradze: Dynamical problems in elasticity. In: *Progress in solid Mechanics.* New York: North-Holland. 1963.

[12] L. Landweber: The axially symmetric potential flow about elongated bodies of revolution. David W. Taylor Model Basin, Rept. no, 761, 1951.

[13] J. Lavoine: *Calcul symbolique, distributions et pseudo-fonctions.* Paris: CNRS. 1959.

[14] J. Moran: Line source distributions and slender-body theory. *J. Fluid. Mech.* **17**, 285-303, 1963.

[15] L. Schwartz: *Théorie des distributions.* Paris: Hermann. 1966.

[16] A. Sellier: Asymptotic expansion of a general integral. *Proc. R. Soc. Lond. A* **452**, 2655-269, 1996.

[17] A. Sellier: Hadamard's finite part concept in dimension $n \geq 2$; definition and change of variables, associated Fubini's theorem, derivation. *Math. Proc. Camb. Phil. Soc.* **122**, 131-148, 1997.

[18] A. Sellier: A slender dielectric body embedded in an arbitrary external potential. Submitted.

[19] J. P. K. Tillett: Axial and transverse Stokes flow past slender axisymmetric bodies. *J. Fluid. Mech.* **44**, 401-417, 1970.

[20] M. Van Dyke: *Perturbation methods in fluid mechanics.* Stanford, CA: Parabolic Press. 1975.

[21] R. P. Zabreyko: *Integral equations.* Noordhoff: Leyden. 1975.

Dr. A. Sellier Ladhyx. Ecole Polytechnique. 91128 Palaiseau Cedex. France.
Email: `sellier@ladhyx.polytechnique.fr`
Fax: (33) 01 69 33 30 30

T. SHAPOSHNIKOVA

Sobolev Multipliers in the Theory of Integral Convolution Operators

In this paper an application of the theory of pointwise multipliers in Sobolev spaces to integral convolution operators in weighted L_2-spaces is discussed.

By a multiplier acting from one functional space S_1 into another S_2 one means a function which defines a bounded linear mapping of S_1 into S_2 by pointwise multiplication. Thus, with any pair of spaces S_1, S_2 one can associate the space of multipliers $M(S_1 \to S_2)$. We use the notation MS instead of $M(S \to S)$.

A theory of multipliers in pairs of Sobolev spaces and their generalizations was developed in our joint book with V. Maz'ya [1], where characterizations of multipliers were given and their various properties and applications were presented. In particular, in Sec. 3.6.1 one can find necessary and sufficient conditions for the continuity of convolutions in pairs of weighted L_2 spaces and a description of their spectrum. The results in Chapter 4 of [1] imply two-sided estimates for the essential norm of such operators and conditions for their compactness. In the present article I give two-sided estimates for the norm of the convolution with positive homogeneous kernel (Theorem 2).

Let
$$K : u \to k * u$$
be the convolution operator with the kernel k. We consider this operator as the mapping
$$K : L_2((1 + |x|^2)^{m/2}) \to L_2((1 + |x|^2)^{l/2}), \tag{1}$$
where $m \geq l \geq 0$ and $L_2((1 + |x|^2)^{s/2})$ is the space of measurable functions on $\mathbf{R}^n$ having the finite norm
$$\left(\int_{\mathbf{R}^n} |u(x)|^2 (1 + |x|^2)^s dx \right)^{1/2}, \quad s \in \mathbf{R},$$
or as the mapping
$$K : \quad L_2(|x|^m) \to L_2(|x|^l), \tag{2}$$
where $2m < n$ and $L_2(|x|^s)$, $0 < s < n/2$, is the completion of $C_0^\infty(\mathbf{R}^n)$ with respect to the norm
$$\left(\int_{\mathbf{R}^n} |u(x)|^2 |x|^{2s} dx \right)^{1/2}.$$

We use the equivalence symbol between two quantities $(a \sim b)$ if their ratio is bounded and separated from zero by positive constants. By $B_r(x)$ we shall mean the open ball in $\mathbf{R}^n$ with center at $x \in \mathbf{R}^n$ and radius r. The Fourier transform in $\mathbf{R}^n$ will be denoted by F.

1. Convolutions with positive homogeneous kernels

In this section we describe positive homogeneous multipliers in pairs of Riesz and Bessel potential spaces and give their applications to convolutions with positive homogeneous kernels.

1.1. Multipliers in Riesz and Bessel potential spaces on the sphere

The space of Bessel potentials $H_p^s(\mathbf{R}^n)$, $1 < p < \infty$, $s \geq 0$ is defined as the completion of $C_0^\infty(\mathbf{R}^n)$ in the norm

$$||u; \mathbf{R}^n||_{H_p^s} = ||F^{-1}(1+|\xi|^2)^{s/2} Fu; \mathbf{R}^n||_{L_p}.$$

Each element of $H_p^s(\mathbf{R}^n)$ is the Bessel potential of order l with density in $L_p(\mathbf{R}^n)$ (see [2]). Let

$$\mathcal{S}_l u(x) = |\nabla_l u(x)| \quad \text{if } \{l\} = 0$$

and

$$\mathcal{S}_l u(x) = \left(\int_0^\infty \left(\int_{B_1} |\nabla_{[l]} u(x+\theta y) - \nabla_{[l]} u(x)| d\theta \right)^2 y^{-1-2\{l\}} dy \right)^{1/2} \quad \text{if } \{l\} > 0.$$

According to [3],

$$||u; \mathbf{R}^n||_{H_p^l} \sim ||\mathcal{S}_l u; \mathbf{R}^n||_{L_p} + ||u; \mathbf{R}^n||_{L_p}.$$

The spaces $H_2^s(\mathbf{R}^n)$ and $W_2^s(\mathbf{R}^n)$ coincide for all $s > 0$.

Replacing $F^{-1}(1+|\xi|^2)^{s/2} Fu$ in the definition of $H_p^s(\mathbf{R}^n)$ by $F^{-1}|\xi|^s Fu$, we obtain the space $h_p^s(\mathbf{R}^n)$ which for $ps < n$ coincides with the space of Riesz potentials of order s with density from $L_p(\mathbf{R}^n)$. The norm in $h_p^s(\mathbf{R}^n)$ is equivalent to $||\mathcal{S}_s u; \mathbf{R}^n||_{L_p}$.

Let S^{n-1} denote the boundary of the n-dimensional unit ball centered at the origin. We introduce the space $H_p^s(S^{n-1})$ in a standard way as follows. Let $\{U_i\}$ be a finite covering of S^{n-1} by open sets with small diameter and let $\{\phi_i\}$ be a family of diffeomorphisms: $U_i \to \mathbf{R}^n$. Further, let $\{\nu_i\}$ be a smooth partition of unity on S^{n-1} subordinate to the covering $\{U_i\}$. We say that a function v on S^{n-1} is contained in the space $H_p^s(S^{n-1})$ if $(\nu_i \circ v) \circ \phi_i^{-1} \in H_p^s(\mathbf{R}^{n-1})$ for all i. The norm in $H_p^s(S^{n-1})$ is introduced by

$$||v; S^{n-1}||_{H_p^s} = \left(\sum_i ||(\nu_i v) \circ \phi_i^{-1}; \mathbf{R}^{n-1}||_{H_p^s}^p \right)^{1/p}.$$

It is well-known that the passage from one collection $\{U_i, \phi_i, \nu_i\}$ to another leads to an equivalent norm.

Proposition 1. *A function f belongs to the space $M(H_p^m(S^{n-1}) \to H_p^l(S^{n-1}))$ if and only if*

$$(\nu_i f) \circ \phi_i^{-1} \in M(H_p^m(\mathbf{R}^{n-1}) \to H_p^l(\mathbf{R}^{n-1}))$$

for all i. Moreover,

$$||f; S^{n-1}||_{M(H_p^m \to H_p^l)} \sim \max_i ||(\nu_i f) \circ \phi_i^{-1}; \mathbf{R}^{n-1}||_{M(H_p^m \to H_p^l)}. \tag{3}$$

Proof. Let ζ_i be a function in $C_0^\infty(U_i)$ such that $\zeta_i \nu_i = \nu_i$. For any $v \in H_p^m(S^{n-1})$ we have

$$||fv; S^{n-1}||_{H_p^l} \leq (\sum_{i,j} ||(\nu_i f \nu_j \zeta_i v) \circ \phi_i^{-1}; \mathbf{R}^{n-1}||_{H_p^l}^p)^{1/p}$$

$$\leq \sup_i ||(\nu_i f) \circ \phi_i^{-1}; \mathbf{R}^{n-1}||_{M(H_p^m \to H_p^l)} (\sum_{i,j} ||(\nu_j \zeta_i v) \circ \phi_i^{-1}; \mathbf{R}^{n-1}||_{H_p^m}^p)^{1/p}.$$

Since the mapping $\phi_j \phi_i^{-1} : \phi_i(U_j \cap U_i) \to \phi_j(U_j \cap U_i)$ is infinitely differentiable, then

$$||(\nu_j \zeta_i v) \circ \phi_i^{-1}; \mathbf{R}^{n-1}||_{H_p^m} = ||(\nu_j \zeta_i v) \circ \phi_j^{-1} \phi_j \phi_i^{-1}; \mathbf{R}^{n-1}||_{H_p^m} \leq c ||(\nu_j v) \circ \phi_j^{-1}; \mathbf{R}^{n-1}||_{H_p^m}$$

and so the required upper estimate for the norm in $M(H_p^m(S^{n-1}) \to H_p^l(S^{n-1}))$ follows.

On the other hand, for any $w \in H_p^m(\mathbf{R}^{n-1})$,

$$||((\nu_i f) \circ \phi_i^{-1}) w; \mathbf{R}^{n-1}||_{H_p^l} = ||(\nu_i f(w \circ \phi_i)) \circ \phi_i^{-1}; \mathbf{R}^{n-1}||_{H_p^l}$$

$$\leq ||\nu_i f(w \circ \phi_i); S^{n-1}||_{H_p^l} \leq ||f; S^{n-1}||_{M(H_p^m \to H_p^l)} ||\nu_i (w \circ \phi_i); S^{n-1}||_{H_p^m}.$$

It is clear that the last norm does not exceed

$$(\sum_j ||(\nu_j \nu_i (w \circ \phi_i)) \circ \phi_j^{-1}; \mathbf{R}^{n-1}||_{H_p^m}^p)^{1/p}$$

$$\leq c (\sum_j ||(\nu_j \nu_i (w \circ \phi_i)) \circ \phi_i^{-1}; \mathbf{R}^{n-1}||_{H_p^m}^p)^{1/p} \leq c ||w; \mathbf{R}^{n-1}||_{H_p^m},$$

and the lower estimate for the norm in $M(H_p^m(S^{n-1}) \to H_p^l(S^{n-1}))$ also follows. □

We note that one can supply $H_p^s(S^{n-1})$ with an equivalent norm using the operator $(1-\delta)^{s/2}$, where δ is the Beltrami operator on the unit sphere. Namely,

$$||v; S^{n-1}||_{H_p^s} \sim ||(1-\delta)^{s/2} v; S^{n-1}||_{L_p}. \tag{4}$$

It is essentially a consequence of the following property established by R.T. Seeley [4]: $(1-\delta)^{s/2}$ is the pseudodifferential operator with principal symbol $|\xi|^s$.

1.2. Another normalization of the spaces H_p^m and h_p^m

Lemma 1. Let $0 \leq l \leq m$, $p > 1$, $pm < n$. Then

$$|| |x|^{l-m} u; \mathbf{R}^n||_{h_p^l} \leq c ||u; \mathbf{R}^n||_{h_p^m}. \tag{5}$$

Proof. We shall show that the inclusion $|x|^{l-m} \in M(h_p^m(\mathbf{R}^n) \to h_p^l(\mathbf{R}^n))$ holds which is equivalent to the statement (5). By $h_{p,loc}^s(\mathbf{R}^n)$ we denote the space $\{\eta u \in h_p^l(\mathbf{R}^n)$ for all $\eta \in C_0^\infty(\mathbf{R}^n)\}$. Further, let $\text{cap}(e, h_p^s)$ be the capacity of a compact $e \subset \mathbf{R}^n$ generated by the norm in $h_p^s(\mathbf{R}^n)$, i.e

$$\operatorname{cap}(e; h_p^s) = \inf\{||u; \mathbf{R}^n||_{h_p^s}^p : u \in C_0^\infty(\mathbf{R}^n), u \geq 1 \text{ on } e\}.$$

In Ch. 2 [1] it was shown that $\gamma \in M(h_p^m(\mathbf{R}^n) \to h_p^l(\mathbf{R}^n))$, $mp < n$, if and only if $\gamma \in (h_{p,loc}^l \cap L_{1,unif})(\mathbf{R}^n)$ and

$$||\mathcal{S}_l\gamma; e||_{L_p}^p \leq c \operatorname{cap}(e, h_p^m) \qquad (6)$$

for any compact $e \subset \mathbf{R}^n$. Moreover,

$$||\gamma; \mathbf{R}^n||_{M(h_p^m \to h_p^l)} \sim \sup_{e \subset \mathbf{R}^n} \frac{||\mathcal{S}_l\gamma; e||_{L_p}}{(\operatorname{cap}(e; h_p^m))^{1/p}} + \sup_{x \in \mathbf{R}^n} ||\gamma; B_1(x)||_{L_1}. \qquad (7)$$

From the easily verified inequality

$$\int_{B_1} ||x + \theta y|^{l-m} - |x|^{l-m}| d\theta \leq c \min\{y, |x|\} |x|^{l-m-1}$$

it follows that $(\mathcal{S}_l|z|^{l-m})(x) \leq c|x|^{-m}$. Besides,

$$\int_e \frac{dx}{|x|^{mp}} \leq c(\operatorname{mes}_n(e))^{1-mp/n} \leq c \operatorname{cap}(e; h_p^m).$$

Therefore,

$$||\mathcal{S}_l|x|^{l-m}; e||_{L_p}^p \leq c \operatorname{cap}(e; h_p^m),$$

and, obviously,

$$\sup_{x \in \mathbf{R}^n} || |y|^{l-m}; B_1(x)||_{L_1} < \infty.$$

Thus, $|x|^{l-m} \in M(h_p^m(\mathbf{R}^n) \to h_p^l(\mathbf{R}^n))$ which completes the proof. □

Using (5) we may equip the space $h_p^m(\mathbf{R}^n)$, $mp < n$, with an equivalent norm. Let $G_k = \{x : 2^{k-1} < |x| < 2^{k+1}\}$, $k = 0, \pm 1, \ldots$, and let $\{\psi_k\}$ be a partition of unity on $\mathbf{R}^n \setminus \{0\}$ subordinate to the covering $\{G_k\}$. Suppose, $|D^\alpha \psi_k| \leq c_\alpha 2^{-k|\alpha|}$.

Lemma 2. *Let $p > 1$, $mp < n$. The relation*

$$||u; \mathbf{R}^n||_{h_p^m} \sim \left(\sum_{k=-\infty}^{\infty} ||\psi_k u; \mathbf{R}^n||_{h_p^m}^p\right)^{1/p} \qquad (8)$$

is valid.

Proof. First we show that the verification of (8) can be easily reduced to the case $0 \leq m < 1$. Suppose that $m \geq 1$ and that the assertion is proved for $m-1$. One can easily check that for integer s the norms of functions $|x|^\alpha \partial^\alpha \psi_k$ in $M h_p^s(\mathbf{R}^n)$ are uniformly bounded with respect to k. The same is true for fractional s by interpolation. Therefore,

$$| ||\psi_k \nabla u; \mathbf{R}^n||_{h_p^{m-1}} - ||\nabla(\psi_k u); \mathbf{R}^n||_{h_p^{m-1}}| \leq ||\zeta_k u \nabla \psi_k; \mathbf{R}^n||_{h_p^{m-1}}$$

$$\leq c||\,|x|^{-1}\zeta_k u; \mathbf{R}^n||_{h_p^{m-1}} \tag{9}$$

where $\zeta_k \in C_0^\infty(G_k)$, $\zeta_k \psi_k = \psi_k$, $|\partial^\alpha \zeta_k| \leq c_\alpha 2^{-k|\alpha|}$. Consequently,

$$||\nabla u; \mathbf{R}^n||_{h_p^{m-1}}^p \leq c \sum_{k=-\infty}^{\infty} ||\psi_k \nabla u; \mathbf{R}^n||_{h_p^{m-1}}^p$$

$$\leq c \sum_{k=-\infty}^{\infty} (||\nabla(\psi_k u); \mathbf{R}^n||_{h_p^{m-1}}^p + ||\,|x|^{-1}\zeta_k u; \mathbf{R}^n||_{h_p^{m-1}}^p).$$

This and (5) imply

$$||u; \mathbf{R}^n||_{h_p^m}^p \leq c \sum_{k=-\infty}^{\infty} (||\psi_k u; \mathbf{R}^n||_{h_p^m}^p + ||\zeta_k u; \mathbf{R}^n||_{h_p^m}^p).$$

Since $\{\psi_k\}$ is the partition of unity and $||\zeta_k; \mathbf{R}^n||_{Mh_p^m} \leq$ const, the upper estimate for the norm $||u; \mathbf{R}^n||_{h_p^m}$ follows.

Next we derive the lower bound. By (9) we have

$$\sum_{k=-\infty}^{\infty} ||\psi_k u; \mathbf{R}^n||_{h_p^m}^p \leq c \sum_{k=-\infty}^{\infty} (||\psi_k \nabla u; \mathbf{R}^n||_{h_p^{m-1}}^p + ||\,|x|^{-1}\zeta_k u; \mathbf{R}^n||_{h_p^{m-1}}^p).$$

Replacing ζ_k by ψ_k in the last norm and using the induction hypothesis we obtain that the right-hand side does not exceed

$$c(||u; \mathbf{R}^n||_{h_p^m}^p + ||\,|x|^{-1} u; \mathbf{R}^n||_{h_p^{m-1}}^p).$$

It remains to make use of (5).

In the case $m = 0$ the relation (8) is trivial. Let $0 < m < 1$. It is clear that

$$||\mathcal{S}_m u; \mathbf{R}^n||_{L_p}^p \sim \sum_{k=-\infty}^{\infty} ||\psi_k \mathcal{S}_m u; \mathbf{R}^n||_{L_p}^p.$$

The definition of the operator $\mathcal{S}_m$ and Minkowski's inequality imply

$$|\psi_k(x)\mathcal{S}_m u(x) - \mathcal{S}_m(\psi_k u)(x)|$$

$$\leq \left(\int_0^\infty (\int_{B_y} |u(x+z)||\psi_k(x+z) - \psi_k(x)|dz)^2 y^{-1-2n-2m} dy\right)^{1/2}$$

$$\leq c \int_{\mathbf{R}^n} |u(z)| \frac{|\psi_k(z) - \psi_k(x)|}{|z-x|^{n+m}} dz.$$

Let $A(x)$ denote the last integral and let $g_k = G_{k-1} \cup G_k \cup G_{k+1}$. Since $\operatorname{supp} \psi_k \subset G_k$, we have

$$A(x) \leq \begin{cases} c\, 2^{-k(n+m)} \int_{G_k} |u(z)|\, dz & \text{for } |x| < 2^{k-2}, \\ c|x|^{-(n+m)} \int_{G_k} |u(z)|\, dz & \text{for } |x| > 2^{k+2}, \end{cases}$$

and

$$A(x) \leq c \left(2^{-k} \int_{B_{2|x|}} \frac{|u(z)|\, dz}{|z-x|^{n+m-1}} + \int_{\mathbf{R}^n \setminus B_{2|x|}} \frac{|u(z)|\, dz}{|z|^{n+m}} \right) \quad \text{for } x \in g_k.$$

Therefore,

$$\|\psi_k \mathcal{S}_m u - \mathcal{S}_m(\psi_k u); \mathbf{R}^n\|_{L_p}^p \leq c\, 2^{k(n-pm-pn)} \left(\int_{G_k} |u(z)|\, dz \right)^p$$

$$+ c\, 2^{-kp} \int_{g_k} \left(\int_{B_{2|x|}} \frac{|u(z)|\, dz}{|z-x|^{n+m-1}} \right)^p dx + c \int_{g_k} \left(\int_{\mathbf{R}^n \setminus B_{2|x|}} \frac{|u(z)|\, dz}{|z|^{n+m}} \right)^p dx.$$

The first term on the right does not exceed $c\, 2^{-kpm} \|u; G_k\|_{L_p}^p$. The second is majorized by

$$c\, 2^{-kp} \left(\max_x \int_{B_{2^{k+3}}} \frac{d\zeta}{|\zeta - x|^{n+m-1}} \right)^p \int_{B_{2^{k+3}}} |u(z)|^p\, dz \leq c\, 2^{-kpm} \int_{B_{2^{k+3}}} |u(z)|^p\, dz.$$

The third term can be rewritten as

$$c \int_{2^{k-2}}^{2^{k+2}} r^{n-1} dr \left(\int_{2r}^{\infty} \frac{v(\rho)\, d\rho}{\rho^{m+1}} \right)^p,$$

where $v(\rho)$ is the mean value of $|u|$ on S^{n-1}. Summing over k and using the one-dimensional Hardy inequality we arrive at

$$\sum_{k=-\infty}^{\infty} \|\psi_k \mathcal{S}_m u - \mathcal{S}_m(\psi_k u); \mathbf{R}^n\|_{L_p}^p$$

$$\leq c \| |x|^{-m} u; \mathbf{R}^n \|_{L_p}^p \sim \sum_{k=-\infty}^{\infty} \| |x|^{-m} \psi_k u; \mathbf{R}^n \|_{L_p}^p. \tag{10}$$

This and (5) imply

$$\|\mathcal{S}_m u; \mathbf{R}^n\|_{L_p}^p \leq c \sum_{k=-\infty}^{\infty} \|\mathcal{S}_m(\psi_k u); \mathbf{R}^n\|_{L_p}^p.$$

So the upper bound for the norm in $h_p^m(\mathbf{R}^n)$ follows.

On the other hand, by (10) we have

$$\sum_{k=-\infty}^{\infty} ||\mathcal{S}_m(\psi_k u); \mathbf{R}^n||_{L_p}^p \leq ||\mathcal{S}_m u; \mathbf{R}^n||_{L_p}^p + |||x|^{-m} u; \mathbf{R}^n||_{L_p}^p.$$

Applying (5) once again we complete the proof.

The relation (8) and the equivalence

$$||u; \mathbf{R}^n||_{H_p^m} \sim ||u; \mathbf{R}^n||_{h_p^m} + ||u; \mathbf{R}^n||_{L_p}$$

immediately imply the result similar to Lemma 2 for the space H_p^m. □

Corollary 1. *Let $p > 1$, $mp < n$. Then*

$$||u; \mathbf{R}^n||_{H_p^m} \sim \Big(\sum_{k=-\infty}^{\infty} ||\psi_k u; \mathbf{R}^n||_{H_p^m}^p \Big)^{1/p}. \qquad (11)$$

As a corollary of Lemma 2 we arrive at the following assertion on the normalization of the space $M(h_p^m(\mathbf{R}^n) \to h_p^l(\mathbf{R}^n))$, $pm < n$, which will be used in the next subsection.

Corollary 2. *If $p > 1$, $mp < n$, then*

$$||\gamma; \mathbf{R}^n||_{M(h_p^m \to h_p^l)} \sim \sup_{-\infty < k < \infty} ||\psi_k \gamma; \mathbf{R}^n||_{M(h_p^m \to h_p^l)}. \qquad (12)$$

1.3. Positive homogeneous multipliers

Theorem 1. *Let $p > 1$, $pm < n$. The function $x \to \gamma(x) = |x|^{l-m} f(x/|x|)$ is contained in both $M(h_p^m(\mathbf{R}^n) \to h_p^m(\mathbf{R}^n))$ and $M(H_p^m(\mathbf{R}^n) \to H_p^m(\mathbf{R}^n))$ if and only if $f \in M(H_p^m(S^{n-1}) \to H_p^l(S^{n-1}))$. Moreover,*

$$||\gamma; \mathbf{R}^n||_{M(h_p^m \to h_p^l)} \sim ||\gamma; \mathbf{R}^n||_{M(H_p^m \to H_p^l)} \sim ||f; S^{n-1}||_{M(H_p^m \to H_p^l)}.$$

Proof. We begin with the space $M(h_p^m(\mathbf{R}^n) \to h_p^l(\mathbf{R}^n))$. Let $\zeta \in C_0^\infty((-1/2, 2))$, $\zeta(t) = 1$ for $t \in (1, 3/2)$, $\zeta_k(x) = \zeta(2^{-k}|x|)$. Since the norms of the functions ζ_k and ψ_k in $Mh_p^l(\mathbf{R}^n)$ are uniformly bounded with respect to k, it follows from (12) that

$$||\gamma; \mathbf{R}^n||_{M(h_p^m \to h_p^l)} \sim \sup_{-\infty < k < \infty} ||\zeta_k \gamma; \mathbf{R}^n||_{M(h_p^m \to h_p^l)}. \qquad (13)$$

Using the homogeneity of γ and that of the norm in $h_p^s(\mathbf{R}^n)$ with respect to the dilation we conclude that the norm on the right in (3) does not depend on k. Therefore,

$$||\gamma; \mathbf{R}^n||_{M(h_p^m \to h_p^l)} \sim ||\zeta \gamma; \mathbf{R}^n||_{M(h_p^m \to h_p^l)}. \qquad (14)$$

Let U_i be any of coordinate neigbourhoods on S^{n-1} and let $\{\phi_i\}$ be the family of diffeomorphisms used in the definition of the space $H_p^l(S^{n-1})$. Since $\mathcal{S}_l \nu_i(x) = O(|x|^{-l})$, we have
$$\|\mathcal{S}_l \nu_i; e\|_{L_p}^p \leq (\mathrm{mes}_n e)^{1-lp/n} \leq c\, \mathrm{cap}(e; h_p^l).$$
So $\nu_i \in Mh_p^l(\mathbf{R}^n)$ and (14) implies
$$\|\gamma; \mathbf{R}^n\|_{M(h_p^m \to h_p^l)} \sim \max_i \|\nu_i \zeta \gamma; \mathbf{R}^n\|_{M(h_p^m \to h_p^l)}. \tag{15}$$

The set $\{x : 1/2 < |x| < 2,\ x/|x| \in U_i\}$ can be supplied with local coordinates $y = (y', y_n)$, where $y_n = |x|$, $y' = \phi_i(x/|x|)$. Since the mapping $\Phi_i : x \to y$ is a diffeomorphism,
$$\|\nu_i \zeta \gamma; \mathbf{R}^n\|_{M(h_p^m \to h_p^l)} \sim \|(\nu_i \zeta \gamma) \circ \Phi_i^{-1}; \mathbf{R}^n\|_{M(h_p^m \to h_p^l)}. \tag{16}$$

Next we note that
$$((\nu_i \zeta \gamma) \circ \Phi_i^{-1})(y) = \zeta(y_n) y_n^{l-m} ((\nu_i f) \circ \phi_i^{-1})(y')$$
and that the function $y \to \zeta(y_n) y_n^{l-m}$ is smooth and has a compact support. Therefore,
$$\|\nu_i \zeta \gamma; \mathbf{R}^n\|_{M(h_p^m \to h_p^l)} \leq c \|(\nu_i f) \circ \Phi_i^{-1}; \mathbf{R}^n\|_{M(h_p^m \to h_p^l)}$$
$$\leq c \|(\nu_i f) \circ \phi_i^{-1}; \mathbf{R}^{n-1}\|_{M(h_p^m \to h_p^l)}. \tag{17}$$

On the other hand, because of the equivalence
$$\|u; \mathbf{R}^n\|_{h_p^l} \sim \sum_{i=1}^n \| |\partial/\partial y_i|^l u; \mathbf{R}^n\|_{L_p},$$
any function $y \to v(y')$ with $v \in C_0^\infty(\mathbf{R}^{n-1})$ satisfies the inequality
$$\|(\nu_i f) \circ \phi_i^{-1} v; \mathbf{R}^{n-1}\|_{h_p^l} \leq c \|(\nu_i \zeta \gamma) \circ \Phi_i^{-1} v \zeta; \mathbf{R}^n\|_{h_p^l}.$$

The right-hand side, obviously, does not exceed
$$c \|(\nu_i \zeta \gamma) \circ \Phi_i^{-1}; \mathbf{R}^n\|_{M(h_p^m \to h_p^l)} \|v\zeta; \mathbf{R}^n\|_{h_p^m}$$
$$\leq c \|(\nu_i \zeta \gamma) \circ \Phi_i^{-1}; \mathbf{R}^n\|_{M(h_p^m \to h_p^l)} \|v; \mathbf{R}^{n-1}\|_{h_p^m}.$$

Thus,
$$\|(\nu_i f) \circ \phi_i^{-1}; \mathbf{R}^{n-1}\|_{M(h_p^m \to h_p^l)} \leq c \|(\nu_i \zeta \gamma) \circ \Phi_i^{-1}; \mathbf{R}^n\|_{M(h_p^m \to h_p^l)}$$
which together with (16) and (17) leads to
$$\|\nu_i \zeta \gamma; \mathbf{R}^n\|_{M(h_p^m \to h_p^l)} \sim \|(\nu_i f) \circ \phi_i^{-1}; \mathbf{R}^{n-1}\|_{M(h_p^m \to h_p^l)}. \tag{18}$$

Comparing (18), (3) and (15) we complete the proof for the space $M(h_p^m(\mathbf{R}^n) \to h_p^l(\mathbf{R}^n))$.

Now we pass to the space $M(H_p^m(\mathbf{R}^n) \to H_p^l(\mathbf{R}^n))$. For all $u \in C_0^\infty(\mathbf{R}^n)$ we have

$$||\gamma u; \mathbf{R}^n||_{h_p^l} \leq c\, ||\gamma; \mathbf{R}^n||_{M(H_p^m \to H_p^l)} (||u; \mathbf{R}^n||_{h_p^m} + ||u; \mathbf{R}^n||_{L_p}).$$

Putting here $u(x) = v(ax)$, where a is an arbitrary positive number and using the positive homogeneity of γ, we find, for any $v \in C_0^\infty(\mathbf{R}^n)$,

$$||\gamma v; \mathbf{R}^n||_{h_p^l} \leq c\, ||\gamma; \mathbf{R}^n||_{M(H_p^m \to H_p^l)} (||v; \mathbf{R}^n||_{h_p^m} + a^{-m}||v; \mathbf{R}^n||_{L_p}).$$

Passing to the limit as $a \to \infty$, we arrive at

$$||\gamma; \mathbf{R}^n||_{M(h_p^m \to h_p^l)} \leq c\, ||\gamma; \mathbf{R}^n||_{M(H_p^m \to H_p^l)}. \tag{19}$$

To derive the opposite estimate we need the following result due to Strichartz [3] on the uniform localization for the space $H_p^l(\mathbf{R}^n)$: Let $\{\mathcal{B}^{(j)}\}_{j \geq 0}$ be a finite covering of $\mathbf{R}^n$ by balls with unit diameters. Further, let $O^{(j)}$ be the center of $\mathcal{B}^{(j)}$, $O^{(0)} = O$, and let $\eta_j(x) = \eta(x - O^{(j)})$, where $\eta \in C_0^\infty(2\mathcal{B}^{(0)})$, $\eta = 1$ on $\mathcal{B}^{(0)}$. Then

$$||u; \mathbf{R}^n||_{H_p^l} \sim \Big(\sum_{j \geq 0} ||u\eta_j; \mathbf{R}^n||_{H_p^l}^p\Big)^{1/p}. \tag{20}$$

Since

$$M(h_p^m(\mathbf{R}^n) \to h_p^l(\mathbf{R}^n)) \subset M(h_p^{m-l}(\mathbf{R}^n) \to L_p(\mathbf{R}^n))$$

(see Ch. 2 [1]), we get

$$||\gamma u\eta_j; \mathbf{R}^n||_{H_p^l} \leq c(||\gamma u\eta_j; \mathbf{R}^n||_{h_p^l} + ||\gamma u\eta_j; \mathbf{R}^n||_{L_p})$$

$$\leq c(||\gamma; \mathbf{R}^n||_{M(h_p^m \to h_p^l)}||u\eta_j; \mathbf{R}^n||_{h_p^m} + ||\gamma; \mathbf{R}^n||_{M(h_p^{m-l} \to h_p^l)}||u\eta_j \mathbf{R}^n||_{h_p^{m-l}})$$

$$\leq c\, ||\gamma; \mathbf{R}^n||_{M(h_p^m \to h_p^l)}||u\eta_j; \mathbf{R}^n||_{H_p^m}.$$

This and (20) imply

$$||\gamma u; \mathbf{R}^n||_{H_p^l} \leq c\, ||\gamma; \mathbf{R}^n||_{M(h_p^m \to h_p^l)}||u; \mathbf{R}^n||_{H_p^m}.$$

Thus,

$$||\gamma; \mathbf{R}^n||_{M(H_p^m \to H_p^l)} \leq c\, ||\gamma; \mathbf{R}^n||_{M(h_p^m \to h_p^l)}$$

and by (19) the norm of γ in $M(H_p^m(\mathbf{R}^n) \to H_p^l(\mathbf{R}^n))$ and in $M(h_p^m(\mathbf{R}^n) \to h_p^l(\mathbf{R}^n))$ are equivalent. The theorem is proved. $\square$

1.4. Convolutions with positive homogeneous kernels

Theorem 1 with $p = 2$ can be interpreted as a theorem on two-sided estimates for the norm of the convolution operator K. Let the kernel k of the operator K be a distribution positive homogeneous of degree $m - l - n$, $m > l$, i.e.

$$k(x) = |x|^{m-l-n} f(x/|x|), \tag{21}$$

where f is a distribution on S^{n-1}. In the case $m = l$ we consider the kernels

$$k(x) = |x|^{-n} f(x/|x|) + \text{const } \delta(x), \tag{22}$$

where δ is the Dirac function and

$$\int_{S^{n-1}} f(\omega) ds_\omega = 0.$$

The kernels of the form (21) correspond to singular integral operators.
Using the representation of the symbol Fk in the form

$$(Fk)(\xi) = |\xi|^{l-m} \sigma(\xi/|\xi|)$$

and the inclusion $Fk \in M(H_2^m(\mathbf{R}^n) \to H_2^l(\mathbf{R}^n))$, from Theorem 1 we obtain

Theorem 2. *Let $m \geq l \geq 0$ and $2m < n$. Then the operator* (1) *with the kernel* (21) *is continuous if and only if*

$$\sigma \in M(H_2^m(S^{n-1}) \to H_2^l(S^{n-1})). \tag{23}$$

Moreover,

$$\|K\| \sim \|\sigma; S^{n-1}\|_{M(H_2^m \to H_2^l)}.$$

The same assertion holds for the operator (2).

Condition (23) can be restated in analytic terms using the results of Chapters 2, 3, 6 [1].
Other applications of multipliers to singular integral operators can be found in [5], [6].

References

[1] V.G. Maz'ya, T.O. Shaposhnikova: *Theory of multipliers in spaces of differentiable functions.* Pitman, London, 1985.

[2] D.R. Adams, L.I. Hedberg: *Function spaces and potential theory.* Springer–Verlag, Berlin–Heidelberg–New York, 1996.

[3] R.S. Strichartz: Multipliers on fractional Sobolev spaces. *J. Math. and Mech.* **16** (1967), no. 9, 1031-1060.

[4] R.T. Seeley: Complex powers of an elliptic operator. *Proc. Symp. Amer. Math. Soc.* Jan. 1967, Boston, 1967, 288-307.

[5] T. Shaposhnikova: On continuity of singular integral operators in Sobolev spaces. *Math. Scand.* **76** (1995), 85-97.

[6] T. Shaposhnikova: On the boundedness of singular integral operators in the Sobolev space. *IABEM Symposium on Boundary Integral Methods for Nonlinear Problems*, Kluwer, 1997, 191-195.

Linköping University, Institute of Technology, Dept. of Mathematics,
S-581 83 Linköping, Sweden
Email: tasha@math.liu.se

O. STEINBACH

Stable Boundary Element Approximations of Steklov–Poincaré Operators

1. Introduction

In this paper we consider boundary element methods for the numerical solution of the mixed boundary value problem

$$Lu(x) = f \quad \text{for } x \in \Omega, \qquad u_{|\Gamma_D} = g, \qquad t = Tu_{|\Gamma_N} = h. \tag{1}$$

In (1) L is a second order elliptic partial differential operator and T is the corresponding conormal derivative operator. The Lipschitz continuous boundary Γ of the bounded domain $\Omega \subset \mathbb{R}^n$ ($n = 2, 3$) is supposed to be split into two disjoint parts Γ_D with Dirichlet boundary conditions and Γ_N with Neumann conditions, respectively. If a fundamental solution $U^*(\cdot, \cdot)$ is known, the solution of (1) can be described by the representation formula

$$u(x) = \int_\Gamma U^*(x,y) t(y) ds_y - \int_\Gamma u(x) T^*(x,y) ds_y + \int_\Omega U^*(x,y) f(y) dy \quad \text{for } x \in \Omega, \tag{2}$$

where $T^*(\cdot, \cdot) = T_y U^*(\cdot, \cdot)$. Hence we have to use some boundary integral method to find the unknown Cauchy data $u_{|\Gamma_N}$ and $t_{|\Gamma_D}$, respectively. Following the direct approach we derive from (2) the boundary integral equation

$$(Vt)(x) = (\tfrac{1}{2}I + K)u(x) - (Nf)(x) \quad \text{for } x \in \Gamma \tag{3}$$

using the standard notations for the single layer potential V, the double layer potential K and the Newton potential N,

$$(Vt)(x) = \int_\Gamma U^*(x,y) t(y) ds_y,$$

$$(Ku)(x) = \int_\Gamma u(y) T^*(x,y) ds_y,$$

$$(Nf)(x) = \int_\Omega U^*(x,y) f(y) dy.$$

The boundary integral equation (3) can be reformulated as a system of boundary integral equations according to the given boundary conditions in (1), see e.g. [3, 5]. Here we will use (3) to define the Steklov–Poincaré operator involved in the Dirichlet–Neumann map defined by the solution of a Dirichlet boundary value problem. The corresponding Galerkin variational formulation can be analysed using standard techniques. However, due to the implicit definition of the Steklov–Poincaré operator, one has to introduce a computable approximation to realize the Galerkin scheme. To ensure solvability and stability of the resulting approximate variational problem, an appropriate stability condition has to be satisfied. We will discuss several possible choices to define such a stable approximation of the Steklov–Poincaré operator based on the boundary integral equation (3). Note that (3) can be discretized either by a Galerkin or collocation scheme. Moreover, there exist some other approaches to discretize the Steklov–Poincaré operator by boundary element methods, e.g. by using the symmetric formulation [7] or hybrid discretization schemes [4, 9].

Steklov–Poincaré operators describing a Dirichlet–Neumann map corresponding to a partial differential operator can also be used in variational formulations for domain decomposition methods, see e.g. [1, 6]. By using different discretization schemes such as finite and boundary elements to approximate the Steklov–Poincaré operator in different subdomains locally one can derive efficient parallel numerical schemes to handle coupled partial differential equations or transmission problems; for a discussion of such formulations based on boundary elements see [11].

In this paper we will not discuss the mapping properties of all boundary integral operators involved; for this see for example [2].

2. Steklov–Poincaré boundary integral formulation

Throughout the paper the single layer potential V is supposed to be invertible; in some cases one has to assume a suitable scaling condition to ensure this. Hence we can solve (3) to derive the Dirichlet–Neumann map

$$t(x) = (Su)(x) - V^{-1}(Nf)(x) \quad \text{for } x \in \Gamma \qquad (4)$$

using the Steklov–Poincaré operator S defined as

$$(Su)(x) = V^{-1}(\frac{1}{2}I + K)u(x) \quad \text{for } x \in \Gamma. \qquad (5)$$

In what follows we will assume that the Steklov–Poincaré operator S is bounded in $H^{1/2}(\Gamma)$ and positiv definite, i.e., that there holds

$$\langle Sv, v \rangle_{L^2(\Gamma)} \geq c \cdot ||v||^2_{H^{1/2}(\Gamma)} \quad \text{for all } v \in W := \{H^{1/2}(\Gamma) : v_{|\Gamma_D} = 0\}. \qquad (6)$$

Let $\tilde{g} \in H^{1/2}(\Gamma)$ be an arbitrary but fixed extension of the given Dirichlet data g. From (4) there follows the variational formulation of the mixed boundary value problem (1) to find $u \in W$ such that

$$\int_\Gamma (S(u + \tilde{g}))(x) \cdot v(x)\, ds_x = \int_\Gamma [h(x) + V^{-1}(Nf)(x)] \cdot v(x)\, ds_x \qquad (7)$$

holds for all test functions $v \in W$. Note that (7) is unique solvable due to the assumptions made above.

If the variational problem (7) is solved, we consider the boundary integral equation (3) with given Dirichlet data $u + \tilde{g}$ to find the corresponding Neumann data $t(x)$ for $x \in \Gamma$. Afterwards we can use the representation formula (2) to compute the solution of the mixed boundary value problem (1) inside Ω.

For a Galerkin discretization of (7) we introduce a boundary triangulation Γ_H of mesh size H and a conforming finite–dimensional trial space

$$W_H = \operatorname{span}\{\psi_k^\mu\}_{k=1}^M \subset W$$

spanned by smoothest splines of polynomial degree μ and satisfying the approximation property

$$\inf_{v_H \in W_H} ||v - v_H||_{H^{1/2}(\Gamma)} \leq c \cdot H^{s-\frac{1}{2}} \cdot ||v||_{H^s(\Gamma)} \qquad (8)$$

for all $v \in H^s(\Gamma)$ and $s \leq \mu + 1$. The Galerkin formulation of (7) reads to find $u_H \in W_H$ such that

$$\langle Su_H, v_H \rangle_{L^2(\Gamma)} = \langle h + V^{-1}Nf - S\tilde{g}, v_H \rangle_{L^2(\Gamma)} =: b(v_H) \quad \text{for all } v_H \in W_H. \qquad (9)$$

Note that due to (6) we have stability and unique solvability of (9) and the solution u_H satisfies the quasi–optimal error estimate

$$||u - u_H||_{H^{1/2}(\Gamma)} \leq c \cdot \inf_{v_H \in W_H} ||u - v_H||_{H^{1/2}(\Gamma)} \qquad (10)$$

and using (8), this gives convergence for $H \to 0$. To realize the Galerkin scheme (9) we have to compute the Galerkin weights

$$S_H[\ell, k] = \langle S\psi_k^\mu, \psi_\ell^\mu \rangle_{L^2(\Gamma)} = \langle V^{-1}(\frac{1}{2}I + K)\psi_k^\mu, \psi_\ell^\mu \rangle_{L^2(\Gamma)} \quad \text{for } k, \ell = 1, \ldots, M,$$

which in general can not be done directly, since V^{-1} is not known explicitly. Hence we have to introduce a computable approximation of S. Note that we have to approximate the right-hand side in (9) as well. This can be done in a similar manner as described in the next section. But we do not need any kind of stability, only some approximation properties are needed. For a discussion of resulting error estimates see [8].

3. Approximate Galerkin discretization

Instead of (9) we consider an approximated Galerkin variational formulation to find $\tilde{u}_H \in W_H$ such that

$$\langle \tilde{S}\tilde{u}_H, v_H \rangle_{L^2(\Gamma)} = b(v_H) \quad \text{for all } v_H \in W_H. \qquad (11)$$

In (11) $\tilde{S}$ is a suitable approximation of the Steklov–Poincaré operator S. To ensure the solvability of (11) we will assume the stability of $\tilde{S}$, i.e., that there holds

$$\langle \tilde{S}v_H, v_H \rangle_{L^2(\Gamma)} \geq c \cdot ||v_H||^2_{H^{1/2}(\Gamma)} \quad \text{for all } v_H \in W_H. \qquad (12)$$

Moreover, we assume that $\tilde{S}$ is bounded,

$$||\tilde{S}v_H||_{H^{-1/2}(\Gamma)} \leq c \cdot ||v_H||_{H^{1/2}(\Gamma)} \quad \text{for all } v_H \in W_H. \qquad (13)$$

Using the Galerkin orthogonality (subtracting (11) from (9))

$$\langle Su_H - \tilde{S}\tilde{u}_H, v_H \rangle_{L^2(\Gamma)} = 0 \quad \text{for all } v_H \in W_H,$$

and (12),

$$\begin{aligned}
c \cdot ||\tilde{u}_H - u_H||^2_{H^{-1/2}(\Gamma)} &\leq \langle \tilde{S}(\tilde{u}_H - u_H), \tilde{u}_H - u_H \rangle_{L^2(\Gamma)} \\
&= \langle (S - \tilde{S})u_H, \tilde{u}_H - u_H \rangle_{L^2(\Gamma)} \\
&\leq ||(S - \tilde{S})u_H||_{H^{-1/2}(\Gamma)} ||\tilde{u}_H - u_H||_{H^{1/2}(\Gamma)},
\end{aligned}$$

we get the error estimate

$$||u - \tilde{u}_H||_{H^{1/2}(\Gamma)} \leq c_1 \cdot ||u - u_H||_{H^{1/2}(\Gamma)} + c_2 \cdot ||(S - \tilde{S})u||_{H^{-1/2}(\Gamma)} \tag{14}$$

when using the triangle inequality twice and the boundedness of $\tilde{S}$. Hence we have to construct a bounded approximate Steklov–Poincaré operator $\tilde{S}$ in such a way that $\tilde{S}$ satisfies the stability condition (12) as well as an appropriate error estimate to be inserted in (14). This can be done as follows:

Let us define a second boundary triangulation Γ_h of mesh size h and a corresponding finite-dimensional trial space

$$Z_h = \text{span}\{\varphi_i^\nu\}_{i=1}^N \subset H^{-1/2}(\Gamma) \tag{15}$$

of discontinuous splines of polynomial degree ν, where we assume the approximation property

$$\inf_{w_h \in Z_h} ||w - w_h||_{H^{-1/2}(\Gamma)} \leq c \cdot h^{\sigma + \frac{1}{2}} \cdot ||w||_{H^\sigma(\Gamma)} \tag{16}$$

for all $w \in H^\sigma(\Gamma)$ with $\sigma \leq \nu + 1$.

Let $v \in W$ arbitrary but fixed. Due to the definition (5) of the Steklov–Poincaré operator S we find that $w = Sv \in H^{-1/2}(\Gamma)$ is a solution of the boundary integral equation

$$(Vw)(x) = (\frac{1}{2}I + K)v(x) \quad \text{for } x \in \Gamma,$$

or in weak form,

$$\langle Vw, \tau \rangle_{L^2(\Gamma)} = \langle (\frac{1}{2}I + K)v, \tau \rangle_{L^2(\Gamma)} \quad \text{for all } \tau \in H^{-1/2}(\Gamma). \tag{17}$$

If we consider a Galerkin variational formulation of (17) to find $w_h \in Z_h$ such that

$$\langle Vw_h, \tau_h \rangle_{L^2(\Gamma)} = \langle (\frac{1}{2}I + K)v, \tau_h \rangle_{L^2(\Gamma)} \quad \text{for all } \tau_h \in Z_h, \tag{18}$$

which is unique solvable due to the invertibility of the single layer potential V, we may define the approximate Steklov–Poincaré operator as

$$\tilde{S}v := w_h \in Z_h, \qquad (19)$$

which is bounded due to the mapping properties of the boundary integral operators involved. Using this definition, the stiffness matrix of the approximated Steklov–Poincaré operator $\tilde{S}$ resulting from (11) can be written as

$$\tilde{S}_H = M_h^\top V_h^{-1}(\tfrac{1}{2}M_h + K_h) \qquad (20)$$

with

$$V_h[j,i] = \langle V\varphi_i^\nu, \varphi_j^\nu \rangle_{L^2(\Gamma)}, \quad K_h[j,k] = \langle K\psi_k^\mu, \varphi_j^\nu \rangle_{L^2(\Gamma)}, \quad M_h[j,k] = \langle \psi_k^\mu, \varphi_j^\nu \rangle_{L^2(\Gamma)}$$

for all $k = 1, \ldots, M; i, j = 1, \ldots, N$.

Now we can formulate a sufficient condition to ensure (12), i.e., we assume that the trial space Z_h is taken in such a way that the stability condition

$$c \cdot \|v_H\|_{H^{1/2}(\Gamma)} \leq \sup_{\tau_h \in Z_h} \frac{|\langle v_H, \tau_h \rangle_{L^2(\Gamma)}|}{\|\tau_h\|_{H^{-1/2}(\Gamma)}} \qquad (21)$$

is satisfied for all $v_H \in W_H$ [10].

On the other hand, there holds a quasi-optimal error estimate for the Galerkin solution of (18),

$$\|w - w_h\|_{H^{-1/2}(\Gamma)} \leq c \cdot \inf_{\tau_h \in Z_h} \|w - \tau_h\|_{H^{-1/2}(\Gamma)},$$

which is obviously the same as

$$\|(S - \tilde{S})v\|_{H^{-1/2}(\Gamma)} \leq c \cdot \inf_{\tau_h \in Z_h} \|Sv - \tau_h\|_{H^{-1/2}(\Gamma)}. \qquad (22)$$

Hence, using (10) and (22), we get from (14) the quasi-optimal error estimate

$$\|u - \tilde{u}_H\|_{H^{1/2}(\Gamma)} \leq \tilde{c}_1 \cdot \inf_{v_H \in W_H} \|u - v_H\|_{H^{1/2}(\Gamma)} + \tilde{c}_2 \cdot \inf_{\tau_h \in Z_h} \|Su - \tau_h\|_{H^{-1/2}(\Gamma)}. \qquad (23)$$

Let us assume $u \in H^s(\Gamma)$ for some $s \leq \mu+1$. Then $Su \in H^{s-1}(\Gamma)$. Therefore, applying (8) and (16),

$$\|u - \tilde{u}_H\|_{H^{1/2}(\Gamma)} \leq c_1 \cdot H^{s-\frac{1}{2}} \cdot \|u\|_{H^s(\Gamma)} + c_2 \cdot h^{\sigma+\frac{1}{2}} \cdot \|Su\|_{H^\sigma(\Gamma)} \qquad (24)$$

with $s \leq \mu + 1$ and $\sigma \leq \min\{s - 1, \nu + 1\}$. >From this we conclude that to get the asymptotic order of convergence it is sufficient to choose

$$\nu = \mu - 1. \qquad (25)$$

To define the trial space Z_h as introduced in (15) we have to take into account the following aspects to get an almost optimal algorithm.

i. the stability condition (21);

ii. the asymptotic order of convergence (25);

iii. the dimension N in (15).

Even the last point is of practical interest, since we have to discretize in (18) all boundary integral operators using Z_h as test space, i.e. the matrix V_h is of dimension N^2 and K_h is of dimension NM. Therefore, an unfavourable choice of N would increase the numerical amount of the discretization process as well as the storage requirements in a dramatical way.

In what follows we discuss three different strategies to define the trial space Z_h satisfying the stability condition (21). Here we assume that the trial space W_H is given with respect to a boundary triangulation Γ_H (see Figure 1) of η_H boundary elements with a mesh size H and W_H is spanned by pieceweise linear continuous splines, i.e. $\mu = 1$.

Mesh refinement

First we define the trial space Z_h by piecewise constant trial functions ($\nu = 0$) with respect to a triangulation Γ_h with η_h boundary elements which is constructed by an additional uniform refinement of Γ_H (see Figure 2), i.e. we assume

$$h = c \cdot H \qquad (26)$$

with an appropriate constant $c < 1$. For a sufficiently small c the stability condition (21) is satisfied. This was proved, e.g. for the coupling of boundary and finite elements in [4, 12]. This definition of Z_h enables us to use the combination of piecewise linear trial functions in W_H and piecewise constant trial functions in Z_h yielding an asymptotically optimal order of convergence, see (24). But, since $M \sim \eta_H$ and $N = \eta_h$ the number N of degrees of freedom in Z_h grows up rapidly. In practical computations it seems to be sufficient to do one additional refinement to define Γ_h from Γ_H where the number of boundary elements grows up with a factor of $2(n-1)$ and n is the space dimension.

Iso–parametric trial functions

A second possibility to satisfy the stability condition (21) is to use iso–parametric trial functions with $\nu = \mu = 1$. Let $\Gamma_h = \Gamma_H$. Note that in this case we do not get any higher order in the asymptotic convergence, see (24), but one may have some higher order of convergence initially [8]. Since the trial space Z_h is defined by discontinuous trial functions, we first introduce a trial space $\tilde{Z}_h$ spanned by piecewise linear continuous splines. Then the stability condition

$$c \cdot \|v_H\|_{H^{1/2}(\Gamma)} = \sup_{\tau_h \in \tilde{Z}_h} \frac{|\langle v_H, \tau_h \rangle_{L^2(\Gamma)}|}{\|\tau_h\|_{H^{-1/2}(\Gamma)}} \qquad (27)$$

is satisfied for all $v_H \in W_H$ and since $\tilde{Z}_h \subset Z_h$, (21) follows. Note that also in the case of iso–parametric trial functions the number N of degrees of freedom grows up. Using discontinuous piecewise linear trial functions we get $N = n\eta_H$ when using triangles in three dimensions (n is the space dimension). Compared with the mesh refinement case, there is no difference in two dimenions, but a slight reduction in three dimensions, $N = 3\eta_H$ versus $N = 4\eta_H$. Hence, the use of iso–parametric trial functions, e.g. piecewise linear ones will give a slight reduction in the storage requirements as well as in the computing times yielding some initial higher convergence rates as in the mesh refinement case using piecewise constant trials.

Non–matching triangulations

In both cases described above, the definition of the trial space Z_h requires the use of appropriate trial functions to satisfy the stability condition (21). Moreover, the dimension N grows up significantly compared to the number of boundary elements η_H in Γ_H. Here we want to define a trial space Z_h using piecewise constant trial functions to get an optimal asymptotic order of convergence and using a triangulation Γ_h in such a way that the stability condition (21) does not require a condition as given in the mesh refinement case. For simplicity of the presentation we will restrict ourselves to the two–dimensional case $n = 2$. Let us define a triangulation $\hat{\Gamma}_h$ as dual triangulation of Γ_H by taking the midpoints of Γ_H as new nodes of $\hat{\Gamma}_h$. Note that in the case of corners the corresponding boundary elements of Γ_h are no straight lines anymore, since they are angular ones. If we define the trial space $\tilde{Z}_h$ with respect to $\hat{\Gamma}_h$ using piecewise constant trial functions we have $\tilde{\eta}_h = M$. Moreover, the stability condition (27) is satisfied. Since angular boundary elements across corners are not suitable for a boundary element approximation, we refine all of these elements adding the corner points to $\hat{\Gamma}_h$ and the final triangulation we denote by Γ_h, see Figure 3. Now we define Z_h by piecewise constant trial functions with respect to Γ_h, again we have the inclusion $\tilde{Z}_h \subset Z_h$. Hence, (21) is satisfied. Note that the number N of degrees of freedom in Z_h is M plus the number of corners appearing in Γ. Hence, an essential reduction in storage requirements as well as in computing times can be reached when using this non–matching triangulation compared with both approaches discussed above. Note that a generalisation to $n = 3$ requires a more detailed discussion to generate the dual mesh as well as the smoothing of the triangulation across corners and edges.

4. Numerical example

We consider a mixed boundary value problem for the two–dimensional Laplacian in a L shaped domain as sketched in Figure 1. Dirichlet boundary conditions are given for $\{(x,0), (0,y) : -0.25 \leq x,y \leq 0\}$ and Neumann boundary conditions are given on the remaining part. The boundary data are taken in such a way that the solution of (1) is given by

$$u(x) = -\log|x - x^*| \quad \text{with } x^* = (-0.1, -0.1).$$

All boundary integral operators are discretized by the Galerkin scheme (18). The trial space W_H is defined with respect to a uniform triangulation of Γ with M nodal points and η_H boundary elements, see for example Figure 1 with $\eta_H = M = 16$, and using piecewise linear trial functions. According to the previous section we consider different definitions of the trial space Z_h. By refining Γ_H we get a boundary triangulation Γ_h with $\eta_h = 2\eta_H$ boundary elements as sketched in Figure 2. The results for the mesh refinement case are given in Table 2. Using piecewise linear and discontinuous splines the trial space Z_h can be defined with respect to the triangulation Γ_H as given in

Figure 1. Due to the discontinuity of the trial functions the number of degrees of freedom is doubled again. The results are given in Table 3. Note that in this case one get some higher order convergence results for the Neumann data as described in [8].

In both cases discussed above the dimension of Z_h is doubled compared with the dimension of W_H. This means that the size of matrices grows up as shown in Table 1, N_C is the number of corners of Γ. According to this the numerical amount of work to discretize the boundary integral operators as well as to perform a matrix times vector multiplication inside an iterative solution process to solve (11) with the stiffness matrix given by (20) grows up, too. Using a non–matching triangulation Γ_h as given in Figure 3 to define the trial space Γ_h by piecewise constant trial functions leads only to a slight increasing of the dimension of Z_h according to the number of corners involved. The results for the non–matching approach are given in Table 4. Due to the definition of Z_h it is clear that the error of the Neumann data is larger than in both other cases, however, the error of the Dirichlet data, is slightly smaller. Note that the original aim was to compute the Dirichlet data, afterwards, one can use any boundary element method to compute the remaining Neumann data.

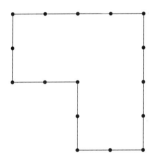

Figure 1. Initial mesh for potential u

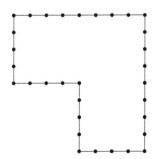

Figure 2. Refined mesh for Steklov–Poincaré approximation

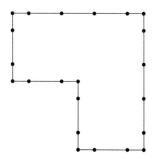

Figure 3. Non-matching mesh for Steklov-Poincaré approximation

	single layer potential V_h	double layer potential K_h
finer mesh	$4 \cdot \eta_H^2$	$2 \cdot \eta_H \cdot M$
iso-parametric	$4 \cdot \eta_H^2$	$2 \cdot \eta_H \cdot M$
non-matching	$(\eta_H + N_C)^2$	$(\eta_H + N_C) \cdot M$

Table 1. Number of elements in BEM matrices

Grid	M	N	$\|\|t - t_h\|\|_{L^2}$	$\|\|u - u_h\|\|_{L^2}$
2	32	64	4.24 −1	2.04 −2
3	64	128	1.84 −1	5.10 −3
4	128	256	8.70 −2	1.28 −3
5	256	512	4.27 −2	3.20 −4
6	512	1024	2.13 −2	8.02 −5

Table 2. Approximation on a finer mesh

Grid	M	N	$\|\|t - t_h\|\|_{L^2}$	$\|\|u - u_h\|\|_{L^2}$
2	32	64	3.20 −1	2.19 −2
3	64	128	9.87 −1	5.41 −3
4	128	256	2.74 −2	1.35 −3
5	256	512	7.33 −3	3.36 −4
6	512	1024	1.99 −3	8.40 −5

Table 3. Approximation with iso-parametric trial functions

Grid	M	N	$\|\|t - t_h\|\|_{L^2}$	$\|\|u - u_h\|\|_{L^2}$
2	32	38	6.41 −1	1.74 −2
3	64	70	3.30 −1	4.37 −3
4	128	134	1.67 −1	1.11 −3
5	256	262	8.42 −2	2.79 −4
6	512	518	4.23 −2	7.02 −5

Table 4. Approximation on a non-matching grid

References

[1] V. I. Agoshkov: Poincaré–Steklov's operators and domain decomposition methods in finite dimensional spaces. In: *First International Symposium on Domain Decomposition Methods for Partial Differential Equations* (R. Glowinski et. al. eds.), SIAM, Philadelphia, 73–112, 1988.

[2] M. Costabel: Boundary integral operators on Lipschitz domains: Elementary results. *SIAM J. Numer. Anal.* **19** (1988) 613–626.

[3] M. Costabel, E. P. Stephan: Boundary integral equations for mixed boundary value problems in polygonal domains and Galerkin approximation. *Banach Centre Publications*, Warsaw, **15** (1985) 175–251.

[4] G. C. Hsiao, E. Schnack, W. L. Wendland: Hybrid coupled finite–boundary element methods for elliptic systems of second order. Preprint 98/13, SFB 404, Universität Stuttgart, 1998.

[5] G. C. Hsiao, E. P. Stephan, W. L. Wendland: On the integral equation method for the plane mixed boundary value problem of the Laplacian. *Math. Meth. Appl. Sci.* **1** (1979) 265–321.

[6] A. Quarteroni, A. Valli: *Numerical Approximation of Partial Differential Equations.* Springer, Berlin, 1994.

[7] G. Schmidt: Boundary element discretization of Poincaré–Steklov operators. *Numer. Math.* **69** (1994) 83–101.

[8] O. Steinbach: A note on initial higher order convergence results for boundary element methods with approximated boundary conditions. Submitted, 1996.

[9] O. Steinbach: On a hybrid boundary element method. *Numer. Math.*, accepted, 1998.

[10] O. Steinbach: Mixed approximations for boundary elements. Submitted, 1999.

[11] O. Steinbach, W. L. Wendland: Domain Decomposition and Boundary Elements. In: *Proceedings of NMA98*, Sofia, 1998.

[12] W. L. Wendland: On asymptotic error estimates for combined BEM and FEM. In: *Finite Element and Boundary Element Techniques from Mathematical and Engineering Point of View* (E. Stein, W. L. Wendland eds.), Springer, Berlin, 273–333, 1988.

Mathematisches Institut A, Universität Stuttgart,
Pfaffenwaldring 57, D 70569 Stuttgart.
Email: steinbach@mathematik.uni-stuttgart.de